Protein Folding

ACS SYMPOSIUM SERIES **526**

Protein Folding

In Vivo and In Vitro

Jeffrey L. Cleland, EDITOR
Genentech, Inc.

Developed from a symposium sponsored
by the Division of Biochemical Technology
at the 203rd National Meeting
of the American Chemical Society,
San Francisco, California,
April 5–10, 1992

American Chemical Society, Washington, DC 1993

Library of Congress Cataloging-in-Publication Data

Protein folding: in vivo and in vitro / Jeffrey L. Cleland, editor.

   p.   cm.—(ACS symposium series, ISSN 0097–6156; 526)

   Developed from a symposium sponsored by the Division of Biochemical Technology at the 203rd National Meeting of the American Chemical Society, San Francisco, California, April 5–10, 1992.

   Includes bibliographical references and index.

   ISBN 0–8412–2640–7

   1. Protein folding—Congresses.

   I. Cleland, Jeffrey L., 1964–   . II. Series.

QP551.P6958213   1993
574.19′245—dc20   93–9740
                                                       CIP

The paper used in this publication meets the minimum requirements of American National Standard for Information Sciences—Permanence of Paper for Printed Library Materials, ANSI Z39.48–1984.

Copyright © 1993

American Chemical Society

All Rights Reserved. The appearance of the code at the bottom of the first page of each chapter in this volume indicates the copyright owner's consent that reprographic copies of the chapter may be made for personal or internal use or for the personal or internal use of specific clients. This consent is given on the condition, however, that the copier pay the stated per-copy fee through the Copyright Clearance Center, Inc., 27 Congress Street, Salem, MA 01970, for copying beyond that permitted by Sections 107 or 108 of the U.S. Copyright Law. This consent does not extend to copying or transmission by any means—graphic or electronic—for any other purpose, such as for general distribution, for advertising or promotional purposes, for creating a new collective work, for resale, or for information storage and retrieval systems. The copying fee for each chapter is indicated in the code at the bottom of the first page of the chapter.

The citation of trade names and/or names of manufacturers in this publication is not to be construed as an endorsement or as approval by ACS of the commercial products or services referenced herein; nor should the mere reference herein to any drawing, specification, chemical process, or other data be regarded as a license or as a conveyance of any right or permission to the holder, reader, or any other person or corporation, to manufacture, reproduce, use, or sell any patented invention or copyrighted work that may in any way be related thereto. Registered names, trademarks, etc., used in this publication, even without specific indication thereof, are not to be considered unprotected by law.

PRINTED IN THE UNITED STATES OF AMERICA

1993 Advisory Board

ACS Symposium Series

M. Joan Comstock, *Series Editor*

V. Dean Adams
University of Nevada—
Reno

Robert J. Alaimo
Procter & Gamble
Pharmaceuticals, Inc.

Mark Arnold
University of Iowa

David Baker
University of Tennessee

Arindam Bose
Pfizer Central Research

Robert F. Brady, Jr.
Naval Research Laboratory

Margaret A. Cavanaugh
National Science Foundation

Dennis W. Hess
Lehigh University

Hiroshi Ito
IBM Almaden Research Center

Madeleine M. Joullie
University of Pennsylvania

Gretchen S. Kohl
Dow-Corning Corporation

Bonnie Lawlor
Institute for Scientific Information

Douglas R. Lloyd
The University of Texas at Austin

Robert McGorrin
Kraft General Foods

Julius J. Menn
Plant Sciences Institute,
U.S. Department of Agriculture

Vincent Pecoraro
University of Michigan

Marshall Phillips
Delmont Laboratories

George W. Roberts
North Carolina State University

A. Truman Schwartz
Macalaster College

John R. Shapley
University of Illinois
at Urbana—Champaign

L. Somasundaram
E. I. du Pont de Nemours and Company

Peter Willett
University of Sheffield (England)

Foreword

THE ACS SYMPOSIUM SERIES was first published in 1974 to provide a mechanism for publishing symposia quickly in book form. The purpose of this series is to publish comprehensive books developed from symposia, which are usually "snapshots in time" of the current research being done on a topic, plus some review material on the topic. For this reason, it is necessary that the papers be published as quickly as possible.

Before a symposium-based book is put under contract, the proposed table of contents is reviewed for appropriateness to the topic and for comprehensiveness of the collection. Some papers are excluded at this point, and others are added to round out the scope of the volume. In addition, a draft of each paper is peer-reviewed prior to final acceptance or rejection. This anonymous review process is supervised by the organizer(s) of the symposium, who become the editor(s) of the book. The authors then revise their papers according to the recommendations of both the reviewers and the editors, prepare camera-ready copy, and submit the final papers to the editors, who check that all necessary revisions have been made.

As a rule, only original research papers and original review papers are included in the volumes. Verbatim reproductions of previously published papers are not accepted.

M. Joan Comstock
Series Editor

Contents

Preface xi

1. Impact of Protein Folding on Biotechnology 1
   Jeffrey L. Cleland

   IN VIVO PROTEIN FOLDING

2. Amino Acid Sequence Determinants of Polypeptide Chain
   Folding and Inclusion Body Formation 24
   Jonathan King, Cameron Haase-Pettingell, Carl Gordon,
   Susan Sather, and Anna Mitraki

3. Duplicated Segment of *Clostridium thermocellum* Cellulases 38
   K. Tokatlidis, Sylvie Salamitou, Tsuchiyoshi Fujino,
   P. Béguin, Prasad Dhurjati, J. Millet, and J.-P. Aubert

4. Probing the Role of Protein Folding in Inclusion
   Body Formation 46
   Boris A. Chrunyk, Judy Evans, and Ronald Wetzel

5. Experimental Investigation of the In Vivo Kinetics
   of Inclusion Body Formation 59
   Jim Klein and Prasad Dhurjati

6. Molecular Chaperones and Their Role in Protein Assembly 72
   Saskia M. van der Vies, Anthony A. Gatenby,
   Paul V. Viitanen, and George H. Lorimer

7. Role of Molecular Chaperones in Transport of Proteins
   into the Mammalian Endoplasmic Reticulum:
   Solubilization-Achieving Proteins, Folding-Accelerating
   Proteins, and Translocation-Mediating Proteins 84
   H. Wiech and R. Zimmermann

8. Folding of Recombinant Human Insulin-Like Growth
   Factor-1 in Yeast 102
   Bhabatosh Chaudhuri and Christine Stephan

9. Role of the Protein-Folding Chaperone BiP in Secretion
   of Foreign Proteins in Eukaryotic Cells 121
   Anne S. Robinson and K. Dane Wittrup

10. GroEL-Mediated Protein Folding 133
    François Baneyx and Anthony A. Gatenby

11. Prolyl Isomerizations as Rate-Determining Steps
    in the Folding of Ribonuclease T1 142
    Lorenz M. Mayr, Thomas Kiefhaber, and Franz X. Schmid

12. Kinetic Control of Protein Folding by Detergent Micelles,
    Liposomes, and Chaperonins 156
    P. M. Horowitz

IN VITRO PROTEIN FOLDING

13. Comparison of Amino Acid Helix Propensities (s-values)
    in Different Experimental Systems 166
    A. Chakrabartty and R. L. Baldwin

14. Single-Step Solubilization and Folding of IGF-1
    Aggregates from *Escherichia coli* 178
    Judy Y. Chang and James R. Swartz

15. Role of Disulfide Bonds in Folding of Recombinant
    Human Granulocyte Colony Stimulating Factor Produced
    in *Escherichia coli* 189
    Hsieng S. Lu, Christi L. Clogston, Lee Ann Merewether,
    Linda O. Narhi, and Thomas C. Boone

16. In Vivo Expression of Correctly Folded Antibody
    Fragments from Microorganisms 203
    Marc Better and Arnold H. Horwitz

17. Characterization of Humanized Anti-p185^{HER2} Antibody
    Fab Fragments Produced in *Escherichia coli* 218
    R. F. Kelley, M. P. O'Connell, P. Carter, L. Presta,
    C. Eigenbrot, M. Covarrubias, B. Snedecor, R. Speckart,
    G. Blank, D. Vetterlein, and C. Kotts

18. Spectral Analysis of Site-Directed Mutants
    of Human Growth Hormone 240
    Michael G. Mulkerrin and Brian C. Cunningham

19. **Altering the Self-Association and Stability of Insulin by Amino Acid Replacement** 254
   David N. Brems, Patricia L. Brown, Christopher Bryant,
   Ronald E. Chance, Richard D. DiMarchi, L. Kenney Green,
   Daniel C. Howey, Harlan B. Long, Alita A. Miller,
   Rohn Millican, Allen H. Pekar, James E. Shields,
   and Bruce H. Frank

INDEXES

Author Index 272

Affiliation Index 272

Subject Index 273

Preface

THE UNDERSTANDING OF PROTEIN FOLDING IS CRITICAL to the production of protein pharmaceuticals, and protein folding plays a vital role in cellular processes. By developing fundamental theories of protein folding, we can design new therapies to affect cellular regulation of protein folding, and we can rationally engineer proteins to display the desired characteristics. In addition, knowledge of protein folding pathways and intracellular regulation will improve methods of producing recombinant proteins for research and commercial use. With the growth of the biotechnology industry, interest in protein folding has increased dramatically since improving processes for the commercial production of recombinant proteins often requires an efficient, high-yield protein folding process.

Recent research has resulted in a rapidly changing perception of protein folding fundamentals. To keep abreast of the many new theories and the growing knowledge of protein folding, seminars and publications such as this text are critical. This book provides an updated review of protein folding research and its impact on future technologies. Leading scientists in protein folding, protein chemistry, and protein design have contributed their extensive experience to the development of this text.

The book is divided into two sections, which focus on the in vivo and in vitro aspects of protein folding. The first section focuses on in vivo protein folding and includes studies of the mechanisms of inclusion body formation and molecular chaperonins (in vivo folding aids). Recent studies of inclusion body formation have led to important insights into the relationship between primary sequence and protein folding, as discussed in the second chapter. When chaperonins fail to interact with the folding protein of interest, inclusion body formation can occur. Studies of the role of the protein sequence in inclusion body formation are described (Chapters 3–4), and the kinetics of inclusion body formation are also analyzed (Chapter 5). In addition, a review of the current knowledge of molecular chaperones is presented (Chapter 6), followed by several investigations into the mechanisms of these unique molecules (Chapters 7–12).

Although many of the studies presented in this section are performed in vitro, the results can be applied to the actual functioning of these proteins in vivo. The role of chaperonins in transport and cellular processing is also presented (Chapter 7), and alterations in the cellular production of different chaperonins in yeast are discussed (Chapters 8–9). The rate-limiting steps during folding may determine the formation of inclusion bodies and the extent to which a protein interacts with chaperonins. To address this problem, the rate-limiting folding of proteins was investigated in terms of both the intrinsic protein properties, such as prolyl isomerization (Chapter 11), and the interaction of proteins with chaperonins (Chapters 10 and 12).

The second section of the text focuses on in vitro protein folding and includes the role of primary sequence in determining structure, the improved recovery of recombinant therapeutic proteins, and the impact of protein engineering on protein folding. For the first time, a comprehensive list of helical propensities is presented for a number of different systems (Chapter 13), and researchers may apply these results to their understanding of the elements of protein conformation. Significant advances in the folding of commercially valuable therapeutic proteins are also described (Chapters 14–15), as the understanding of folding pathways is critical to the success of commercial proteins that require a folding step in the recovery process. In the development of improved therapeutic proteins, protein folding can be a major factor in determining the success of the product. Recent research in protein engineering and antibody technology has resulted in the ability to produce recombinant humanized antibodies for therapeutic use, as discussed in Chapters 16 and 17. Mutagenesis of known therapeutic proteins has been studied (Chapters 18–19) to determine the role of specific residues in protein conformation and stability. These studies will aid the development of improved therapeutic protein products.

Jonathan King was essential in the organization of the symposia, and his guidance was greatly appreciated. The Divisions of Biochemical Technology and Biological Chemistry and the Biotechnology Secretariat were vital for the support of this program. The program coordinators for this ACS meeting, Charles Goochee, Sharon Shoemaker, and Judith Klinman, made the organization and planning of the symposia an efficient process. George Georgiou and Paul Horowitz were kind enough to chair a session, and their contributions to the symposium as well as reviews of the text are greatly appreciated. The ACS acquisitions editor, Barbara Tansill, was instrumental in the timely publication of the book. The successful completion of this text was also made possible by many understanding colleagues at Genentech: Andrew J. S. Jones, Michael F. Powell, Rodney Pearlman, and Jessica Burdman. Finally, the authors of each chapter provided excellent reports of innovative new research or updated reviews, and they should be commended for taking the time and effort to contribute to this text.

JEFFREY L. CLELAND
Genentech, Inc.
South San Francisco, CA 94080

January 22, 1993

Chapter 1

Impact of Protein Folding on Biotechnology

Jeffrey L. Cleland

Pharmaceutical Research and Development, Genentech, Inc., 460 Point San Bruno Boulevard, South San Francisco, CA 94080

>Protein folding has played a major role in both academic and industrial research and development in biotechnology. The biotechnology industry has relied on both *in vivo* and *in vitro* protein folding to successfully produce the currently approved therapeutic proteins. In this article, the methods used to produce therapeutic proteins are discussed along with the rationale for choosing an expression system. An industrial perspective of the protein folding problem and potential solutions is provided with an emphasis on the practical aspects of improving *in vivo* or *in vitro* folding based on current research. In addition, the future directions of research in protein folding are projected focusing on the potential for improving *in vivo* folding, predicting protein structure, and ultimately developing new therapeutics.

Over the past fifteen years, the biotechnology industry has developed from a fledgling group of research start-ups to companies with marketed products. To bring products to market, biotechnology companies have not only designed and discovered new drugs, they have also developed new processes for the large scale production of biopharmaceuticals. A wide range of therapeutic proteins are approved or in clinical trials. Each of the proteins in clinical trials as well as the United States Food and Drug Administration (FDA) approved therapeutic proteins are produced by different methods (see Table 1). The major processes employed for the production of these therapeutic proteins involve either *Escherichia coli* or mammalian cell expression systems.

When considering the appropriate expression system for production of the desired protein, the manufacturer must consider the advantages and disadvantages of each method. Table 2 lists the characteristics of the major expression systems: bacteria, yeast, mammalian cells, and insect cells. The three host cell types used in approved therapeutic proteins are *E. coli* (eg. insulin and

TABLE 1: Therapeutic Proteins: Characteristics and Expression Systems (Listed in Order of Approval)

Protein	Indication (*I*)	FDA Status[a] (*I*)	Expression System	Approx. M.W.	Glycosylation (# of Sites)	Disulfide Bonds
insulin	diabetes	Approved 1982	*E. coli*	6,000	No	Yes (3)
human growth hormone	growth hormone deficiency in children	Approved 1985	*E. coli*	23,000	No	Yes (2)
alpha interferon	hairy cell leukemia, genital warts, AIDS-related Kaposi's sarcoma, non-A, non-B hepatitis	Approved 1986-1991	*E. coli*	18,000	No	Yes
Antibody (OKT 3)	reversal of acute kidney transplant rejection	Approved 1986	Hybridoma (animal cells)	150,000	Yes	Yes
tissue plasminogen activator	acute myocardial infarction, acute pulmonary embolism	Approved 1987, 1990	Chinese Hamster Ovary cells	70,000	Yes (2 or 3)	Yes (17)
erythropoietin	anemia form chronic renal failure or AIDS therapy	Approved 1989	Chinese Hamster Ovary cells	34,000	Yes (3)	Yes (2)
gamma interferon	chronic granulomatous disease	Approved 1990	*E. coli*	34,000 (Dimer)	No	No

Protein	Indication	Status	Host	MW		
granulocyte colony stimulating factor	chemotherapy-induced neutropenia	Approved 1991	E. coli	16,000	No	Yes (3)
granulocyte macrophage colony stimulating factor	autologous bone marrow transplatation	Approved 1991	E. coli	16,000	No	Yes (2)
interleukin-2	chemotherapy	Approved 1992	E. coli	17,000	No	Yes (1)
antibody (Centoxin and E5 MAb)	sepsis and septic shock	Approved 1992	Hybridoma	150,000	Yes	Yes
human basic fibroblast growth factor	venuous stasis, diabetic leg and foot ulcers	Phase III	E. coli	18,000	No	Yes
human deoxyribonuclease	cystic fibrosis	Phase III	Chinese Hamster Ovary cells	31,000	Yes (2)	Yes (2)
beta interferon	multiple sclerosis	Phase III	E. coli	20,000	No	Yes

[a] Approval date may vary with indication. See reference *1* for details.

TABLE 2: Comparison of Different Expression Systems

Host	Expression Level[a]	Secretion[b]	Glycosylation[c]	Redox. Potential
E. coli	High	Periplasm	None	cytoplasm - reduced periplasm - oxidized
Yeast	High-Moderate	Yes	High mannose only	cytosol - reduced secretory pathway[d] - oxidized
Mammalian cells (2, 3)	Moderate-Low	Yes	Complex, high mannose or phosphorylated[e]	cytosol - reduced secretory pathway[d] - oxidized
Insect cells (baculovirus, 4)	Moderate-Low	Yes	Complex, high mannose or phosphorylated[f]	cytosol - reduced secretory pathway[d] - oxidized

[a] Expression levels can vary for each host and are strongly dependent upon the desired protein and the promoter as well as the fermentation conditions.
[b] Eukaryotic expression systems can be designed to express the protein into the culture media. Expression in E. coli can result in either perplasmic or cytoplasmic accumulation of the protein.
[c] The level of glycosylation and the glycosylation pattern varies for both CHO and insect cells. In addition, these cells will often produce the desired protein with a high level of microheterogenity (carbohydrate branching, sialation, etc.).
[d] Secretory pathway includes the endoplasmic reticulum, Golgi, and secretory vesicles.
[e] Glycosylation in mammalian cells varys among different cell types and is dependent upon the desired protein and the culture conditions (3).
[f] Glycosylation pattern in insect cells is usually not similar to the pattern in mammalian cells. Insect cells tend to produce proteins with more mannose and less total carbohydrate (4).

human growth hormone), Chinese Hamster Ovary (CHO) cells (eg. tissue plasminogen activator (rhtPA) and erythropoietin (EPO)), and hybridoma cells (eg. OKT3 and anti-sepsis antibodies). It is interesting to note that the approved therapeutics expressed in *E. coli* are all small proteins that do not require glycosylation for activity. In contrast, both CHO derived proteins, rhtPA and EPO, are large and extensively glycosylated. Often, glycosylation is required for the protein to retain its bioactivity, reduce its antigenicity, maintain its native conformation, and prolong its serum half-life. If glycosylation is required, then either mammalian cell, insect cell, or yeast expression systems must be used.

In addition to glycosylation, the host organism dictates the *in vivo* protein folding efficiency. Eukaryotes have a different reduced-to-oxidized potential (redox. potential) within the various cellular components. Since most proteins are synthesized and processed in an oxidizing environment in eukaryotes, disulfide bond formation is enhanced during the initial folding events. Eukaryotes also possess a catalytic protein, protein disulfide isomerase (PDI), to assure correct disulfide bond formation. The importance of this intracellular protein is displayed in the production of rhtPA by CHO cells. This protein has 17 disulfide bonds and one free thiol. Therefore, the *in vitro* folding efficiency of *E. coli* derived rhtPA is very low (5). There are also several potentially misfolded and disulfide-scrambled intermediates that could form during rhtPA folding. In contrast, CHO cells are able to efficiently produce native rhtPA by using their intrinsic cellular machinery which includes PDI and other chaperones. Another catalytic chaperone protein, peptidyl prolyl isomerase (PPI), also exists in many eukaryotes and catalyzes the conversion of proline residues between the *cis* and *trans* states. Several other assistant proteins also referred to as chaperones exist in both eukaryotes and prokaryotes as well as plants (see references 6 and 7 for reviews).

Based on their protein processing characteristics, animals cells would appear to be the natural choice for production of most proteins of therapeutic interest. However, as noted in Table 1, the majority of approved proteins are produced in *E. coli*. The original rationale for choosing an *E. coli* expression system arose from the lack of knowledge regarding animal cells. At the initial development of the biotechnology industry, animal cell culture was underdeveloped and the FDA had just begun to investigate some of the issues (DNA, viruses, etc.) surrounding the use of CHO cells. Thus, *E. coli* had an initial lead over CHO cells in both technology development and FDA approval. Besides their late development, eukaryotic expression systems have several potential limitations that reduce their utility for protein production. First of all, animal cells require longer growth times (~7 days) to achieve maximum cell density and achieve a much lower cell density ($1-10 \times 10^6$ cells/ml) than *E. coli* (1×10^8 cells/ml). The same proteins that facilitate folding in eukaryotes, molecular chaperones, can also interfere with high levels of protein production. When a cell is forced to overexpress the desired protein, chaperones can become overwhelmed and the protein may then accumulate as partially folded structures inside the cell. This event has been observed in both eukaryotes and prokaryotes (yeast, *8*; mammalian cells, *9*; *E. coli*, *10, 11*). In addition, animal cells are often grown in expensive media, containing several growth components that may interfere with the recovery of the desired protein. The high fermentation costs, technical challenges, and often low expression level of animal cells have resulted in the decision to proceed with *E. coli*-based products. For many companies, the yields obtained during *in vitro* folding of *E. coli*-based proteins are often the limiting factor in the successful development of the product. In many cases, it is possible to use an animal cell system to obtain reasonable yields at costs approaching an *E. coli*-based process. If the FDA approved therapeutic proteins are any indication, animal cell systems are primarily used when the final protein product (rhtPA, EPO, and antibodies) can command a premium price. These prices are necessary to support the more expensive production as well as recover the investment in research and development.

To avoid some of the problems with animal cell expression, bacterial systems are chosen because they usually provide a high level of expression of the desired protein. The expression of the recombinant protein can constitute 5 to 20% of the total cellular protein. Bacterial systems such as *E. coli* lack the ability to secrete large proteins into the culture media and the cellular machinery to facilitate proper folding. Therefore, the high expression levels usually result in the formation of inclusion bodies, which are large insoluble aggregates of the desired protein. Several recombinant proteins are produced as inclusion bodies in *E. coli* (see reference *12* for a review). These insoluble protein aggregates must be solubilized and refolded to form the native protein. A major advantage of this production method is the ability to easily isolate the product from the cellular components by centrifugation or microfiltration. However, if the refolding yields are low, some of the benefits of inclusion body formation are negated. Inclusion bodies can be formed by proteins at different states of folding. Several studies have shown that inclusion bodies result from the association of folding intermediates (*11,13*). The intermediates form intermolecular bonds (van der Waals interactions, hydrogen bonds, disulfide bonds, etc.). These bonds are typically difficult to perturb since the intermolecular interactions are analogous to intramolecular bonds formed in the core of the native protein. To disrupt these aggregates, denaturants such as guanidine hydrochloride or urea are utilized and, if intermolecular disulfide bonds are formed, reducing agents must be used. The denaturants must then be removed to allow the protein to regain its native structure. During the refolding process, the protein undergoes a series of conformational changes which may or may not result in correctly folded protein. Potential pathways for the protein during refolding include the formation of misfolded intermediates, native-like intermediates, aggregates, or the native protein as shown in Figure 1. Each of the off-pathway reactions can dramatically reduce the recovery of native protein. These off pathway structures can also be difficult to remove from the final refolding mixture since their chromatographic properties are often similar to the native protein.

To solve the protein folding problem, one must develop an understanding of both the *in vivo* and *in vitro* folding events. Each individual study on a single protein can yield insight into the general mechanisms of protein folding. Several approaches have been applied to the study of protein folding and many of these

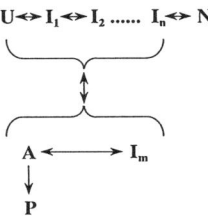

Figure 1: The potential folding pathways for an unfolded protein (U) are shown assuming that it is refolded *in vitro* by dilution with a simple buffer (no additives). The unfolded protein will fold to form an intermediate structure (I_1) which has some secondary structure. This intermediate as well as others in the folding pathway ($I_2...I_n$) often associate to form soluble aggregates (A). These soluble aggregates can agglomerate to form large irreversible precipitates (P) that must be resolubilized in denaturants. All intermediates as well as the unfolded protein can form misfolded or off-pathway intermediates (I_m) that can reduce the yield of native protein by becoming a kinetic trap for the preceding species. Some intermediates on the later portion of the pathway obtain a native-like conformation (I_n) and eventually assume the native state (N).

approaches will yield insight into the design of improved processes as well as improved therapeutics.

Solving the Protein Folding Problem for Industrial Production

The success of therapeutic proteins may depend upon the ability to produce proteins inexpensively on an industrial scale. In light of the possible price fixing for drugs in the United States and elsewhere, pharmaceutical companies must produce their products at low costs to recover the large development expenditures required for pharmaceuticals. To achieve this goal, companies have encouraged both internal and external development of improved methods for the production of proteins. If low yields in protein folding from *E. coli* can be overcome or high levels of properly folded protein can be achieved in the periplasm of *E. coli*, the inexpensive production of proteins is possible. To achieve this goal, researchers have studied two different methods. The first method is the optimization of *in vivo* production of properly folded protein in *E. coli*. The second technique involves the development of improved methods to increase the yields from *in vitro* protein folding. Both of these approaches have provided some improvement in the yield of native protein.

In Vivo **Folding in *E. coli*.** To enhance the recovery of *in vivo* folded proteins in *E. coli*, several researchers have modified the growth conditions. Modifications in the growth conditions, if successful, are less difficult than other potential alternatives such as coexpression of chaperones or mutations in the desired protein. Several studies by King and coworkers have shown that temperature can play a critical role in the formation of inclusion bodies (*14, 15*). For example, in the model system of p22 tailspike protein, a mutant protein that aggregated *in vivo* at high temperatures was produced (*14*). This mutant protein formed a temperature labile intermediate during both *in vivo* and *in vitro* folding (*16,17*). At 40°C, the intermediate aggregated to form insoluble inclusion bodies and folded to form the native protein at 37°C (*17*). These studies indicate the potential role of temperature in the formation of inclusion bodies. In addition, if inclusion body formation is primarily driven by endothermic processes such as hydrophobic interactions, high temperatures should be avoided during the fermentation. In many industrial fermentations, heat transfer and temperature control can be quite variable resulting in periods of global or local temperature increases above the desired growth temperature. These differences could be significant enough to result in the increased formation of insoluble aggregates. The effect of temperature on both *in vivo* and *in vitro* folding should be further studied to address this potential problem.

Another important factor in the *in vivo* production of folded proteins in *E. coli* is the fermentation media composition. Studies by Georgiou and coworkers have shown that inclusion body formation can be dramatically affected by the addition of nonmetabolizable sugars such as sucrose and raffinose to the culture media (*18*). In these studies, the formation of periplasmic inclusion bodies of β-lactamase was dramatically reduced through the addition of these sugars to the culture media. These sugars could have either stabilized the formation of the native structure or prevented the association of β-lactamase intermediates. The use of sucrose during *in vitro* folding of β-lactamase also resulted in improved recovery of native protein (*19*). It may then be possible to perform an *in vitro* screening of media components that will enhance the formation of soluble protein. These components must be able to enter the periplasm of *E. coli* and prevent aggregation. Further studies of the effect of culture media components on protein folding are needed to assess the potential for this method of producing high concentrations of soluble protein in *E. coli*.

If changes in the culture environment are not successful in enhancing the recovery of soluble protein, alterations in the cellular components or mutations in the desired protein may be required. One potential change at the cellular level would involve the overexpression of the chaperones that inhibit protein aggregation. In *E. coli*, these proteins are DnaK, DnaJ, and the GroE complex, GroEL and GroES. It has been hypothesized that these proteins work in concert to produce soluble correctly folded protein (*20*). The proposed mechanism involves DnaK binding to an unfolded polypeptide chain followed by binding of DnaJ to the complex. The protein then forms an intermediate structure that may then bind to GroEL. Finally, after a series of ATP hydrolysis steps and interaction with GroES, the native protein is released (*20*). An initial attempt at increasing the production of soluble protein in *E. coli* has been performed by overexpression of DnaK in *E. coli* overproducing recombinant human growth hormone (*21*). The overexpression of DnaK appeared to reduce the amount of insoluble protein accumulated in the cells (*21*). However, if the chaperones work together in a concerted manner, it will be necessary to overexpress all of them to eliminate intracellular aggregation. Overexpression of the chaperone cascade may have adverse effects on cell growth and yield. Chaperones interact with cellular proteins during their synthesis and, therefore, chaperone overexpression may inhibit proper folding by altering the distribution of protein bound to the chaperones. In addition, the *in vivo* overproduction of the chaperone proteins at high levels will inevitably reduce the overall fermentation yield, desired protein mass per unit cell mass or media component. Therefore, chaperone overexpression may not be a feasible alternative for large scale protein production.

Another possible method for increased production of soluble protein in *E. coli* could be based on recent studies comparing mutant and wild-type protein folding. As mentioned previously, King and coworkers produced a temperature-sensitive folding mutant of the phage p22 tailspike protein (*14*). This mutant formed a temperature labile intermediate that aggregated at elevated temperatures (40°C). The temperature labile nature of this folding intermediate has been suppressed by further changes in the protein sequence. By introducing single amino acid substitutions, the temperature sensitive mutant was able to fold to the native state at 40°C without forming aggregates (*22, 23*). In contrast, a point mutation in bovine growth hormone (bGH) resulted in the stabilization of an intermediate that had a greater propensity to aggregate (*24*). Bovine growth hormone is difficult to renature since it forms a stable hydrophobic intermediate which readily aggregates (*25*). To avoid this problem, the section of bGH that had been shown to cause aggregation was removed and replaced by the homologous sequence from human growth hormone. The resulting mutant protein followed the same folding pathway as the wild type, but did not form off-pathway aggregates (26). These results indicate that it may be possible to make aggregation-suppressing alterations in a protein's primary sequence without dramatically altering the folding pathway or the native conformation. Either the use of single site mutations or homology within protein classes should provide methods for producing proteins that do not aggregate during folding. Another approach may involve the choice of expression vectors and expression of only the bioactive portion of the desired protein. For example, Carter and coworkers were able to achieve high levels of expression of an antibody fragment, $F(ab')_2$, in *E. coli* (27). In this case, the fragment was the only portion of the native protein needed for activity. To make these changes industrially relevant, they must be considered at the very early stages of product development so that the correct molecule is chosen for preclinical and clinical studies.

In Vitro **Protein Folding.** A more common approach to solving the protein folding problem has been the study of *in vitro* folding or "refolding." In addition, this approach has been widely applied in the production of therapeutic proteins. Over the past twenty years, several researchers have performed *in vitro* folding studies on different proteins in an attempt to understand the underlying mechanisms as well as improve the efficiency of the refolding process. Typically, the initial studies performed to characterize refolding investigate the effects of denaturant, redox reagent for disulfide containing proteins, and protein concentration. The solution conditions such as pH, buffer components, and salts used in the refolding step are then analyzed for their effects on refolding. Unfortunately, each protein is likely to fold by a different pathway for each condition and it is therefore difficult to develop general rules for refolding. However, intuition based on previous refolding studies can provide general insight into the appropriate conditions to assess. Each additional protein refolding study on an individual protein provides further information into the rules associated with the folding process.

Several methods (see Table 3) have been utilized to either study *in vitro* folding or enhance recovery of native protein. One method uses molecular chaperones, intracellular proteins, to facilitate folding. Researchers have performed *in vitro* folding studies with different chaperones as shown in Table 3 primarily to assess their mechanism of action. However, these molecules might also be useful in facilitating *in vitro* folding. For example, by using both catalytic chaperones, PDI and PPI, the refolding rate of RNase T1 was greatly enhanced (31) and these chaperones could also be applied to other proteins that have disulfide bonds and proline residues. The increased folding rate could result in a reduced accumulation of unstable intermediates which are prone to aggregate. Noncatalytic chaperones have also been studied for their effect on *in vitro* folding. In particular, the GroE complex has been shown to inhibit aggregation and therefore facilitate folding in several studies (32-34). If these approaches are considered on an industrial scale, the costs resulting from the amount of protein required ([Aide]/[Protein] mass ratio in Table 3) to observe a significant effect would likely outweigh the benefit. In most cases, one would have to produce

TABLE 3: *In Vitro* Folding Aids: Effects on Recovery Yield of Native Protein and Refolding Rate

Folding Aid	Protein(s) Refolded	Recovery Yield[a]	Refolding Rate[a]	[Aid] /[Protein][b] (mass ratio)	Reference
Catalytic Chaperones:					
Protein Disulfide Isomerase (PDI)	RNase A	0	+	1.4	28
Peptidyl-Prolyl Isomerase (PPI)	Immunoglobulin light chain	0	+ +	0.02 - 0.4	29, 30
PDI + PPI	RNase T1	0	+ + +	3.2 (2.7 PDI/ 0.5 PPI)	31
Noncatalytic Chaperones:					
GroE Complex[c]	Citrate synthase	+ + +	0	10 - 100	32
	Rhodanese	+ +	0	5 - 30	33
	Rhodanese	+ +	0	600	34
Sugars:					
Glycerol	Citrate synthase	+	0	>1000	35
	Lysozyme	+ + +	0	4000	36
Sucrose	Beta-lactamase	+ + +	0	9 - 108	19
BSA	Citrate synthase	+	0	2	35
BSA + Glycerol	Citrate synthase	+ +	0	>1000	35

Continued on next page

TABLE 3: *In Vitro* Folding Aids: Effects on Recovery Yield of Native Protein and Refolding Rate (continued)

Folding Aid	Protein(s) Refolded	Recovery Yield[a]	Refolding Rate[a]	[Aid] /[Protein][b] (mass ratio)	Reference
Micelles:					
Lauryl maltoside/ cardiolipin	Rhodanese	++	0	2 - 6	37
aerosal OT	RNase A	+	0	36	38
Surfactants:					
Lauryl maltoside	Rhodanese	++	0	2 - 100	39
Cetyltrimethyl- ammonium bromide	Rhodanese	+++	0	20	40
Cetyltrimethyl- ammonium chloride	Porcine growth hormone	++	0	2	41
	Interleukin-1beta	++	0	2	41
	Insulin-like growth factor II fusion	++	0	2	41
N-decyl-N,N-dimethyl- 3-ammonio-1-propane sulfonate	Rhodanese	+	0	200	40
N-dodecyl-N,N-dimethyl- 3-ammonio- 1-propanesulfonate	Rhodanese	++	0	200	40
Nonidet P-40	Rhodanese	+	0	200	40

Additive	Protein				Ref
Triton X-100	Rhodanese	+	0	200	40
Polymers:					
Polyethylene glycol (3350 MW)	Carbonic anhydrase	+++	0	0.2	42
	recombinant human Deoxyribonuclease	+++	0	0.6	43
	recombinant human Interferon gamma	++	0	0.4	43
	recombinant human tissue plasminogen activator	++	0	1	43
Polyalanine (3000 MW)	Carbonic anhydrase	+++	0	3	44
Polylysine (2500 MW)	Carbonic anhydrase	+++	+?	3	44
Polyglycine (3500 MW)	Carbonic anhydrase	+++	0	1	44

Continued on next page

TABLE 3: *In Vitro* Folding Aids: Effects on Recovery Yield of Native Protein and Refolding Rate (continued)

Folding Aid	Protein(s) Refolded	Recovery Yield[a]	Refolding Rate[a]	[Aid] /[Protein][b] (mass ratio)	Reference
Ligands and Inhibitors:					
Oxalacetate	Citrate synthase	+	0	0.3	35
BSA + Oxalacetate	Citrate synthase	++	0	3	35
Phenylalanine	Pyruvate kinase (unfolding)	++	0	5	45
Monoclonal antibody (anti-S-Protein) + S-Peptide from RNase A	RNase A	+++	0	26	46

[a] Scoring for yield and refolding rate: 0 = no effect, + = minor effect (10 - 30% increase), ++ = moderate effect (30 - 70% increase), +++ = major effect (>70% increase); See references for actual changes from controls.
[b] Mass ratios were determined from references directly or by calculation of mass concentrations by using the molecular weights provided in the reference.
[c] GroE Complex consists of a 14-mer of GroEL (58 kDa) and 7-mer of GroES (10 kDa).

more of the chaperone than the desired protein. This problem could be overcome if the chaperones are reused or recycled.

If an optimal folding aide could be designed, it would have several critical properties. It would be cost effective, meaning that the improved recovery would outweigh the reagent costs or the reagent would be reusable without a significant reduction in efficiency. The folding aide would inhibit protein aggregation without adversely affecting the formation of native protein. In addition, it must be easily separated from the native protein after completion of folding. If the aide could operate at low concentrations and provide catalysis of folding, it would also greatly reduce process costs. Although a single molecule has not yet been discovered that can provide these properties for the folding of all proteins, there are several folding aides that have been somewhat successful (see Table 3). In particular, sugars, surfactants, and polymers have shown some utility in increasing the recovery yield of native protein during refolding. Sugars have been observed to stabilize native proteins (47) and this effect has been considered for their choice as a folding aide (19). Usually, sugars are used at high concentrations (>10% w/w) to improve refolding yields (19, 35, 36). At these concentrations, the protein will become preferentially hydrated and a compact state such as the native protein will be favored (47). In contrast, surfactants have been used to solubilize inclusion bodies and enhance refolding by binding to the protein (39-41). The surfactant concentrations are typically lower than those used for sugars, but the concentrations often exceed the critical micelle concentration and it is unclear whether micelle formation is necessary to achieve an improvement in the refolding yield. A few studies have assessed the effects of micelles on protein folding and these studies indicate that micelles are useful for membrane-associated proteins (37) or hydrophilic proteins which could partition into a single reverse micelle (38). It appears likely that surfactants interact with unfolded or partially folded proteins in a stoichiometric fashion and, therefore, do not depend on micellar formation for their folding aide function. The stoichiometric binding of polymers during refolding has been studied in detail (42). In these studies, a folding intermediate of bovine carbonic anhydrase B was observed to bind to polyethylene glycol (PEG). Additional studies with PEG also revealed the apparent stoichiometric relationship between the polymer and the refolding protein. The amount of PEG required for enhanced refolding of other proteins was also correlated to a specific optimal reaction stoichiometries and the observed stoichiometries were probably dependent upon the physicochemical properties of each protein (43). If this same hypothesis is applied to previous studies with surfactants, the surfactant concentrations may be reduced dramatically depending upon their ability to bind folding intermediates and inhibit their aggregation. Sugars, surfactants, and polymers do not appear to provide any enhancement in the rate of refolding, but merely act as inhibitors of the off-pathway aggregation reaction. The only exception could be the use of polylysine (2500 MW) in the refolding of bovine carbonic anhydrase B where an increase in the initial rate of folding was observed (44). The unique ability of polyamino acids to potentially act as folding templates warrants additional studies. Further investigations of the effects of sugars, surfactants, and polymers should lead to the development of general rules for their application in protein refolding.

In addition to nonspecific folding aides, several researchers have attempted to use compounds that bind to the native protein with high affinity. In particular, ligands and inhibitors have been used to enhance refolding (35,46) or stabilize the native protein (45). In each case, the conceptual approach focussed on the stabilization of the native protein by shifting the refolding equilibrium toward the native-inhibitor complex. Some success has also been achieved through the use of monoclonal antibodies to the native protein (46). Antibodies will facilitate the formation of the correct native-like epitope and, therefore, guide the protein along the proper folding pathway. From an industrial standpoint, it may be interesting to consider a process involving immobilized antibodies that contact the denatured protein and then slowly equilibrate with refolding buffer. This process would allow the reuse of the reagents and provide easier separation as well as increased recovery of native protein. Again, additional studies are needed to assess the validity of this approach for the general refolding of proteins.

Clearly, a great deal of interesting and insightful research has been performed on the *in vitro* folding of several proteins. These analyses have just begun to provide important insight into the design of better refolding schemes. Future studies on this aspect of protein folding may provide the industrial process researcher with the tools needed to achieve high yields during refolding.

Protein Folding from a Research and Development Perspective

Although the applied research on protein folding has a direct and significant impact on the biotechnology industry, fundamental studies of protein folding are also critical to the future success of biotechnology (for recent reviews see 48 and 49). Continuing research into the role of chaperones and *in vivo* folding will yield insight into the cellular processing of proteins and, perhaps, provide some clues about improving protein expression and folding at the cellular level. This research would also complement studies on the influence of the protein's primary sequence on protein folding. With an understanding of the folding rules obtained from primary sequence information, structure prediction for any sequence of amino acids could be performed. Finally, if one had a complete understanding of protein folding rules and structure prediction, new drugs could be developed with improved properties such as higher affinity or activity. To achieve this level of understanding, the phenomena involved in protein folding will clearly require the continuation of fundamental research.

Analysis of *In Vivo* Folding. Major strides have recently been made to understand the cellular mechanisms of protein folding. In particular, the understanding of chaperone function has improved significantly over the past few years. The cooperative nature of molecular chaperone action has recently been discovered (20) and these results indicate that the concerted action of the chaperones may be necessary to insure proper *in vivo* folding of the protein. The *E. coli* chaperones apparently operate in sequential order, DnaK, DnaJ, and

GroE complex, on the unfolded protein (20). The last chaperone set in the cascade, the GroE complex, has been the focus of several studies. GroEL, the 14-mer of the 58 kDa heat shock protein, has been observed to function with positive cooperativity. More than one unfolded protein may bind to the GroEL double donut structure (two 7-mer rings) and, thereby, reduce the number of unfolded or partially folded proteins in the cytoplasm (50). Therefore, GroEL can both reduce intracellular aggregation by decreasing the concentration of folding intermediates and enhance the assembly of multimeric proteins by bringing the subunits together. Further research on GroEL has revealed that it is regulated by phosphorylation at high temperatures. Upon induction by heat shock, GroEL becomes reversibly phosphorylated and the phosphorylated form can bind and release denatured proteins without assistance from its usual cofactor, GroES (51). The cellular process for recovery from heat shock seems to indicate the possibility for improving the function of the chaperone, GroEL. With subsequent knowledge of the GroEL monomer conformation, improvements in GroE complex efficiency may be possible through genetic engineering methods.

Primary Sequence Analysis and Structure Prediction. In addition to gaining knowledge on the function of cellular machinery in protein folding, researchers have made progress in developing methods for prediction of protein conformation from the primary sequence. One method of structure prediction is based on the use of homology within a protein family. This technique brought together the evolutionary knowledge within a protein family and several advanced computer sorting methods, and successfully predicted the structure of a protein domain (52). This research indicates that there is hope for developing a methodology to determine *a priori* the conformation of all known proteins. Computer simulations have also been used to predict protein folding and the effect of single site mutations on the protein. By modelling a protein as a block copolymer with chain segments consisting of hydrophobic or hydrophilic monomers, researchers have been able to assess the effect of a single mutation on the stability of the native protein (53). This approach may also provide an explanation for the effect of point mutations on suppression of temperature-sensitive folding observed in the phage p22 tailspike protein double mutants (suppressor-temperature sensitive folding) where a single residue change resulted in temperature-insensitive folding (22,23). Other mutagenesis studies have revealed the importance of internal residue packing (54, 55) and potential repetition in sequences (56). The internal packing of a protein can apparently tolerate some minor changes in sequence, but the general consensus of these regions must be conserved to maintain the integrity and stability of the protein (54, 55). Therefore, the hydrophobic residues clustered in protein cores may be a common feature of protein structure which would reduce the complexity of structure prediction methodologies. Further simplification of structure prediction is elucidated from the studies of alanine replacements in T4 lysozyme. These studies have shown that solvent-exposed residues on alpha helices do not contribute to protein folding or stability (56). If there are indeed segments in the amino acid sequence that do not contribute to protein folding, then the prediction of structure would be much simpler. Future research will focus on the applications of mutagenesis, x-ray crystallography, structure prediction algorithms, and sequence homology to elucidate the role of a protein's sequence in determining its conformation and folding.

Application of Folding Knowledge to the Development of Future Products

With a growing knowledge of the effects of changes in protein sequence on folding and conformation, more proteins will be altered to provide increased stability, specificity, or activity. Several structural alterations in enzymes have been performed to increase stability toward heat (57) or organic solvents (58). The primary purpose of these studies was to develop enzymes for use as catalysts in industrial processes. The structure of enzymes has also been analyzed for the development of inhibitors. A recent review by Kuntz describes the use of knowledge of the protein's structure and function to design HIV protease inhibitors as well as receptor antagonists or agonists (59). One might consider taking this knowledge one step further and developing small molecule analogs or mimetics to replace proteins. These structural mimetics often have many advantages over the original protein (see reference 60 for a review of mimetics). The small molecules can be delivered orally, unlike proteins which are usually delivered subcutaneously or intravenously. These molecules can also be made resistant to attack by proteases and can be engineered to have a greater specificity and activity than the original protein. In general, protein engineering coupled with an understanding of protein structure prediction will ultimately lead to the development of additional small-molecule therapeutics.

In the future, biotechnology may become a tool for the development of small organic molecules and the understanding of biological functions. Each of these tasks will require additional knowledge of protein folding and its impact on protein stability and design. In many cases, it may not be possible to design small molecule agonists to replace proteins as observed for human growth hormone (61). The biotechnology industry will then be required to produce these proteins for therapeutic use. Therefore, both *in vivo* and *in vitro* protein folding will continue to have an impact on the biotechnology industry.

Acknowledgments

This manuscript and text was made possible by the assistance and patience of my colleagues at Genentech. In particular, I would like to thank Dr. Andrew J. S. Jones and Jessica Burdman for their thoughtful review of this manuscript. Also, Drs. Jones, Michael F. Powell, and Rodney Pearlman were instrumental in providing me with the time to complete this text. Jessica Burdman played a vital role in all correspondence and should be commended for her efforts.

Literature Cited

1. "Biotechnology Medicines in Development", *Genetic Engineering News* **Jan. 1992**, pp. 27-28.

2. Hwang, C.; Sinkskey, A. J.; Lodish, H. F.; *Science* **1992**, *257*, pp. 1496-1502.
3. Goochee, C. F.; Monica, T.; *Biotechnology* **1990**, *8*, pp. 421-427.
4. Luckow, V. A.; In *Recombinant DNA Techniques and Applications*; Prokop, A; Bajpai, R. K.; Ho, C. S., Eds.; McGrawHill Inc., St. Louis, 1992, pp. 97-152.
5. Sarimientos, P.; Duchesne, M.; Denéfle, P.; Boiziau, J.; Fromage, N.; Delporte. N., Parker, F.; Leliévre, Y.; Mayaux, J.-F.; Cartwright T.; *Biotechnology* **1989**, *7*, pp. 495-501.
6. Ellis, R. J.; van der Vies, S. M.; *Annu. Rev. Biochem.* **1991**, *60*, pp. 321-347.
7. Rothman, J. E.; *Cell* **1989**, *59*, pp. 591-601.
8. Cousens, L. S.; Shuster, J. R.; Gallegos, C.; Ku, L.; Stempien, M. M.; Urdea, M. S.; Sanchez-Pescador, R.; Taylor, A.; Tekamp-Olson, P.; *Gene* **1987**, *61*, pp. 265-275.
9. Marquardt, T; Helenius, A.; *J. Cell Biol.* **1992**, *117*, pp. 505-513.
10. Kane, J. F., Hartley, D. L.; *TIBTECH* **1988**, *6*, pp. 95-101.
11. Mitraki, A.; Haase-Pettingell, C.; King, J.; In: *Protein Refolding*; Georgiou, G., DeBernardez-Clark, E., Eds.; A.C.S. Symposium Series, Vol. 470, American Chemical Society, Washington, D.C., 1991, pp. 35-49.
12. Marston, F. A. O.; *Biochem. J.* **1986**, *240*, pp. 1-12.
13. Mitraki, A.; King, J.; *Biotechnology* **1989**, *7*, pp. 690-697.
14. Goldenberg, D. P.; Smith, D. H.; King, J.; *PNAS* **1983**, *80*, pp. 7060-7064.
15. Haase-Pettingell, C.; King, J.; *J. Biol. Chem.* **1988**, *263*, pp. 4977-4983.
16. Sturtevant, J. M.; Yu, M.; Haase-Pettingell, C.; King, J.; *J. Biol.Chem.* **1989**, *264*, pp. 10693-10698.
17. Seckler, R.; Fuchs, A.; King, J.; Jaenicke, R.; *J. Biol. Chem.* **1989**, *264*, pp. 11750-11753.
18. Bowden, G. A.; Georgiou, G. *Biotech. Prog.* **1988**, *4*, pp. 97-101.
19. Valax, P.; Georgiou, G. In *Protein Refolding*; Georgiou, G.; DeBernardez-Clark, E., Eds.; ACS Symposium Series, Vol. 470; American Chemical Society, Washington, D.C., 1991; pp. 97-109.
20. Langer, T.; Lu, C.; Echols, H.; Flanagan, J.; Hayer, M. K.; Hartl, F. U.; *Nature* **1992**, *356*, pp. 683-689.
21. Blum, P.; Velligan, M.; Lin, N.; Matin, A.; *Biotechnology* **1992**, *10*, pp.301-309.
22. Mitraki, A.; Fane, B.; Haase-Pettingell, C.; Sturtevant, J.; King, J.; *Science* **1991**, *253*, pp. 54-58.
23. Fane, B.; Villafane, R.; Mitraki, A.; King, J.; *J. Biol. Chem.* **1991**, *261*, pp. 11640-11648.
24. Brems, D. N.; Plaisted, S. M.; Havel, H. A.; Tomich, C.-S. C. *PNAS* **1988**, *85*, pp. 3367-3371.
25. Brems, D. N.; *Biochemistry* **1988**, *27*, pp. 4541-4546.
26. Lehrman, S. R.; Tuis, J. L.; Havel, H. A.; Haskell, R. J.; Putnam, S. D.; Tomich, C.- S. C.; *Biochemistry* **1991**, *30*, pp. 5777-5784.
27. Carter, P.; Kelley, R. F.; Rodriques, M. L.; Snedecor, B.; Covarrubias, M.; Velligan, M. D.; Wong, W. L. T.; Rowland, A. M.; Kotts, C. E.; Carver, M. E.; Yang, M.; Bourell, J. H.; Shepard, H. M.; Henner, D.; *Biotechnology* **1992**, *10*, pp. 163-167.

28. Lyles, M. M.; Gilbert, H. F.; *Biochemistry* **1991**, *30*, pp. 619-625.
29. Lang, K.; Schmid, F. X.; Fischer, G.; *Nature* **1987**, *329*, pp. 268-270.
30. Lang, K.; Schmid, F. X.; *Nature* **1988**, *331*, pp. 453-455.
31. Schönbrunner, E. R.; Schmid, F. X.; *PNAS* **1992**, *89*, pp. 4510-4513.
32. Buchner, J.; Schmidt, M.; Fuchs, M.; Jaenicke, R.; Rudolph, R.; Schmid, F. X.; Kiefhaber, T.; *Biochemistry* **1991**, *30*, pp. 1586-1591.
33. Martin, J.; Langer, T.; Boteva, R.; Schramel, A.; Horwich, A L.; Hartl, F.-U.; *Nature* **1991**, *352*, pp. 36-42.
34. Miller, D. M.; Kurzban, G. P.; Mendoza, J. A.; Chirgwin, J. M.; Hardies, S. C.; Horowitz, P. M.; *Biochim. Biophys. Acta* **1992**, *1121*, pp. 286-292.
35. Zhi, W.; Landry, S. J.; Gierasch, L. M.; Svere, P. A.; *Protein Sci.* **1992**, *1*, pp. 522-529.
36. Sawano, H.; Kuomoto, Y.; Ohta, K.; Sasaki, Y.; Segawa, S.; Tachibana, H.; *FEBS Lett.* **1992**, *303*, pp. 11-14.
37. Zardeneta, G.; Horowitz, P. M.; *J. Biol. Chem.* **1992**, *267*, pp. 5811-5816.
38. Hagen, A.; Hatton, T. A.; Wang, D. I. C. *Biotech. Bioeng.* **1990**, *35*, pp. 966-975.
39. Tandon, S.; Horwitz, P. M. *J. Biol. Chemistry* **1986**, *261*, pp. 15615-15618.
40. Tandon, S.; Horwitz, P. M. *J. Biol. Chemistry* **1987**, *262*, pp. 4486-4491.
41. Puri, N. K.; Crivelli, E.; Cardamone, M.; Fiddes, R.; Bertolini, J.; Ninham, B.; Brandon, M. R.; *Biochem. J.* **1992**, *285*, pp. 871-879.
42. Cleland, J. L.; Hedgepeth, C.; Wang, D. I. C.; *J. Biol. Chem.* **1992**, *267*, pp. 13327-13334.
43. Cleland, J. L.; Builder, S. E.; Swartz, J. E.; Winkler, M.; Chang, J. Y.; Wang, D. I. C.; *Biotechnology* **1992**, *10*, pp. 1013-1019.
44. Cleland, J. L.; Wang, D. I. C.; In *Biocatalyst Design for Stability and Specificity*; Himmel, M., Ed.; ACS Symposium Series; American Chemical Society, Washington, D.C., Chapter 12, *in press*.
45. Consler, T. G.; Lee, J. C.; *J. Biol. Chem.* **1988**, *263*, pp. 2787-2793.
46. Carlson, J. D.; Yarmush, M. L.; *Biotechnology* **1992**, *10*, pp. 86-91.
47. Arawaka, T.; Timasheff, S. N.; *Biochemistry* **1982**, *21*, pp. 6536-6544.
48. DeBernardez-Clark, E.; Georgiou, G.; In *Protein Refolding*; Georgiou, G.; DeBernardez-Clark, E., Eds.; ACS Symposium Series, Vol. 470; American Chemical Society, Washington, D.C., 1991; pp. 1-20.
49. Cleland, J. L.; Wang, D. I. C.; In *Biotechnology, Second Edition, Volume 3: Bioprocessing (Reaction and Process Engineering)*; Stephanopoulos, G. N., Ed.; VCH Publishers, Germany, Chapter 23, *in press*.
50. Bochkareva, E. S.; Lissan, N. M.; Flynn, G. C.; Rothman, J. E.; Girshovich, A. S.; *J. Biol. Chem.* **1992**, *267*, pp. 6796-6800.
51. Sherman, M. Y.; Goldberg, A. L.; *Nature* **1992**, *357*, pp. 167-169.
52. Benner, S. A.; *Curr. Opin. Struct. Biol.* **1992**, *2*, pp. 402-412.
53. Shortle, D.; Chan, H. S.; Dill, K. A.; *Protein Sci.* **1992**, *1*, pp. 201-215.
54. Lim, W. A.; Sauer, R. T.; *J. Mol. Biol.* **1991**, *219*, pp. 359-376.
55. Eriksson, A. E.; Baase, W. A.; Zhang, X.-J.; Heinz, D. W.; Blaber, M.; Baldwin, E. P.; Matthews, B. W.; *Science* **1992**, *255*, pp. 178-183.
56. Heinz, D. W.; Baase, W. A.; Matthews, B. W.; *PNAS* **1992**, *89*, pp. 3751-3755.

57. Quax, W. J.; Mrabet, N. T.; Luiten, R. G. M.; Schuurhuizen, P. W.; Stanssens, P.; Lasters, I.; *Biotechnology* **1991**, *9*, pp. 738-742.
58. Dordick, J. S.; *Biotechnol. Prog.* **1992**, *8*, pp. 259-267.
59. Kuntz, I. D.; *Science* **1992**, *257*, pp. 1078-1082.
60. Saragovi, H. U.; Greene, M. I.; Chrusciel, R. A.; Kahn, M.; *Biotechnology* **1992**, *10*, pp. 773-778.
61. DeVos, A. M.; Ultsch, M.; Kossiakoff, A. A.; *Science* **1992**, *255*, pp. 306-312.

RECEIVED January 4, 1993

IN VIVO PROTEIN FOLDING

Chapter 2

Amino Acid Sequence Determinants of Polypeptide Chain Folding and Inclusion Body Formation

Jonathan King, Cameron Haase-Pettingell, Carl Gordon, Susan Sather, and Anna Mitraki

Department of Biology, Massachusetts Institute of Technology, Cambridge, MA 02139

Newly synthesized polypeptide chains within cells pass through a series of partially folded intermediates in reaching their native state. Association of these intermediates into the aggregated inclusion body state often competes with productive folding into the native state, in both prokaryotic and eukaryotic cells. For the thermostable tailspike of phage P22, a thermolabile early folding intermediate switches from the productive pathway to the inclusion body pathway with increasing temperature. Two classes of amino acid substitutions influence the behavior of these intracellular folding intermediates. Temperature sensitive folding (tsf) mutations destabilize the critical intermediates in the chain folding pathway at elevated temperatures, resulting in their polymerization into the aggregated inclusion body state, a kinetic trap for the chains. Global suppressors of the tsf mutations inhibit chain entry into the inclusion body pathway. Neither the tsf or global suppressor substitutions alter the activity or stability of the native state once correctly folded. For tsf mutations in the major coat protein, overexpression of the GroEL/S chaperonin efficiently suppresses the tsf folding defects. The isolation of global suppressor mutations, and overexpression of chaperonins, may be general strategies for solving problems of protein expression and protein recovery in biotechnology.

The passage of newly synthesized polypeptide chains from the poorly defined conformation on the ribosome to the fully native functional state is emerging as a complex multi-step pathway (*1, 2, 3*). In eukaryotic cells, chains destined for export pass through a variety of cellular compartments including the endoplasmic reticulum, golgi, and secretion vesicles. Other chains must pass from the cytoplasm into cellular organelles such as the mitochondria with its double membrane, the nucleus, and chloroplasts in plants.

Folding intermediates have to solve a variety of problems; they must pass through channels, avoid aggregated states, avoid proteolysis, reach particular cellular compartments, transiently bind to chaperones and other auxiliary proteins, and finally reach the native state. In fact, folding intermediates generally have properties quite distinct from their own native states, in terms of solubility, stability, half lives, interactions with cellular components such as chaperones or signal recognition particles (4, 5, 6). As a result there is no a priori reason to assume that the conformation of folding intermediates will represent simply a subset of the native conformation.

The generation of misfolded, generally aggregated, chains has emerged as a practical problem both in the biotechnology industry and in biomedical research in general. It has been particularly, though not exclusively, associated with expression of cloned genes in heterologous hosts (7, 8, 9). The aggregated inclusion body state resulting from the association of folding intermediates differs from the precipitated state of native proteins (8, 9, 10). Both appear to be insoluble, but precipitates generated by exceeding solubility limits or salting out return to solution upon dilution, whereas inclusion bodies cannot be solubilized by dilution.

The inclusion body problem can often be circumvented by exporting the newly synthesized chains outside the cytoplasm, limiting their spatial concentration. However, this is less feasible for oligomeric proteins, where formation of the product may require maintenance of high chain concentrations (11, 12, 13).

Protein Folding Intermediates and Inclusion Body Formation

The aggregation of polypeptide chains during *in vitro* refolding was initially carefully studied by Michel Goldberg (14) and Rainer Jaenicke and their colleagues, (11, 12), and more recently by Brems (15, 16) and Cleland and Wang (17, 18). They found that aggregates were not derived from the native or denatured form of the protein, but from folding intermediates in the pathway. The aggregated state represents a kinetic trap for a folding intermediate, which otherwise would proceed to its native state.

In vitro aggregation and *in vivo* inclusion body formation are similar phenomena, originating from the association or polymerization of specific folding intermediates (6). For cloned genes expressed in a heterologous intracellular environment, the failure to obtain native protein in the heterologous environment may derive from the instability of folding intermediates which have evolved for a particular intracellular environment. The association reaction seems to be highly selective, since aggregates contain largely the overexpressed protein, despite the presence of numerous species of partially folded polypeptide chains in the intracellular environment. This selectivity must reflect an aggregation mechanism that proceeds through specific interactions between folding intermediates and therefore is in some sense encoded in the primary amino acid sequence.

Chaperones and Inclusion Body Formation

In prokaryotic cells many of the chaperones were identified due to their induction under heat shock or other stress conditions (3, 6, 19-23). However, they are essential proteins for the organisms and, thus, in some cases are not auxiliary functions that can be dispensed with. In eukaryotic cells many of the chaperones play essential functions, such as mitochondrial import (3, 6), maturation of tubulin (24) and visual pigment formation (25). Recent studies of chaperone function *in vitro* indicate that GroEL recognizes folding intermediates that channel into the aggregation pathway (21, 22). The chaperonins transform these structures to a conformation that is past the off pathway junction, releasing them in a state that proceeds efficiently to the native state.

Amino Acid Sequences Influencing Chain Folding and Aggregation

Many sources of information -prosequences, signal sequences, cofactors, auxilliary proteins - are drawn upon during polypeptide chain folding. Nonetheless, the primary amino acid sequence remains the predominant determinant of native protein structure. We have been interested in amino acid sequence information controlling the conformation and properties of the folding intermediates, as opposed to those residues and sequences which stabilize the native state.

The production of structural subunits of virions in heterologous hosts has been a critical aspect in the development of subunit vaccines. Two model systems for investigating the folding of structural proteins of viruses in bacteria are the tailspike and coat proteins of phage P22. The thermostable tailspike endorhamnosidase of bacteriophage P22 has provided an experimental system for investigating intracellular folding intermediates and inclusion body formation (26 - 41). More recently, we have extended the methodology developed for the study of the tailspike to the major coat protein of P22, the product of gene 5 (42, 43, 44).

Folding Pathways for the P22 Tailspike Endorhamnosidase

Each tailspike is a trimer of the 666 amino acid chain encoded by gene 9 of P22. Six tailspikes assemble onto the phage capsid to form the cell recognition and attachment apparatus of the phage. The native protein is thermostable, with a T_m of 88° C, and is resistant to SDS and proteases (29, 30). The newly synthesized chains fold and assemble in the host cytoplasm without undergoing any known covalent modifications. During the *in vivo* folding pathway (Figure 1), the chain passes through single chain and triple-chain defined folding intermediates that are sensitive to heat, SDS, and proteases (4, 31, 32). Both species can be trapped in the cold and detected by native gel electrophoresis (4, 32). The SDS-resistance of the native state allows unambiguous identification of the mature polypeptide chains in SDS gels and thus makes it possible to follow the intracellular state of newly synthesized tailspike chains as shown in Figure 2 (4, 27, 32). No intermediate in the tailspike pathway corresponds to a native monomer. Because interchain interactions are required for the native structure, the kinetic species are in different conformations from their own native state.

Robert Seckler and coworkers have studied the refolding of urea-denatured tailspike chains by fluorescence and other direct measurements (34, 35). The folding pathway followed by chains refolding from the fully denatured state proceeds similarly to that *in vivo*, passing through single chain and triple chain partially folded

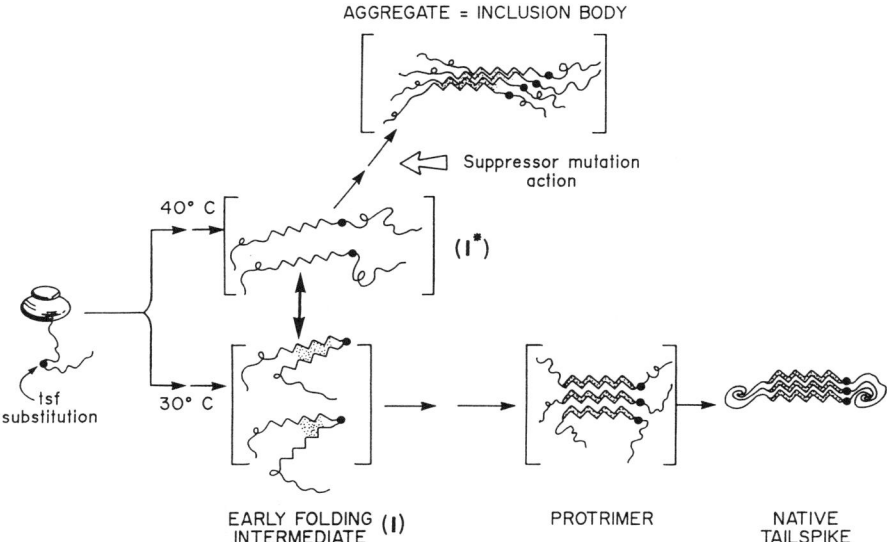

Figure 1: The folding and association pathway of the P22 tailspike. After release from the ribosome, an early single chain folding intermediate [I] is formed, which further proceeds to a species competent for chain-chain association. These chains associate to form the protrimer, in which the chains are associated but not fully folded. The protrimer folds further to form the very thermostable and SDS resistant native tailspike. The critical folded intermediate is thermolabile even in the wild type folding pathway, melting to an [I*] species which proceeds to the aggregated inclusion body state at the higher temperature range of growth. The dark dot symbolizes a temperature sensitive folding mutation, which prevents the intermediates from proceeding through the productive pathway at higher temperatures *(26)*. Adapted from *(40)*.

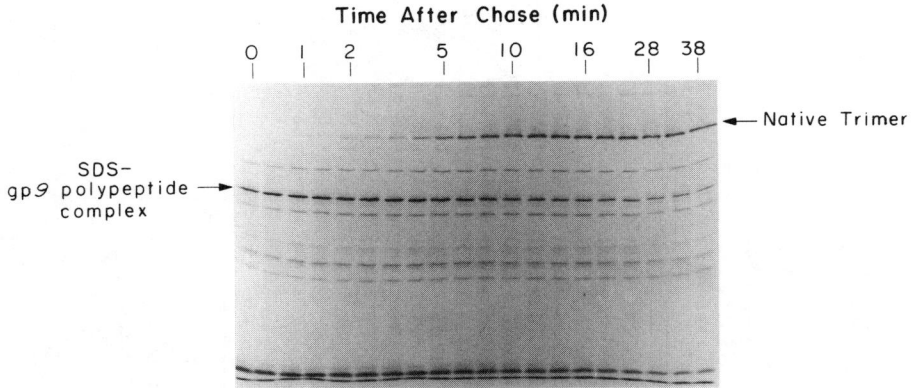

Figure 2: The time course of the in vivo folding and assembly of newly synthesized tailspike polypeptide chains into the SDS resistant native tailspike. Salmonella typhimurium cells were infected at permissive temperature with P22 phage defective in capsid assembly to allow accumulation of higher than normal amounts of soluble native tailspike protein. The culture was labeled 60 min after infection with a two minute pulse of ^{14}C amino acids, and samples were taken at various times and frozen in the presence of SDS. The sample were thawed, promoting lysis, and the proteins were separated by electrophoresis through a SDS polyacrylamide gel. The gel was dried and applied to X-ray film. All the partially folded non-native species are denatured in SDS without heating and migrate as SDS polypeptide complexes. The native trimeric tailspike binds little detergent and, thus, migrates more slowly through the gel *(26,27,32)*. Adapted from *(4)*.

intermediates (*34, 35*). Thus, the tailspike provides one of the very few systems where the *in vivo* pathway off the ribosome can be compared with the *in vitro* refolding pathway out of denaturant.

At low temperature, (30° C) almost 100% of the newly synthesized folding intermediates form the native trimer. As the temperature of folding increases, the early partially folded intermediate partitions between the native pathway and an aggregation pathway leading to inclusion bodies (*27, 31*). At 39° C, about 20% of the newly synthesized polypeptide chains reach the mature form, while the remainder partitions to the inclusion body. The temperature dependence of this partitioning suggests that an early intermediate in the pathway is thermolabile and is being partially denatured within the cytoplasmic environment (Figure 1). The thermal denaturation of folding intermediates, which are generally much less stable than their own native states, is probably one of the major problems to which the inducible heat shock chaperones respond.

If cells containing native tailspikes produced at permissive temperatures *in vivo* are shifted up to restrictive temperatures, the mature tailspikes stay SDS resistant (*30*). Thus once the native form is attained, its stability within the cell is not altered at higher temperatures. The aggregated state is not the product of the intracellular denaturation of the native protein, but rather the result of polymerization of a folding intermediate. If chains that have been synthesized at high temperatures are shifted to low temperature early enough, they can reenter the productive pathway (*28, 31*). However, once aggregated the chains cannot be recovered by lowering the temperature. These results indicate that aggregation is a conformational trap for folding intermediates that can be kinetically avoided. In the scheme below the step induced by higher temperature is modeled as the conversion from the productive intermediate I to the species I* which prefers the aggregation pathway. Temperature and chain conformation, rather than chain concentration, are critical in determining the pathway tanek by the nascent chain.

$$[I^*] \rightarrow \text{aggregates}$$
$$\cdot$$
$$\cdot$$

Nascent chain --> [newly synthesized chain] <--> [I] <--> [Ipt] <--> pt -->Trimer

where pt refers to the protrimer.

Temperature Sensitive Folding Mutations Identifying Critical Residues for Tailspike Folding

Temperature sensitive folding (tsf) mutations alter the folding pathway without influencing the properties of the native form (*26, 27*). The tsf polypeptide chains reach the native state at permissive temperatures *in vivo*, but fail to reach it at high temperatures (*28, 32*). The native state of the mutant proteins, once formed at permissive temperature, has similar thermal stability and biological activity to the wild type (24, 28, 29). Thus, the residues at the sites of these mutations make very little if any contribution to the stabilization or function of the native state. They behave as if they destabilize the already thermolabile single chain folding intermediate. They represent a class of residues within polypeptide chains, not crucial for the maintenance of the native form of the protein, but making a significant contribution to the stability and/or kinetic fates of folding intermediates.

Since the native state of the tsf mutant polypeptide chains is fully stable at elevated temperatures, the failure of the chains to fold into that state at elevated temperatures must reflect a pathway that is kinetically rather than thermodynamically controlled (30).

The mutations are located at more than 30 sites in the central region of the polypeptide chain, with the local sequences resembling those associated with surface beta turns (36, 37, 42). These sites apparently are kinetically important in directing turns and/or stabilizing a critical folding intermediate in the beta sheet fold at high temperature. The surface location of these sites explains the tolerance for the mutant substitutions in the native conformation, for example arginines for glycines (37). Thus, selection of tsf mutations detects surface sites that are kinetically important in chain folding.

Temperature Sensitive Folding Mutants of the Major Coat Protein.

The folding of another protein, the major coat protein of phage P22, can also be studied using a mutational approach. The major coat protein of P22 is a 47,000 D chain encoded by gene 5 of P22. About 420 coat molecules copolymerize with approximately 300 scaffolding molecules to form the procapsid of the phage (43). This packages the DNA and transforms to the mature capsid. Temperature sensitive mutations have been mapped to 17 sites in the chain (44). At low temperature these chains fold up and form the procapsid and mature capsid of infectious virons. At high temperature the chains fail to reach the folded monomer state as indicated by the lack of recognition by the scaffolding subunits (44). The chains accumulate in an inclusion body state presumably derived from a folding intermediate.

Global Suppressors of Protein Folding Defects and Inclusion Body Formation

Starting with temperature sensitive folding mutants of tailspike, a second set of mutants was selected, which alleviated the folding defects of the starting alleles (38). A group of these second site suppressor mutations mapped within the gene coding for the tailspike. Two nearby substitutions in the center of the polypeptide chains, Val331>Ala and Ala334>Val, were able to suppress diverse tsf and absolutely defective amino acid substitutions spanning over 200 residues in the primary sequence (39). Figure 3 shows the intracellular folding of wild type, tsf, and su/tsf polypeptide chains. Whereas native tailspikes were not detected in the tsf infected cells, a significant fraction of the chains matured to the native trimer in the double mutant infected cells.

The strains containing the suppressor mutations by themselves were not defective in any of their physiological functions. The analysis of the purified

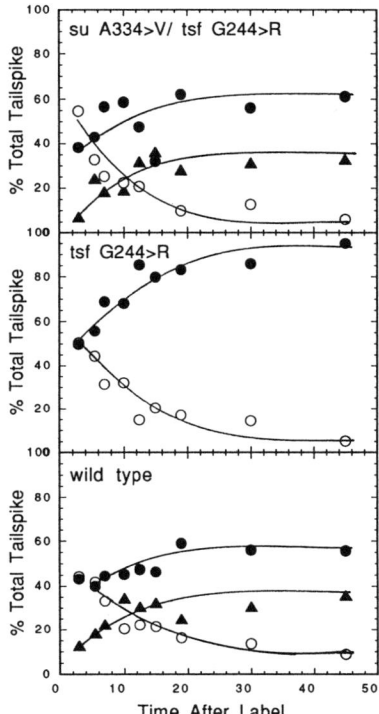

Figure 3: Kinetics of the intracellular folding and aggregation at 39°C of the tsfG244>R, suA334>V/tsfG244>R and wild type polypeptide chains. Infected cells were pulse labeled with ^{14}C amino acids at 39°C. Samples were taken at different times, concentrated, and lysed by freezing and thawing. Samples were centrifuged, and pellet and supernatant fractions were mixed with SDS sample buffer and separated by SDS-PAGE. The gel was dried and applied to X-ray film. The resulting tailspike bands were quantified with a microdensitometer. (▲) Native tailspike, (O) Folding intermediates, (•) Aggregated species or Inclusion body.

suppressor proteins, alone or in combination with a tsf mutation, showed that their thermal stability and activity were not altered with respect to wild type. The chains carrying only the global suppressor mutations (Figure 4) mature more efficiently than wild type at high temperatures, suggesting that they might be "super-folding" substitutions. However, examination of the kinetics of intracellular chain folding revealed that the suppressors substitutions did not change the rate of the intermediate to mature steps in chain folding. Rather they inhibited the loss of folding intermediates to the kinetically trapped inclusion body state (40). Of additional practical significance is the absence of effects on native function or stability of the suppressor substitutions

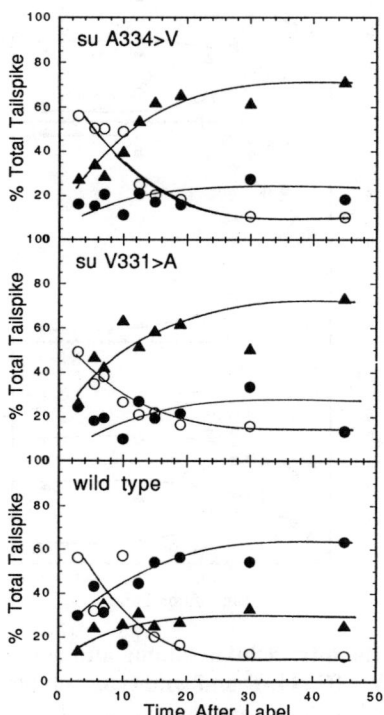

Figure 4: Kinetics of intracellular folding and aggregation at 39°C of the su V331>A, su A334>V, and wild type polypeptide chains. The protocol was the same as in figure 4. (▲) Native tailspike; (○) Folding intermediates; and (•) Aggregated species = Inclusion bodies

The local sequence surrounding the suppressor region is as follows:

Ser-Tyr-Gly-Ser-**Val**-Ser-Ser-**Ala**-Gln-Phe-Leu-Arg
331 334

It is difficult to imagine how a single amino acid substitution could alter the aggregation behavior of a 666 amino acid chain, if that process were "nonspecific". Site specific mutagenesis studies by Myeong-Hee Yu and coworkers (41) show that the suppressor phenotype is very specific to the substitution with only Ala or Gly at site 331 and Val or Ile at 334 functioning as suppressors. The observations support models in which inclusion body formation is a rather specific off-pathway polymerization process involving domains of folding intermediates that would normally productively interact in the intradomain mode (Figure 1).

Three general models seem possible: 1) the suppressors identify the site on the folding intermediate that initiates the association reaction, 2) the mutations stabilize the productive conformation of the folding intermediate(s) that the tsf mutations are destabilizing or, 3) the suppressors could create a site for chaperone recognition or binding that is absent from the wild type chain.

Regardless of the mechanism this kind of suppressor sequences identifies an additional class of sequence information for the folding of polypeptide chains: sequences that ensure passage of the chain through the productive pathway, by preventing off-pathway traps.

Mutational Suppression of Inclusion Body Formation in Eukaryotic Proteins

Mutations that alter inclusion body formation have been described recently for a number of systems. Ronald Wetzel and colleagues have reported that subtle amino acid substitutions can either decrease or increase inclusion body formation for interferon and interleukin 1, without affecting biological activity (45). The authors have also developed screening methods that allow systematic mutational analysis. The same kind of mutations were found by Rinas et al. (46) on basic fibroblast growth factor. The existence of a global second site suppressor of temperature sensitive mutants has also been reported for the human receptor-like protein tyrosine phosphatase (47). This mutation reduced the amount of protein that was intracellularly accumulating in inclusion bodies.

Suppression of Inclusion Body Formation by Overexpression of Chaperonins.

LaRossa, Gatenby and Van Dyk reported that introduction of plasmids over-expressing groE chaperonins suppressed temperature sensitive mutants in a variety of Salmonella genes, as well as P22 tailspike mutants (48). The effects reported for the tailspike tsf mutants were quite small. Further investigation of this suppression has led us to conclude that the tailspike tsf chains are not being rescued by overexpression of the groE chaperonin. During *in vitro* refolding the groE chaperonin binds to an early tailspike intermediate but does not suppress the folding defect (49).

In contrast, the tsf mutations in gene 5 are efficiently rescued. As shown in Table 1, plating the gene 5 tsf mutants on strains overproducing the chaperonin brings the yield of phage almost back to wild type levels. Though the heat shock operon is induced at 39°C in Salmonella without the plasmid, we suspect that these normal chaperonin levels are inadequate to cope with the very high concentration of misfolded coat chains in the ts mutant infected cells. The chaperonin likely acts on the same intermediate affected by the coat protein tsf mutations (Figure 5).

Much more work will be needed to identify the actual conformational features of intermediates that are critical in the choice to proceed down the productive pathway, or to switch to the aggregation pathway. These features are encoded in the amino acid sequence, and therefore subject to mutational modification.

Table 1. Effect of Overexpressing GroEL and GroES on P22 Phage Carrying Temperature-Sensitive Mutations in the Coat Protein and Tailspike Protein

P22 Strain	S. typhimurium Cell Host with Plasmid	Efficiency of Plating 30 °C[1]	Efficiency of Plating 39 °C[1]
Wild-type	pBR322	1.0	0.9
	pGroELS	1.0	0.9
Coat ts: Phe353›Leu	pBR322	1.1	3.6x10^{-6}
	pGroELS	1.1	1.0
Coat ts: Gly282›Asp	pBR322	1.0	9.1x10^{-6}
	pGroELS	1.0	1.0
Tailspike ts: Gly435›Glu	pBR322	0.9	6.3x10^{-7}
	pGroELS	1.1	6.3x10^{-7}
Tailspike ts: Glu309›Val	pBR322	1.1	1.9x10^{-2}
	pGroELS	0.9	1.4x10^{-2}

All host cells were derived from *S. typhimurium* strain DB7136. Cells used in the plating experiments were grown to 5x10^8/ml and concentrated ten-fold, before being mixed with phage in soft agar. pGroELS cells were constructed by transforming DB7136 *Salmonella* cells with pOF39, which contains the *E. Coli groE* operon (Fayet *et al*, 1986). pBR322 cells were transformed with a control pBR322 plasmid.

[1]Efficiency of plating is the ratio of the number of plaques seen when phage were plated on cells carrying either of the two plasmids at the specified temperature to the number of plaques seen when phage were plated at 30 °C on DB7136.

Reference:

Fayet, O., Louarn, J.-M., and Georgopoulos, C. (1986) *Mol Gen Genet* **202**, 435-445

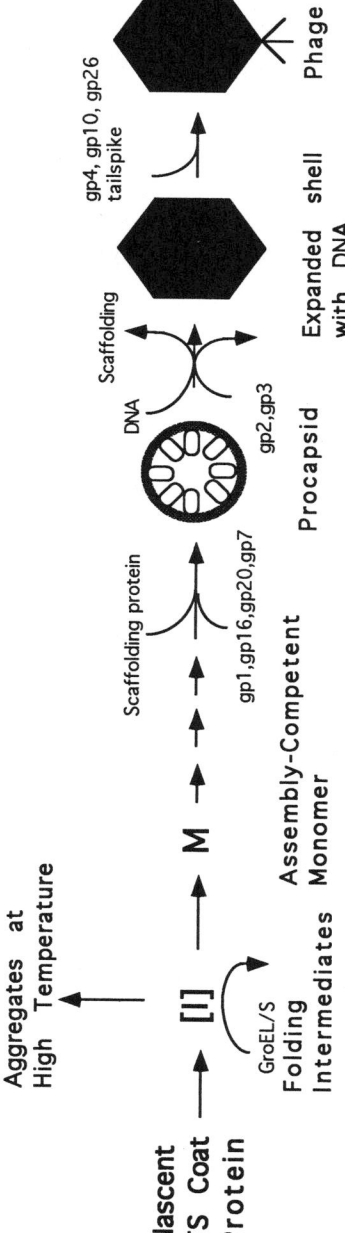

Figure 5: The pathway for the maturation and assembly of the major coat protein into the procapsid and mature virion. The temperature sensitive folding mutations prevent the folding intermediate from proceeding to the native monomer. Instead the chains channel to the inclusion body state. This reaction is blocked by overexpression of the GroE chaperonins.

Acknowledgements: We thank Cynthia Woolley for her skillful assistance with this manuscript. This work was supported by grant GM17,980 from the National Institutes of Health and NSF Grant ECD 8803014. Susan Sather was supported by an NIH Biophysical Chemistry Training Grant and Carl Gordon by a Howard Hughes Predoctoral Fellowship.

Literature Cited

1. Creighton, T.E. *Prog. Biophys. Mol. Biol.* **1978**, *33*, 231-298.
2. Kim, P.S. and Baldwin, R.L. *Ann. Rev. Biochem.* **1990**, *59*, 631-660.
3. Langer, T., Lu, C., Echols, H., Flanagan, J., Hayer, M.K., and Hartl, F-U. *Nature* **1992**, *356*, 683-6892.
4. Goldenberg, D. and King, J., *Proc. Natl. Acad. Sci. USA* **1982**, *79*, 3403-3407.
5. Brems, D. N. *Biochemistry* **1988**, *27*, 4541-4546.
6. Ostermann, J., Horwich., A., Neupert, W. and Hartl F-U., *Nature* **1989**, *341*, 125-130
7. Marston, F. A. O. *Biochem. J.* **1986**, *240* 1-12.
8. Mitraki A.; King J. *Bio/Technology* **1989**, *7*, 690-697.
9. De Bernardez-Clark, E. and Georgiou, G. In *Protein Refolding*; De Bernardez-Clark and G. Georgiou, Eds.; ACS Symposium Series 470; American Chemical Society: Washington, D.C., 1991; pp 1-20.
10. Schein, C. H. *Bio/Technology* **1990**, *8*, 308-317.
11. Zettlmeissl, G.; Rudolph, R.; Jaenicke, R. *Biochemistry* **1979**, *18*, 5567-5571.
12. Rudolph, R.; Zettlmeissl, G.; Jaenicke, R. *Biochemistry* **1979**, *18*, 5572-5575.
13. Teschke, C.M. & King, J. *Current Opinion in Biotechnology* **1992**, *333*, 468-473.
14. London J.; Skrzynia C.; Goldberg M. 1974. *Eur. J. Biochem.* **1974**, *47*, 409-415.
15. Brems D. N. *Biochemistry* **1988**, *27*, 4541-4546.
16. Brems, D. N.; Plaisted, S. M.; Kauffman, E. W.; Havel, H. A. *Biochemistry* **1986**, *25*, 6539-6543.
17. Cleland, J.L.; Wang, D.I.C. *Bio/Technology* **1990**, *8*, 1274-1278.
18. Cleland, J.L.; Wang, D. *Biochemistry* **1990**, *29*, 11072-11078.
19. Hemmingsen, S. M.; Woolford, C.; van der Vries, S. M.; Tilly, K.; Dennis, D. T.; Georgopoulos, C. P.; Hendrix, R. W.; Ellis, R. J. *Nature* **1988**, *333*, 330-334.
20. Bochkareva, E. S.; Lissin, A. S.; Girshovich, A. S. *Nature* **1988**, *336*, 254-257.
21. Goloubinoff, B.; Gatenby, A.A; Lorimer, G. *Nature* **1989**, *337*, 44-47.
22. Goloubinoff, P.; Christeller, J.T.; Gatenby, A. A.; Lorimer, G. H. *Nature* **1989**, *342*, 884-889.
23. Randall, L. L.; Hardy, S. J. S.; Thom, J. A. *Ann. Rev. Microbiol.* **1987**, *41*, 507-541.

24. Yaffe, M.B.;Farr, G. W.;Miklos, D.; Horwich, A. L.; Sternlicht, M.L.; Sternlicht, H. *Nature* **1992**, *358*, 245-248.
25. Colley, N. J.; Baker, E. K.; Stamnes, M.A.; Zuker, C.S. *Cell* **1991**, *67*, 255-263.
26. King, J.; Fane, B.; Haase-Pettingell, C.; Mitraki, A.; Villafane, R.; Yu, M-H. In *Protein Folding, Deciphering the second half of the genetic code*; Gierasch, L.M. and King, J., Eds.; A.A.A.S.: Washington D.C., 1990; pp. 225-239.
27. Goldenberg, D. P.; Berget, P. B.; King, J. *J. Biol. Chem.* **1982**, *257*, 7864-7871.
28. Smith, D.H.; King, J. *J. Mol. Biol.* **1981**, *145*, 653-676.
29. Goldenberg, D.; King, J. *J. Mol. Biol.* **1981**, *145*, 633-651.
30. Sturtevant, J.; Yu, M-h; Haase-Pettingell, C.; King, J. *J. Biol. Chem.* **1989**, *264*, 10693-10698.
31. Haase-Pettingell, C.; King J. *J. Biol. Chem.* **1988**, *263*, 4977-4983.
32. Goldenberg, D.P.; Smith, D.H.; King, J. *Nat. Acad. Sci. USA* **1983**, *80*, 7060-7064.
33. Thomas, G. J. Jr.; Becka, R.; Sargent, D.; Y, M-H.; King J. *Biochemistry* **1990**, *29*, 4181-4187.
34. Seckler, R. Fuchs, A., King, J. and Jaenicke, R. *J. Biol. Chem.* **1989**, *264*, 11750-117535.
35. Fuchs, A., Seiderer, C., and Seckler, R. *Biochemistry* **1991**, *30*, 6598-6604.
36. Yu, M.-H.; King, J. *Proc. Natl. Acad. Sci. USA.* **1984**, *81*, 6584-6588.
37. Yu, M.-H.; King, J. *J. Biol. Chem.* **1988**, *263*, 1424-1431.
38. Fane, B. and King, J. *Genetics* **1991**, *127*, 263-277.
39. Fane, B.; Villafane, R.; Mitraki, A.; King, J. *J. Biol. Chem.* **1991**, *266*, 11640-11648.
40. Mitraki, A., Fane, B., Haase-Pettingell, C., Sturtevant, J. and King, J. *Science* **1991**, *253*, 54-58.
41. Lee, S.C., Koh, H. and Yu, M.-H. *J. Biol. Chem.* 1991, *266*, 23191-23196.
42. Villafane, R. and King, J. *J. Mol. Biol.* **1988**, *204*, 607-619.
43. Prevelige, P.E.;Thomas, D.; King, J. *J. Mol. Biol.* **1988**, *202*, 743-757.
44. Gordon, C. and King, J. *J. Biol. Chem.* **1993**, in press.
45. Wetzel, R., Perry, L.J. and Vielleux, C. *Bio/Technology* **1991**, *9*, 731-737.
46. Rinas, U.;Tsai, L.B.; Lyons, D.; Gox, G.M.; Stearns, G.; Fieschko, J.; Fenton, D. and Bailey, J.E. *Bio/Technology* **1992**, *10*, 435-440.
47. Tsai, A.Y.M.; Itoh, M.; Streuli, M.; Thai, T. and Saito, H. *J. Biol. Chem.* **1991**, *266*, 10534-10543
48. VanDyk, T.K.; Gatenby, A.A. and LaRossa, T.A. *Nature* **1989**, *342*, 451-453.
49. Brunschier, R.; Danner, M.; Seckler, R. *J. Biol. Chem.* **1993**, in press.

RECEIVED January 22, 1993

Chapter 3

Duplicated Segment of *Clostridium thermocellum* Cellulases

K. Tokatlidis[1,2], Sylvie Salamitou[1], Tsuchiyoshi Fujino[1,3], P. Béguin[1], Prasad Dhurjati[2], J. Millet[1], and J.-P. Aubert[1]

[1]Unité de Physiologie Cellulaire and URA 1300 Centre National de la Recherche Scientifique, Institut Pasteur, 25 rue du Dr. Roux, 75724 Paris Cedex 15, France
[2]Department of Chemical Engineering, University of Delaware, Newark, DE 19716

> All but one of the *Clostridium thermocellum* endoglucanases and xylanases sequenced to date contain a highly conserved, duplicated segment of 22 residues, which is generally located at the COOH terminus of the protein. In *Escherichia coli* clones overproducing endoglucanase CelD, this segment contributes to the formation of cytoplasmic inclusion bodies containing fully active enzyme. In *C. thermocellum*, the duplicated segment appears to anchor the various catalytic components to a large scaffolding component, leading to the formation of the high molecular weight complex, termed cellulosome, that is responsible for crystalline cellulose degradation.

Clostridium thermocellum is a Gram-positive, spore-forming bacterium with an optimal growth temperature of about 60°C. The thermostable cellulase system produced by this bacterium has a very high specific activity towards cotton, a form of cellulose that is most recalcitrant to enzymatic hydrolysis due to its high degree of crystallinity (*1*).

The cellulase system of *C. thermocellum* consists of an extracellular multienzyme complex of about 2-4 MDa (*2, 3*), termed cellulosome (*4*), which is composed of at least 14-18 different types of subunits, ranging from 40 to 250 kDa (*2*). Many of the components are endowed with endoglucanase or xylanase activity. A notable exception is the largest component, termed S1, a 210-250 kDa glycoprotein devoid of catalytic activity.

A number of *C. thermocellum* genes involved in cellulose or hemicellulose hydrolysis have been cloned and expressed in *Escherichia coli*. These include fifteen distinct endoglucanase (*cel*), two xylanase (*xyn*), and two ß-glucosidase (*bgl*) genes (*5-7*). Eight of the *cel* genes, one of the *xyn* genes, and the two *bgl* genes have been sequenced (reviewed in (*8*); (*9-11*), and unpublished data). The encoded endoglucanases and the xylanase contain catalytic domains that are representative of four of the seven broad cellulase and xylanase families defined by hydrophobic cluster analysis (*12*). Furthermore, most of them share a conserved region of about 65 residues, usually located at the COOH end of the proteins. The region contains two homologous segments of 22 residues each linked together by 9-15 amino acids (Figure 1). It is often, but not always, separated from the rest of the protein by a sequence rich in proline and hydroxy amino acids. The duplicated segment is not directly involved in

[3]On leave from Nagoya Seiraku Company Ltd., Nagoya, Japan

```
CelA 416  G DVNGDGNVNSTDLTMLKRYLLK SVTNINREA
CelB 501  G DVNGDGRVNSSDVALLKRYLLG LVENINKEA
CelD 584  G DVNDDGKVNSTDLTLLKRYVLK AVSTLPSSKAEKA
CelE 414  G DVNGDGKINSTDCTMLKRYILR GIEEFPSPSGIIA
CelF 669  G DVNFDGRINSTDYSRLKRYVIK SLEFTDPEEHQKFIAA
CelG 502  G DINSDGNVNSTDLGILKRIIVK NPPASANMDA
CelH 831  G DLNFDNAVNSTDLLMLKRYILK SLELGTSEHEEKFKKA
XynZ 429  G DLNGDGNINSSDLQALKRHLLG ISPLGEALLR
OrfX      G DVNLDGQVNSTDFSLLKRYILK VVDINSINVTN

CelA 448  A DVNRDGAINSSDMTILKRYLIK SIPHLPY-COOH
CelB 533  A DVNVSGTVNSTDLAIMKRYVLR SISELPY-COOH
CelD 620  A DVNRDGRVNSSDVTILSRYLIR VIEKLPI-COOH
CelE 450  A DVNADLKINSTDLVLMKKYLLR SIDKFPAED...
CelF 708  A DVDGNGRINSTDLYVLNRYILK LIEKFPAEQ-COOH
CelG 535  A DVNADGKVNSTDYTVLKRYLLR SIDKLPHTT-COOH
CelH 870  A DLNRDNKVDSTDLTILKRYLLY AISEIPI-COOH
XynZ 463  A DVNRSGKVDSTDYSVLKRYILR IITEFPG...
OrfX      A DMNNDGNINSTDISILKRILLR N-COOH
```

Figure 1. Alignment of the conserved, duplicated segment present in endoglucanases CelA, CelB, CelD, CelE, CelF, CelG, CelH, in xylanase XynZ, and in a putative cellulosome component encoded by an open reading frame (OrfX) located upstream of *celE*. The reiterated region is framed. Amino acids in boldface type are identical or have similar chemical properties. Numbers indicate, for each line, the position of the leftmost residue within the sequence of each protein. Similarity criteria are : V, L, I, M; R, K; D, E; N, Q; Y, F, W; S, T. (Reproduced with permission from Béguin, P., Millet, J. and Aubert, J.-P., *FEMS Microbiol. Lett.*, in press. Copyright 1992 Elsevier Science Publishers B.V.)

the catalytic function of the cellulases, since several experiments have shown that truncated enzymes, encoded by genes deprived of the duplicated segment, retained their catalytic activity towards soluble substrates (*13-15*). Likewise, removal of the duplicated segment from the sequence of endoglucanase CelE did not affect the cellulose binding affinity of the enzyme (*16*). However, the duplicated segment is present in all cellulases and xylanases known to be part of the cellulosome (*17-19*). By contrast, endoglucanase CelC, the only enzyme known not to possess the duplicated segment does not belong to the cellulosome (E.A. Bayer and R. Lamed, personal communication). These observations hinted towards it having a role in the structural organization of the cellulosome.

The first part of this chapter will review evidence that the duplicated segment is responsible for the formation of inclusion bodies containing active endoglucanase CelD when the corresponding gene, *celD*, is overexpressed in *Escherichia coli*. The second part will discuss the role of this segment in the structural organization of the cellulosome.

Role of the Duplicated Segment in the Formation of Inclusion Bodies Containing Active Endoglucanase CelD

Using recombinant DNA technology, high yields of many eukaryotic and prokaryotic proteins have been obtained in foreign hosts. However, it has often been observed, particularly in *Escherichia coli*, that recombinant gene products precipitate and form insoluble cytoplasmic inclusion bodies (*20-24*). Biochemical analysis showed that they contain the recombinant protein as a major constituent. Inclusion bodies can readily be observed by phase contrast microscopy as highly refractile granules usually located at one extremity of the cell. Electron microscopy revealed them to be amorphous masses not surrounded by a membrane (*20-27*). However, recent data by Bowden *et al.* (*28*) indicate that β-lactamase forms two kinds of inclusion bodies, found in the periplasm and in the cytoplasm, respectively. Periplasmic inclusion bodies are amorphous, whereas cytoplasmic granules appear to be more regular.

Cellular factors such as chaperonins can have an important role in controlling inclusion body formation. One recent report showed that overproduction of the molecular chaperonin DnaK reduces the partitioning of human growth hormone into inclusion bodies (*29*). The most plausible mechanism for the formation of inclusion bodies is that they are formed from intermediates of the protein folding pathway that operates *in vivo* (reviewed in (*30*), see also paper by J. King, this volume).

Recovery of the protein in a soluble form usually requires treatment with a chaotropic agent, such as urea, followed by a renaturation/refolding step during which the chaotropic agent is removed (*21, 24*). In many cases, the recovery of native proteins from inclusion bodies has proved quite difficult, due to problems in the refolding of the polypeptides obtained after dissolution of the inclusion bodies. The case of CelD is an exception. By contrast with other proteins found in inclusion bodies, the enzyme can be extracted from the inclusion body fraction in the presence of 5 M urea, and after dialysis of the urea extract, it is easily purified in a highly active form. Furthermore, the protein crystallizes readily, which suggests that it has a homogeneous and presumably native conformation (*27*). Therefore, it was of interest to investigate the basis of such a behaviour.

Activity of Insoluble CelD within Inclusion Bodies. Two explanations could account for the high activity of CelD purified from inclusion bodies. Either the enzyme was very efficiently renatured after dissolving the inclusion bodies in 5 M urea, or the enzyme was not denatured in its insoluble form nor during the urea treatment. To differentiate between these two hypotheses, the specific activity of the protein inside of the inclusion bodies was assayed directly using the small, diffusible substrate *p*-nitrophenyl-β-D-cellobioside (*p*-NPC). The specific activity was calculated after estimating the amount of antigenic CelD present in the inclusion body fraction by Western blotting using an anti-CelD rabbit polyclonal antiserum. The measured specific activity, 1.4 +/- 0.25 U/mg, was very close to that of purified CelD (1.6 +/- 0.4 U/mg) (31). Furthermore, although inclusion bodies are soluble in 5 M urea, the activity of purified CelD at room temperature was not affected by urea concentrations ranging up to 8 M. Studies based on intrinsic fluorescence emission spectra and far-ultraviolet circular dichroism confirmed the absence of detectable structural changes (32). Thus, the high specific activity of the enzyme purified from inclusion bodies is probably due to the fact that it does not acheive a denatured form at any stage.

Role of the Duplicated Segment in the Formation of Inclusion Bodies by CelD. As stated above, measurements of activity show that inclusion body formation does not entail the denaturation of the catalytic core of the enzyme, which comprises about 90 % of the polypeptide chain. However, this does not rule out that small surface perturbations not affecting activity, such as changes in the conformation of the duplicated segment, might enhance the probability of aggregation. Thus, we investigated whether the duplicated segment might be a candidate for mediating intermolecular interactions leading to the precipitation of CelD (31).

Earlier evidence had suggested that inclusion bodies were formed for CelD and XynZ only in *E. coli* clones harboring plasmids with the corresponding genes carrying the duplicated segment. Polypeptide products encoded by truncated genes devoid of the duplicated segment did not form inclusion bodies (15 and unpublished data).

As shown by Western blotting, CelD was present in inclusion bodies almost exclusively as a 68 kDa polypeptide. Truncation of the 68 kDa species occurred during the dialysis step performed after solubilizing the enzyme in urea, yielding a form of 65 kDa, which is lacking part of the duplicated segment (15). The 68 kDa form was absent from the soluble cytoplasmic fraction, which consisted entirely of the 65 kDa form, and some lower M_r degradation products.

The 68 kDa species contains the intact form of the duplicated segment, since it cross-reacted in Western blotting experiments with an antiserum directed against endoglucanase CelA, whose only sequence similarity with CelD lies in the duplicated segment carried by both enzymes. By contrast, anti-CelA antiserum did not react with the 65 kDa CelD species, nor with a 63 kDa species derived from a clone from which the whole sequence encoding the duplicated segment was deleted (31).

All these results support the hypothesis that the duplicated segment is essential for the interactions leading to inclusion body formation by CelD.

Role of the Duplicated Segment in the Structural Organization of the Cellulosome

Purification of Intact CelD from Inclusion Bodies. In order to define the properties conferred on CelD by the duplicated segment, it was essential to obtain the intact form of the enzyme. The technique recommended by Babbit and coworkers for the purification of *Torpedo californica* creatine kinase and bovine pancreatic trypsin inhibitor (33) proved most efficient to prevent proteolysis of the 68 kDa form during the extraction procedure. The 68 kDa form could be obtained at at least 95 % purity (unpublished data) by washing inclusion bodies at least twice in 0.15 M NaCl, followed by incubation for 16 h at 37 °C with agitation in the presence of 0.1 M Tris-HCl buffer, pH 7.7, containing 2.5 % n-octyl-glucoside. Presumably n-octyl glucoside solubilizes membranes cosedimenting with inclusion bodies and eliminates a membrane-bound protease that is dissoved in urea and refolded during the dialysis step. The rest of the purification was carried out as described (27). The preparation obtained had a specific activity similar to those of the 65 and 63 kDa forms of CelD purified previously (15, 27).

Presence of the Duplicated Segment in Several Cellulosome Components. The hypothesis that the duplicated segment plays a role in the organization of the cellulosome implies that it should be specific for cellulosomal components and detectable in many of these. We looked by Western blotting for the presence in cellulosomal components of epitopes cross-reacting with the duplicated segment of CelD (*34*).

Polyclonal rabbit antibodies against the 68 kDa form of CelD were obtained and preadsorbed against an excess of the 63 kDa species in order to block the recognition of epitopes located on the core enzyme. These antibodies, which still reacted with the 68 kDa species, revealed in the culture supernatant of *C. thermocellum* a set of 6-8 bands ranging in M_r between 50,000 and 100,000. All polypeptides recognized were adsorbed on a cellulose affinity column and were associated with high molecular weight complexes larger than 1 MDa, indicating that they corresponded to cellulosomal components.

Binding of CelD and XynZ Carrying the Duplicated Segment to the Largest Component of the Cellulosome. Another set of Western blotting experiments was performed in order to determine whether the duplicated segment conferred on cellulosomal proteins the capacity to bind to other cellulosomal components (*34*). ^{125}I-labelled forms CelD and XynZ carrying the duplicated segment were used as probes against cellulosomal proteins separated by SDS-PAGE and transferred onto nitrocellulose. In both cases, the same set of cellulosome-specific polypeptides ranging between 95 and 250 kDa was revealed. The most prominent band corresponded to the large, non-catalytic subunit S1. Control experiments performed under the same conditions with forms of CelD and XynZ devoid of the duplicated segment failed to reveal any band, even after 5-fold longer exposure of the blots. These results suggest that the duplicated segment serves to anchor the various cellulases and xylanases of the cellulosome to S1, which would act as a scaffolding component of the complex.

Discussion and Perspectives

The duplicated segment behaves as an independent domain that is distinct, in structure and function, from the catalytic core of cellulolytic or xylanolytic enzymes. It mediates binding to S1 of two proteins, CelD and XynZ, which are totally different in sequence, and most likely in tertiary structure. Its presence or absence does not seem to influence the catalytic properties of the enzymes tested to date. In addition, the part of the duplicated segment remaining in the 65 kDa form that was crystallized is not ordered in the crystal lattice as is the rest of the polypeptide (*35*).

The recognition of S1 by ^{125}I-labelled CelD and XynZ suggested that S1 might contain several complementary "receptors", to which the duplicated segment would bind. Cloning and sequencing of a fragment encoding part of the S1 protein showed that the S1 polypeptide does indeed contain at least three segments of about 165 residues each having sequences very similar to each other (Figure 2). These segments are responsible for binding to the duplicated segment of CelD (*36*). The cloning and sequencing of the entire S1 gene indicates that S1 contains six further receptor segments located upstream from the three segments described in (*36*) (A.L. Demain, personal communication).

Enhancement of inclusion body formation, as observed for CelD, is probably due to less specific intermolecular interactions, and may be correlated with increased hydrophobicity. In the case of CelD, ammonium sulphate precipitation of the 68 kDa form of the protein occurs at a lower concentration than for the truncated 65 kDa form (Figure 3). A decrease in the ammonium sulphate concentration required for precipitation was also observed for the intact form as compared to the truncated form of the *Clostridium cellulolyticum* endoglucanase CelCCA, which carries a duplicated

```
  1' DLDAVRIKVDTVNAKPGYTVRIPVRFTGIPSKGIANCDFVYSYDPNVLEIIEIEPGE
166' DLDAVRIKVDTVNAKPGDTVRIPVRFSGIPSKGIANCDFVYSYDPNVLEIIEIEPGD
332'    NKLTLKIGRAEGRPGDTVEIPVNLYGVPQKGIASGDFVVSYDPNVLEIIEIEPGE

 58' LIVDPNPTKSFDTAVYPDRKMIVFLFAEDSGTGAYAITEDGVFATIVRKVKSGAPNG
223' IIVDPNPDKSFDTAVYPDRKIIVFLFAEDSGTGAYAITKDGVFATIVAKVKEGAPNG
387' LIVDPNPTKSFDTAVYPDRKMIVFLFAEDSGTGAYAITEDGVFATIVAKVKEGAPEG

115' LSVIKFVEVGGFANNDLVEQKTQFFDGGVNVGDTTEPAT-PTTPVTTPTTTD
280' LSVIKFVEVGGFANNDLVEQKTQFFDGGVNVGDTTVPTTSPTTTPPEPTITP
444' FSAIEISEFGAFADNDLVEVETDLINGGVLVTN
```

Figure 2. Alignment of the three COOH-terminal segments of S1. Amino acids that are identical or have similar chemical properties are shown on a stippled background. Numbers indicate, for each line, the position of the leftmost residue relative to the beginning of the fragment whose sequence was determined in ref. (*36*).

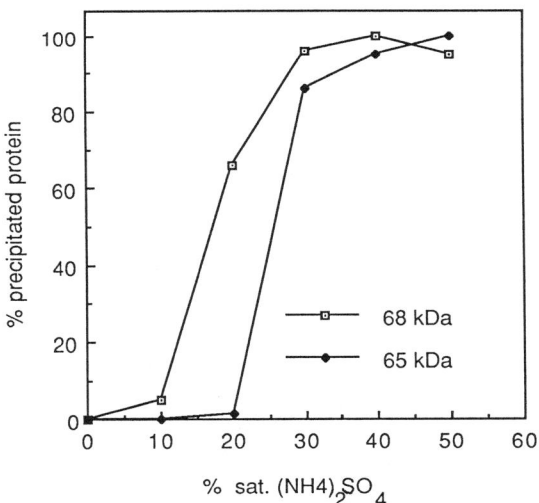

Figure 3. Ammonium sulphate precipitation curves of the 68 and 65 kDa forms of endoglucanase CelD. The proteins were incubated overnight at 4 °C in solutions prepared from saturated ammonium sulphate and an appropriate volume of 50 mM Tris-HCl buffer, pH 7.7. Precipitated material was pelleted for 15 min at 13,000 g_{max}, resuspended in water and assayed for protein using the Coomassie Blue binding assay (*45*). Values are expressed as percentage of the maximal amount precipitated.

segment similar to that found in *C. thermocellum* enzymes (*37*). Thus, the duplicated segment seems to render such proteins more hydrophobic, and so more sensitive to salting-out.

Other polypeptides have been reported to induce the formation of insoluble inclusion bodies when fused to otherwise soluble proteins. As observed for CelD, the protein in the inclusion bodies thus obtained is less susceptible to proteolysis. Among the polypeptides that lead to aggregation are the Rop protein (*38*), the TrpLE protein (*39, 40*), the C-terminal part of diphtheria toxin (*41, 42*) and prochymosin (*43, 44*). TrpLE fusion proteins are now commonly used for the production of antigenic material. However, none of these systems has been reported to produce inclusion bodies containing enzymatically active protein.

A variety of questions remain concerning the duplicated segment. First, determining the structural basis of the interaction with S1 is clearly required to understand the organization of the cellulosome. We are currently trying to crystallize the 68 kDa form of CelD, since the structural information already available on the core enzyme (*35*) should greatly facilitate the interpretation of x-ray diffraction data. Second, although the duplicated segments carried by the catalytic components are quite similar, they are not identical, and the same is true of the corresponding receptors on S1. Does this reflect recognition specificity (i.e. does each of the receptors on S1 bind a defined catalytic component preferentially), or do these components attach to S1 in a more or less random order ? Third, if the duplicated segment is grafted onto other, non-cellulosomal proteins, will these proteins be endowed with S1 binding and inclusion body formation properties ? This question is currently being investigated using CelC, a non-cellulosomal endoglucanase that does not form inclusion bodies in *E. coli*, as a test protein.

Acknowledgments. We are grateful to Pr. A.L. Demain for providing results prior to publication and Pr. M.P. Coughlan for critical reading of the manuscript. K.T. was supported by Presidential Young Investigator Grant ECE-8552492 awarded to P.D. This work was supported by grant CPL C462 from the Commission of the European Communities and by research funds from the University of Paris 7.

Literature cited

1. Johnson, E. A.; Sakajoh, M.; Halliwell, G.; Madia, A.; Demain, A. L. *Appl. Environ. Microbiol.* **1982**, *43*, 1125-1132.
2. Lamed, R.; Setter, E.; Bayer, E. A. *J. Bacteriol.* **1983**, *156*, 828-836.
3. Coughlan, M. P.; Hon-Nami, K.; Hon-Nami, H.; Ljungdahl, L. G.; Paulin, J. J.; Rigsby, W. E. *Biochem. Biophys. Res. Comm.* **1985**, *130*, 904-909.
4. Lamed, R.; Setter, E.; Kenig, R.; Bayer, E. A. *Biotechnol. Bioeng. Symp.* **1983**, *13*, 163-181.
5. Millet, J.; Pétré, D.; Béguin, P.; Raynaud, O.; Aubert, J.-P. *FEMS Microbiol. Lett.* **1985**, *29*, 145-149.
6. Hazlewood, G. P.; Romaniec, M. P. M.; Davidson, K.; Grépinet, O.; Béguin, P.; Millet, J.; Raynaud, O.; Aubert, J.-P. *FEMS Microbiol. Lett.* **1988**, *51*, 231-236.
7. Gräbnitz, F.; Staudenbauer, W. L. *Biotechnol. Lett.* **1988**, *10*, 73-78.
8. Béguin, P. *Annu. Rev. Microbiol.* **1990**, *44*, 219-248.
9. Gräbnitz, F.; Rücknagel, K. P.; Seiß, M.; Staudenbauer, W. L. *Molec. Gen. Genet.* **1989**, *217*, 70-76.
10. Gräbnitz, F.; Seiss, M.; Rücknagel, K. P.; Staudenbauer, W. L. *Eur. J. Biochem.* **1991**, *200*, 301-309.
11. Navarro, A.; Chebrou, M.-C.; Béguin, P.; Aubert, J.-P. *Res. Microbiol.* **1991**, *142*, 927-936.
12. Henrissat, B.; Claeyssens, M.; Tomme, P.; Lemesle, L.; Mornon, J.-P. *Gene* **1989**, *81*, 83-95.

13. Hall, J.; Hazlewood, G. P.; Barker, P. J.; Gilbert, H. J. *Gene* **1988**, *69*, 29-38.
14. Grépinet, O.; Chebrou, M.-C.; Béguin, P. *J. Bacteriol.* **1988**, *170*, 4582-4588.
15. Chauvaux, S.; Béguin, P.; Aubert, J.-P.; Bhat, K. M.; Gow, L. A.; Wood, T. M.; Bairoch, A. *Biochem. J.* **1990**, *265*, 261-265.
16. Durrant, A. J.; Hall, J.; Hazlewood, G. P.; Gilbert, H. J. *Biochem. J.* **1991**, *273*, 289-293.
17. Lamed, R.; Bayer, E. A. In *FEMS Symposium No. 43. Biochemistry and Genetics of Cellulose Degradation;* Aubert, J.-P.; Béguin, P.; Millet, J., eds.; Academic Press, London & New York: 1988, pp. 101-116 .
18. Grépinet, O.; Chebrou, M.-C.; Béguin, P. *J. Bacteriol.* **1988**, *170*, 4576-4581.
19. Hazlewood, G. P.; Davidson, K.; Clarke, J. H.; Durrant, A. J.; Hall, J.; Gilbert, H. J. *Enzyme Microb. Technol.* **1990**, *12*, 656-662.
20. Williams, D.; Van Frank, R. M.; Muth, W.; Burnett, J. *Science* **1982**, *215*, 687-689.
21. Marston, F. A. O. *Biochem. J.* **1986**, *240*, 1-12.
22. Krueger, J. K.; Kulke, M. H.; Schutt, C.; Stock, J. *Biopharm* **1989**, *3*, 40-45.
23. Schein, C. H. *Bio/Technology* **1989**, *7*, 1141-1148.
24. Schein, C. H. *Bio/Technology* **1990**, *8*, 308-317.
25. Schoemaker, J. M.; Brasnett, A. H.; Marston, F. A. O. *EMBO J.* **1985**, *4*, 775-780.
26. Schoner, R. G.; Ellis, L. F.; Schoner, E. F. *Bio/Technology* **1985**, *3*, 151-154.
27. Joliff, G.; Béguin, P.; Juy, M.; Millet, J.; Ryter, A.; Poljak, R.; Aubert, J.-P. *Bio/Technology* **1986**, *4*, 896-900.
28. Bowden, G. A.; Paredes, A. M.; Georgiou, G. *Bio/Technology* **1991**, *9*, 725-730.
29. Blum, P.; Velligan, M.; Lin, N.; Matin, A. *Bio/Technology* **1992**, *10*, 301-304.
30. Mitraki, A.; King, J. *Bio/Technology* **1989**, *7*, 690-697.
31. Tokatlidis, K.; Dhurjati, P.; Millet, J.; Béguin, P.; Aubert, J.-P. *FEBS Lett.* **1991**, *282*, 205-208.
32. Chaffotte, A. F.; Guillou, Y.; Goldberg, M. E. *Eur. J. Biochem.* **1992**, *205*, 369-373.
33. Babbitt, P. C.; West, B. L.; Buechter, D. D.; Kuntz, I. D.; Kenyon, G. L. *Bio/Technology* **1990**, *8*, 945-949.
34. Tokatlidis, K.; Salamitou, S.; Béguin, P.; Dhurjati, P.; Aubert, J.-P. *FEBS Lett.* **1991**, *291*, 185-188.
35. Juy, M.; Amit, A. G.; Alzari, P. M.; Poljak, R. J.; Claeyssens, M.; Béguin, P.; Aubert, J.-P. *Nature* **1992**, *357*, 89-91.
36. Fujino, T.; Béguin, P.; Aubert, J.-P. *FEMS Microbiol. Lett.* **1992**, *94*, 165-170.
37. Fierobe, H.-P.; Gaudin, C.; Belaich, A.; Loutfi, M.; Faure, E.; Bagnara, C.; Baty, D.; Belaich, J.-P. *J. Bacteriol.* **1991**, *173*, 7956-7962.
38. Giza, P. E.; Huang, R. C. C. *Gene* **1989**, *78*, 73-84.
39. Kleid, D. G.; Yansura, D.; Small, B.; Dowbenko, D.; Moore, D. M.; Grubman, M. J.; McKercher, P. D.; Morgan, D. O.; Robertson, B.; Bachrach, H. L. *Science* **1981**, *214*, 1125-1129.
40. Koerner, T. J.; Hill, J. E.; Myers, A. M.; Tzagoloff, A. *Methods in Enzymol.* **1991**, *194*, 477-490.
41. Bishai, W. R.; Rappuoli, R.; Murphy, J. R. *J. Bacteriol.* **1987**, *169*, 5140-5151.
42. Bishai, W. R.; Miyanohara, A.; Murphy, J. R. *J. Bacteriol.* **1987**, *169*, 1554-1563.
43. McCaman, M. T.; Andrews, W. H.; Files, J. G. *J. Biotechnol.* **1985**, *2*, 177-190.
44. McCaman, M. T. *J. Bacteriol.* **1989**, *171*, 1225-1227.
45. Bradford, M. *Anal. Biochem.* **1986**, *72*, 248-254.

RECEIVED October 26, 1992

Chapter 4

Probing the Role of Protein Folding in Inclusion Body Formation

Boris A. Chrunyk[1], Judy Evans, and Ronald Wetzel

Macromolecular Sciences Department, SmithKline Beecham Pharmaceuticals, 709 Swedeland Road, King of Prussia, PA 19406

Point mutations in recombinant human interleukin-1β (IL-1β) have been identified which exhibit dramatic changes in the level of inclusion bodies (IBs) in *E. coli* producing this protein. In both wild type and mutants, a smaller percentage of IL-1β is deposited into IBs when cells are grown at 32° C rather than 42° C. Some of the mutant proteins have been purified and examined for their stabilities and folding kinetics by monitoring the fluorescence of the lone tryptophan. In addition, aggregation associated with thermal treatment of the native protein, and with the refolding of the protein from the denatured state, was monitored. Several lines of evidence suggest that at least one sequence variant, Lys97->Val (K97V), may form IBs due to an alteration in the properties of a folding intermediate. Implications of the structural locations of this and other IB mutants will be discussed.

The expression of heterologous proteins in bacterial hosts sometimes leads to the deposition of the product of interest into amorphous refractile particles known as inclusion bodies (IBs) (*1, 2*). This often poses technical problems in protein recovery, since (a) denaturing conditions are often required for solubilization of IBs, and (b) once solubilized in denaturant, refolding conditions must be identified to allow recovery of active protein. At the same time, in those cases where refolding conditions can be established, inclusion body formation may actually prove advantageous. For instance, inclusion body formation can protect the protein from proteolytic digestion during expression. In addition, since in many cases the isolated inclusion body is highly enriched in the overexpressing protein, separation of IBs from soluble cellular contents can serve as an initial purification step.

It would thus be useful to be able to control inclusion body formation, and acquiring an understanding of the mechanism(s) of IB formation would seem a valuable approach to this goal. Several external factors have been suggested to influence inclusion body deposition in bacterial cells (*2*). To date, however, the only external factor proven to influence the balance between soluble or insoluble expression in a variety of systems is temperature (*3, 4*). Characteristics of the overexpressed protein itself undoubtedly contribute significantly to their deposition into inclusion bodies. For

[1]Current address: Central Research Division, Pfizer, Inc., Groton, CT 06340

0097–6156/93/0526–0046$06.00/0
© 1993 American Chemical Society

example, most proteins are only marginally stable at 25°C (5) and eventually undergo cooperative unfolding at some higher temperature (T_m). The unfolded state that is *populated at equilibrium* at temperatures above this T_m would be expected to be prone to aggregation due to the exposure of normally buried hydrophobic residues (6). Similarly, proteins which require disulfide bonds to achieve net folding stability may *populate at equilibrium* aggregation prone unfolded states under the reducing conditions of the cytoplasm. Alternatively, aggregation of *transiently populated* folding intermediates may be responsible for some IBs, for example if the intermediates are poorly soluble and/or build up to relatively high concentrations as controlled by folding kinetics.

How might sequence determine the fate of a freshly synthesized polypeptide chain? Since many proteins can achieve their native folds *in vitro* without requiring accessory factors (7), the code for folding, in many cases, is determined by the amino acid sequence alone. Alterations in the sequence by as little as a single point mutation are known to produce dramatic effects on the thermodynamic stability of a protein, the kinetics of folding, or both (8), thus leading to the kinds of species discussed above as potential sources of aggregate formation. In discussing the origins and consequences of off pathway reactions it is useful to invoke a generalized folding scheme such as that shown in Figure 1. Here, P represents proteolytic digestion and Agg represents aggregate formation. In this scheme, there exists a potential for aggregate formation or proteolysis at any stage in the folding of the polypeptide chain. Mutations can alter the stability of the native protein or generate a proteolytic site, both of which can lead to decreased levels of soluble product. As suggested for P22 tailspike protein, mutations can also alter IB levels *via* effects on a folding intermediate (9, 10).

In their investigation of mutations in the trimeric Salmonella phage P22 tail spike protein, King and coworkers (9, 10) isolated temperature sensitive folding (*tsf*) mutations. At permissive temperatures (<39°C), WT and the mutant tail spike proteins both folded and assembled properly *in vivo*, while at higher temperatures the *tsf* mutants were driven into inclusion bodies. *In vitro* characterization of the WT and mutant proteins showed that all had similar T_m values, some 40° higher than the non-permissive temperature at which IBs were formed *in vivo*(11). This result suggested that folding or assembly, and not stability, of the *tsf* mutant proteins plays the major role in inclusion body formation. Specifically, a temperature sensitive folding intermediate was implicated (9, 10).

In order to further examine the roles of protein folding and stability in the formation of inclusion bodies, we have initiated studies with recombinant human interleukin 1β (hIL-1β). Interleukin 1β is a small (17kD) monomeric protein, whose X-ray (12) and solution NMR (13, 14) structures are known. This allows for a more structural approach to analysis of mutagenic effects on inclusion body formation than is currently possible with the trimeric P22 tailspike protein. Additionally, previous work of Craig et al. (15) has established conditions for reversible *in vitro* folding of IL-1β. We review here the results of *in vivo/in vitro* comparisons of a collection of mutant IL-1β proteins which exhibit differing extents of inclusion body formation when synthesized in transformed *E. coli* (16-19).

Inclusion Body Screening

We obtained a collection of about 70 point mutants in IL-1β from Dr. Peter Young and his co-workers; these sequence variants were previously prepared by site-directed mutagenesis to explore structure/function relationships in the biological activity of this protein (20, 21). These mutant IL-1β proteins were screened for their tendencies to be deposited into inclusion bodies during growth of *E. coli*. Since inclusion bodies are dense aggregates, centrifugation of cell lysates results in a separation of soluble protein from the cell debris and IBs. The total expression and fraction of protein in the soluble or insoluble phases were determined by laser densitometry of SDS-PAGE gels. The results for several of the mutant proteins are shown in Table I (*18*). Also shown are T_{agg} and $\Delta\Delta G_u$ values, to be discussed below, for each mutant. The T_{agg} value is the temperature for the onset of protein aggregation, as observed under defined buffer and heating conditions and as monitored by scattering of fluorescence (*18*). The $\Delta\Delta G_u$ value is the difference in free energy of unfolding between a mutant protein and the wild type, in kcal/mole, where the free energies are determined by equilibrium unfolding measurements (*18*).

It is immediately evident that there is considerable variability in both expression and in inclusion body percentage for the mutant proteins compared to wild type. Mutations are observed to alter expression level, IB content, or both. Thus, mutational effects in IL-1β are consistent with the general scheme invoked in Figure 1, and with mutational effects on IB formation observed in bacterial expression of P22 tailspike protein (*9, 10*) and several other proteins (*3, 4*).

In Vivo Temperature Effects

As discussed above, growth temperature can be a factor in inclusion body formation. To determine whether growth temperature is a factor in deposition of IL-1β into inclusion bodies, the plasmids carrying the WT, L10T, and K97V genes were transformed into *E. coli* strain AR120. In this strain, the P_L promoter can be induced by nalidixic acid (*22*). Overnight cultures of the AR120 strains containing the respective plasmids were used to inoculate fresh Luria Broth. The strains were grown to an A_{600} of ~ 1 at which time protein synthesis was induced by addition of 40 ug/ml nalidixic acid and grown for 5 hours. The cells were lysed and the soluble and insoluble fractions separated. SDS-PAGE analysis was followed by Western blotting. The results of densitometry scans of the Western blot are shown in Table II.

Table II shows that, as observed for IB formation by some other proteins (*3, 4*), reducing the temperature from 42°C to 32°C causes a reduction in the percentage of protein found in the insoluble fraction for each of the three sequence variants tested. At 32°C, almost none of the WT protein is found in the insoluble fraction, while the two

Table I. Mutational Effects on IB Formation in IL-1β[a]

Mutation	Expression[b]	% IL-1β in IBs	T_{agg} °C	$\Delta\Delta G_u$[c]
WT	100	8	58.4	
T9A	111	8	58.4	-0.4±0.2
T9E	33	47	nd	nd
T9L	47	6	55.4	-0.4±0.1
T9Q	66	24	52.3	-1.9±0.4
T9W	37	4	nd	nd
L10T	79	83	nd	nd
L10N	49	59	nd	nd
K97R	117	1	56.8	-0.1±0.1
K97G	96	24	54.2	-1.1±0.2
K97V	109	61	53.0	+0.7±0.2

[a] Heat induction at 42°C of the λ P_L promoter in *E. coli* strain AR58. [b]Total expression relative to WT. [c]Kcal/mole, errors represent 95% confidence limits. Adapted from reference 18.

mutant proteins show IB formation that is still quite significant, but which is nonetheless reduced from 42° C levels. These results are important for two reasons. They show that IB formation by IL-1β exhibits the temperature-dependence found in many other IB systems (3, 4), including the P22 tailspike protein prototype (23). Second, *in vitro* models for IB formation can eventually be tested for their relevance to the *in vivo* process by their temperature dependence.

Thermodynamic Stability and Inclusion Bodies

In order to determine the importance of protein stability to inclusion body deposition, the guanidine hydrochloride (Gdn-HCl) induced unfolding of the IL-1β proteins was initiated using the fluorescence of the lone tryptophan at position 120 as a probe of structure. The fluorescence monitored denaturation of the IL-1β proteins showed unusual behavior (Figure 2) in that upon increasing denaturant up to approximately 1 M Gdn-HCl, the fluorescence intensity at the emission maximum of 343 nm increases. Above 1 M Gdn-HCL, the fluorescence decreases, consistent with solution quenching of tryptophan fluorescence upon unfolding of the protein. (In previous folding studies on wild type IL-1β, Craig and coworkers observed an initial burst of fluorescence in kinetic unfolding experiments (15). They attributed this burst to the disruption, during unfolding, of an interaction between tryptophan and another group on the protein which quenches the tryptophan fluorescence in the native state.) In order to derive thermodynamic parameters from denaturant induced unfolding curves a model is required, but the observed biphasic behavior in the equilibrium unfolding curve makes it difficult *a priori* to justify a simple two state approximation. However, an analysis of the equilibrium data showed that there may actually be two processes occurring in the initial increases in denaturant. Figure 2 shows there is an immediate large jump in fluorescence intensity from 0 to 0.3 M Gdn-HCl, followed by a more gradual increase in fluorescence with increasing denaturant up to 1 M Gdn-HCl. It is sometimes possible to separate two such phenomena to enable establishment of native baselines for fitting to a two-state model of unfolding (24); we used such techniques to improve the fit of the equilibrium curves obtained for IL-1β sequence variants (18).

Although the behavior of the fluorescence can be handled mathematically, the physical interpretation of the effects is not immediately obvious. Since ionic strength, pH, and denaturants are all observed to produce an increase in fluorescence (15), and since no obvious quenching group could be identified in the X-ray crystal structure, alternatives to the rationale by Craig et al. were considered. Tryptophan 120 is only 60% buried in the native structure (15) and lies in a small pocket near the surface of the protein. There appears to be a salt bridge between lysine 138 and glutamate 111 in the crystal structure, which has also been confirmed by solution NMR studies (14), and this salt bridge appears to be involved in stabilizing one end of the tryptophan pocket. Release of this salt bridge could result in a conformational change in the region of W120, which might further bury the tryptophan and thus increase its fluorescence. Our

Table II. *In Vivo* IB Formation as a Function of Induction Temperature[a]

Induction Temp.	WT IB%	L10T IB%	K97V IB%
32 °C	6.4	41.0	17.0
42 °C	18.0	73.0	47.0

[a]Densitometry of Western blot as described in text. These values express the percentage of total IL-1β produced which is found in the insoluble fraction, and are thus analogous to the second column of Table I. Data from reference 18.

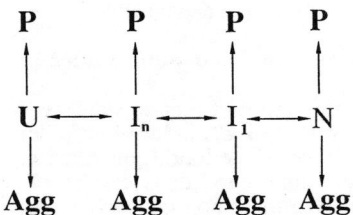

Figure 1. General scheme for protein folding, aggregation, and proteolysis in the cell. Adapted from reference 3.

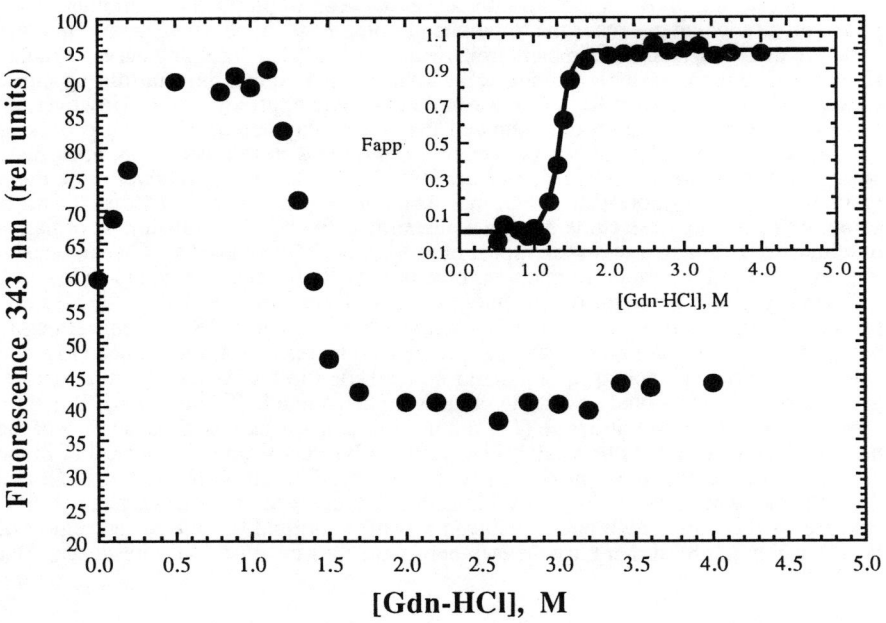

Figure 2. Relative fluorescence of the lone tryptophan in IL-1β WT as a function of guanidine hydrochloride concentration at 25°C, pH 6.5. Inset shows the data fit to a two state model of unfolding as described in the text. Adapted from reference 18.

proposed model for the fluorescence changes observed in this protein is that the initial increase in fluorescence up to 0.3 M Gdn-HCl is due to release of the salt bridge by virtue of an increase in the ionic strength of the solution, and that the subsequent increase in fluorescence results from gradual readjustments in the tryptophan pocket as denaturant strength continues to increase.

With a model in place, the unfolding of a set of IL-1β mutant proteins was initiated. These proteins were chosen to encompass a range of IB contents but at the same time exhibit sufficient expression of IL-1β in the soluble fraction that material for folding studies could be isolated from lysis supernatants. The fluorescence data was converted to a fraction of unfolded protein, F_{app}, and fit to a two state model (24) utilizing linear free energy relationships (25). No evidence for equilibrium stable intermediates was observed in this study (18) or in previous studies of the denaturation of WT IL-1β (15). Non-linear least squares fitting of these curves to a two state model of unfolding results in the thermodynamic parameters shown in Table I (18). A comparison of the free energy of unfolding, calculated at the WT midpoint (26), with IB formation *in vivo* is shown in Figure 3. Since these mutations were chosen for further study based on their overall expression of soluble protein, several of the proteins are clustered near the WT value. Surprisingly and significantly, K97V, which is predominantly deposited into inclusion bodies *in vivo*, was found to be more stable than WT by almost 1 kcal/mol. For the remainder of the proteins there is a fair trend between IB formation and stability. Since many members of this limited set of mutant proteins are similar in stability to the WT, however, this trend is not strong and cannot be viewed as a significant correlation. In addition, the result with the K97V variant is difficult to justify with a model for IB formation depending solely on thermodynamic stability effects.

In Vitro Thermal Studies

Western analysis of SDS-PAGE on cell lysates grown at lower temperature (Table II) suggested that temperature is a factor in IB formation. Since stability against reversible unfolding, described above, did not strongly correlate with IB formation, we turned our attention to studies of stability against irreversible thermal denaturation. Previous calorimetric studies with the WT protein and some mutants showed that IL-1β does not melt reversibly but aggregates upon thermal denaturation (C. Brouilette, personal communication) and we have confirmed these results. The aggregates are not a result of interchain disulfide linkages, since gel analysis shows the absence of disulfide linked oligomers in the aggregates (BAC & RW, unpublished observations). In fact, strongly reducing conditions were implemented in all experiments to eliminate this as a possibility.

Since thermal denaturation of proteins is often associated with unfolding (6, 27), estimates of T_ms can be obtained in these cases by monitoring light scattering, an indicator of aggregation and precipitation, as a function of temperature (3, 6, 27). Figure 4 shows the heating curves of the WT and mutant IL-1β proteins as monitored by scattering of fluorescence. The estimated temperatures for the onset of aggregation, T_{agg}, listed in Table I, are obtained by extrapolation of the linear portion of the curve of fluorescence with respect to temperature to zero relative fluorescence (18). The value for WT based on this analysis is 58.4°C, very close to the 62°C determined calorimetrically (C. Brouilette, personal communication). In fact, we would expect the T_{agg}, which is a measure of the temperature of *onset* of aggregation, to be lower than the calorimetric T_m, which is defined as the temperature of maximal heat absorption. Of the seven proteins surveyed, three showed significantly reduced T_{agg} values. Interestingly, two of these proteins, K97G and K97V, also show increased inclusion body formation. A comparison of the T_{agg} and the amount of protein expressed in inclusion bodies is shown in Figure 5.

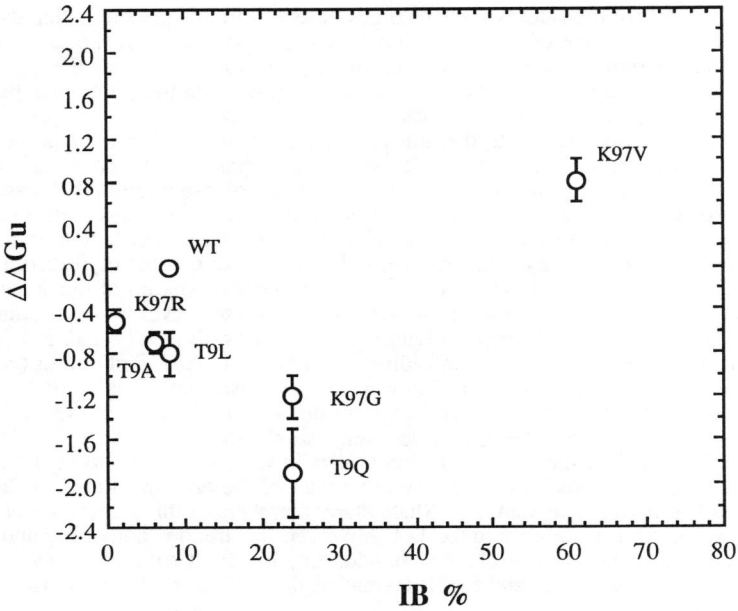

Figure 3. Comparison of the free energy of unfolding relative to WT, calculated at the WT midpoint, to IB formation *in vivo* using values from Table I. Adapted from reference 18.

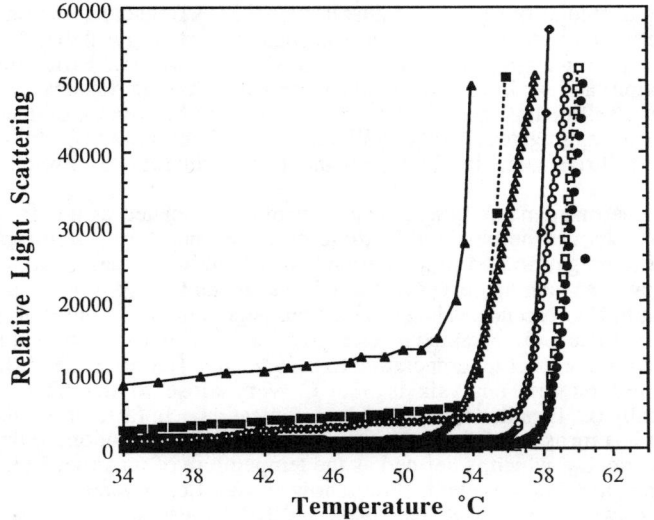

Figure 4. Thermal aggregation curves of WT and mutant proteins in 10 mM MES, 2 mM EDTA, 10 mM DTT, pH 6.5. Key: WT, ● ; K97R, --o-- ; K97G, --■-- ; K97V, --△-- ; T9L, --◇-- ; T9A, --□-- ; T9Q, --▲-- .

The value of the T_{agg} determination is two fold- (a) it measures a property, aggregation, which resembles the *in vivo* property of IB formation, and (b) it measures this as a response to temperature (as opposed, for example, to Gdn-HCl concentration), and can thus be directly related to the *in vivo* response. In the limited set of variants studied, Figure 5 shows there is a trend of increased IB formation with decreased thermal stability. However, Figure 5 also shows that the *in vitro* aggregation observed starting from the native state is not directly relevant to the *in vivo* process, since aggregation *in vitro* requires temperatures over 20°C higher than the lowest growth temperature found to support IB formation *in vivo* by the same sequence variants (Table II). These results are thus qualitatively similar to those obtained in studies of temperature sensitive folding (*tsf*) mutations in the P22 tailspike protein, where T_m values near 90° C were obtained (28) for wild type and *tsf* mutants which form IBs in the 40° C range (23). This temperature difference has been the primary evidence for suggestions that *tsf* mutations influence folding intermediates in the P22 tailspike system (9, 10).

Kinetic Studies

Since the experiments described above failed to demonstrate a strong correlation between either thermodynamic or thermal stability and IB formation, we turned our attention to folding kinetics. Previous work of Craig and coworkers established reversible folding conditions for the WT IL-1β at 25° C, and we utilized modifications of these conditions to evaluate the refolding/unfolding properties of the WT and mutant proteins, using tryptophan fluorescence as a probe (19). Refolding is characterized by a very fast increase in fluorescence, which is complete in the dead time of manual mixing, followed by a slow phase ranging from ~5000 seconds at the midpoint to 50 seconds at 0.2 M Gdn-HCl. A comparison of the WT and two mutant proteins is shown in Figure 6. Although in some of the mutant proteins the unfolding rate was affected, the refolding rates were largely unaffected. The simplest explanation of these results is that differences among the sequence variants in inclusion body formation do not depend on differences in kinetic lifetimes of folding intermediates.

These results do not preclude a role for kinetic intermediates, however. For instance, important differences among the mutants may occur during fast kinetic phase(s) inaccessible to manual mixing techniques. Our observation of a rapid change in fluorescence upon refolding suggests the existence of faster phases, whose further characterization will require fast kinetic techniques. An alternative explanation for the failure to observe differences in folding intermediates is that the tryptophan probe may not be sensitive to the folding event associated with aggregation. In fact, recent studies using covalently attached fluorescent probes (29) suggest that there do exist phases in the unfolding kinetics of IL-1β distinct from those we observed in tryptophan monitored studies. Finally, mutational effects may be more related to changes in the aggregation tendencies of folding intermediates than in their kinetic lifetimes. Although such details are not yet elucidated, recent experiments have added further support for the role of folding intermediates in these mutational effects (see below).

Discussion

It is interesting to consider both the position and the substituting residue when analyzing the structural aspects of inclusion body formation and *in vivo* proteolysis. Figure 7 shows the three dimensional structure of IL-1β (12). Strand 1, which contains both position 9 and 10, is part of the core 6 stranded β-barrel. Position 9 is partially buried with its side chain pointing toward solvent. Position 10 is completely buried with the side chain pointing into the center of the barrel. Mutations in the barrel strands might be expected to perturb the stability of the protein since packing inside the barrel may be critical. In fact, replacement of the wild type Leu at position 10 with

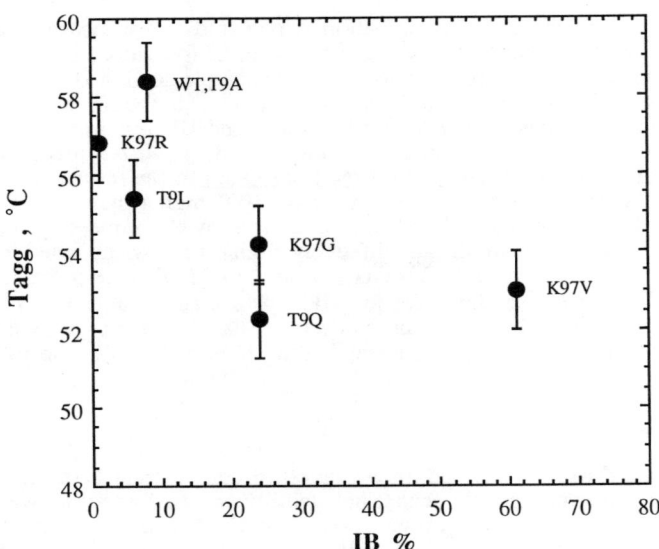

Figure 5. Comparison of the temperature for onset of aggregation, T_{agg}, to IB formation *in vivo* using values from Table I. Adapted from reference 18.

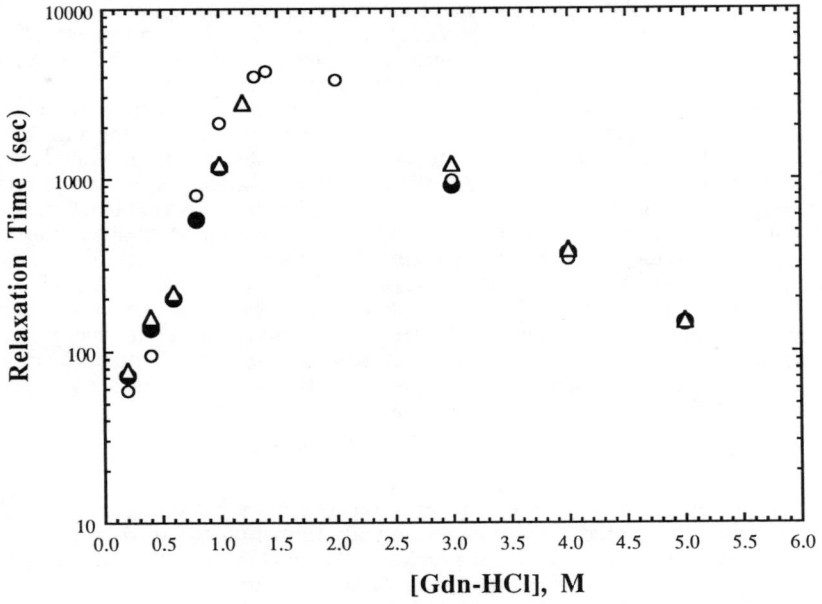

Figure 6. Reversible folding kinetics of WT (o), T9A (●), and K97V (Δ) sequence variants as a function of the concentration of guanidine hydrochloride at 25° C, pH 6.5. Adapted from reference 19.

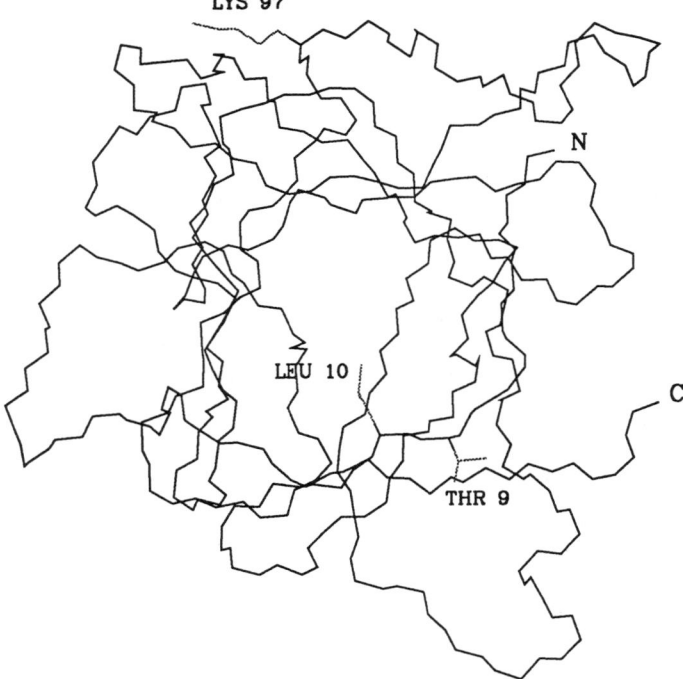

Figure 7. Model of the three-dimensional structure of IL-1β showing the main chain and the N- and C- termini. The side chains of the residues at positions 9, 10 and 97 are shown in mottled lines. Adapted from reference 18.

hydrophilic residues results in either low expression (*18*) or high inclusion body content (Table I). Although the position 9 side chain is pointing toward solvent, residue replacements which increase hydrophobicity are tolerated with respect to IB formation, while hydrophilic side chains increase IB formation. For example, the T9L mutation produces essentially no effect on IB formation, yet (1) it is slightly destabilizing, both in terms of $\Delta\Delta G_u$ and T_{agg} (Table I), and (2) the surface exposure of additional hydrophobicity would be expected to decrease solubility of the native state. It is thus possible that hydrophobic character at this site is especially important in stabilizing or maintaining the solubility of a folding intermediate.

Position 97 is located in a long surface exposed loop between strand 7 and core strand 8 (*30*). Surprisingly, replacements at this position result in significant alterations in thermodynamic stability. One might expect that replacements in a mobile loop region would have little or no effect on stability. However, a replacement of a loop residue in the α subunit of tryptophan synthase from *Salmonella typhimurium* was shown to result in a stabilization of an equilibrium intermediate (*31*). A closer examination of the loop region in IL-1β indicates the presence of a small hydrophobic cluster which reduces the overall mobility of the loop as evidenced by the crystallographic temperature factors (*30*). This cluster may be critical for the folding and stability of the protein. Conservative replacement of the wild type Lys with Arg results in little change, either in stability or IB%. However, removal of the side chain by replacement with Gly results in a strong decrease in stability and increased percentage of IBs over the WT value. Perhaps the removal of the side chain at this position re-orients the cluster and alters the packing forces stabilizing it. The X-ray crystal structure shows that the methylene groups of Lys97 appear to pack on the surface of the protein. Replacement with Gly would remove these packing interactions, as well as increase conformational entropy in the unfolded state. The substitution of Val at position 97 <u>increases</u> thermodynamic stability. This increased stability may be the result of an extra hydrophobic group packing, *via* non-native interactions, into the cluster, or simply by enhancing packing energy in the native contact area by removal of the protonated nitrogen of the Lys. Significantly, this K97V replacement, which enhances thermodynamic stability, also leads to enhanced deposition of the protein into inclusion bodies

As in the case of the P22 tailspike protein (*10*), mutations in recombinant human interleukin-1β lead to variations in the deposition of the protein into inclusion bodies. As found in the tailspike protein studies, the influence of growth temperature on the delicate balance between soluble and insoluble expression does not appear to be related to the thermodynamic stability of the native state. <u>Thermodynamic</u> stabilities of IL-1β sequence variants do not correlate strongly with IB formation (Figure 3). <u>Thermal</u> stability, which is often a reasonable measure of thermodynamic stability (*6*), also does not seem to control inclusion body formation; significantly, temperatures for the onset of thermally induced aggregation *in vitro* (T_{agg} values, Table I) are considerably higher than temperatures which are known to support substantial inclusion body formation in *E. coli* cultures (Table II).

The unusual behavior of the K97V mutant is of some interest. Although the thermal stabilities of most of the sequence variants correlate well with their thermodynamic stabilities, IL-1β(K97V) is a dramatic outlier (*18*). There are other examples of uncoupling of thermodynamic and thermal stability (*6*). For example, T4 lysozyme mutants have been constructed which are less stable than wild type to reversible unfolding but more stable against irreversible denaturation (*32*); this may be due to mutational effects on the aggregation competence of an unfolded form. In contrast, IL-1β(K97V) is more stable against reversible unfolding but less stable against irreversible denaturation; this may be due to a local unfolding event, at temperatures well below those required for cooperative unfolding, which unmasks an aggregation surface on the otherwise folded protein. The location of position 97 in a microdomain-like hydrophobic cluster is consistent with this possibility.

Since mutational effects on inclusion body formation by IL-1β mutants do not seem to be related to the stability of the native state, a reasonable hypothesis is that the mutations influence some property of a folding intermediate. Recent kinetic experiments have shown that the K97V variant aggregates upon refolding at 42° C while the WT does not (*19*). Further, this *in vitro* folding dependent aggregation of K97V mirrors the *in vivo* result, in that refolding of K97V at lower temperatures does not lead to aggregation (*19*). These results provide further, more direct, support for the involvement of a folding intermediate in the formation of IL-1β inclusion bodies.

It is tempting to speculate on how an independently folding subdomain containing Lys 97, as invoked above, might also play a role in the folding-dependent aggregation of the K97V mutant. Our folding kinetics results on the wild type protein support previously published suggestions that folding of IL-1β may begin with a relatively rapid collapse to a "molten globule" type intermediate (*33*), and, consistent with what is known about such intermediates (*34*), subsequent, slower, folding steps would involve adjustments of side chain packing contacts and loop conformations. Surface exposure of the hydrophobic residues of the Lys97 loop in such a relatively long-lived intermediate would be expected to generate an aggregation-prone species, whose aggregation would be expected to be enhanced at higher temperatures (*35*).

We have shown that mutations in recombinant human IL-1β can have a pronounced effect on the expression of soluble protein and that these effects do not correlate with characteristics of the native state such as location in particular secondary structure type, or surface exposure. Subtle changes at a given position in the molecule can tip the balance between soluble and insoluble expression. As discussed above, we have recently obtained direct experimental support for the involvement of a folding intermediate in the deposition of mutant IL-1β protein K97V into inclusion bodies (*19*). We hope to continue to elucidate the *in vivo* mechanism by considering the role of chaperone proteins, and by searching for other genetic *loci* which influence inclusion body formation, including as suppressers of IB formation (*11, 36*).

Acknowledgments. The authors wish to thank Dr. Peter Young and Jay Lillquist for the gift of the IL-1β WT and mutant plasmids.

Literature Cited

1. Marston, R. A. O., *Biochem. J.* **1986**, *240*, 1-12.
2. Schein, C. H., *Biotechnology*. **1989**, 7, 1141-1149.
3. Wetzel, R., in *Protein Engineering - A Practical Approach;* A. R. Rees; Sternberg, M. J. E.; Wetzel, R., Ed.; IRL Press at Oxford University Press: Oxford, 1992; pp. 191-219.
4. Wetzel, R., in *Stability of Protein Pharmaceuticals: In Vivo Pathways of Degradation and Strategies for Protein Stabilization;* T. J. Ahern; Manning, M. C., Ed.; Plenum Press: New York, 1992; In Press.
5. Baldwin, R. L.; Eisenberg, D., in *Protein Engineering;* D. L. Oxender, Ed.; Alan R. Liss: New York, 1987; pp. 127-148.
6. Wetzel, R.; Perry, L. J.; Mulkerrin, M. G.; Randall, M., in *Protein Design and the Development of New Therapeutics and Vaccines; Proceedings of the Sixth Annual Smith, Kline and French Research Symposium;* G. Poste; Hook, J. B., Ed.; Plenum: New York, 1990; pp. 79-115.
7. Anfinsen, C. B., *Science*. **1973**, *181*, 223-230.
8. Beasty, A. M.; Hurle, M. R.; Manz, J. T.; Stackhouse, T.; Onuffer, J. J.; Matthews, C. R., *Biochemistry*. **1986**, *25*, 2965-74.
9. King, J.; Fane, B.; Haase-Pettingell, C.; Mitraki, A.; Villafane, R., in *Protein Design and the Development of New Therapeutics and Vaccines;* J. B. Hook; Poste, G., Ed.; Plenum: New York, 1990; pp. 59-78.

10. King, J.; Fane, B.; Haase-Pettingell, C.; Mitraki, A.; Villafane, R.; Yu, M.-H., in *Protein Folding: Deciphering the Second Half of the Genetic Code;* L. M. Gierasch; King, J., Ed.; American Association for the Advancement of Science: Washington, D.C., 1990; pp. 225-240.
11. Mitraki, A.; Fane, B.; Haase-Pettingell, C.; Sturtevant, J.; King, J., *Science.* **1991**, *253*, 54-58.
12. Finzel, B. C.; Clancy, L. L.; Holland, D. R.; Muchmore, S. W.; Watenpaugh, K. D.; Einspahr, H. M., *J. Mol. Biol.* **1989**, *209*, 779-791.
13. Driscoll, P. C.; Gronenborn, A. M.; Wingfield, P. T.; Clore, G. M., *Biochem.* **1990**, *29*, 4668-4682.
14. Clore, G. M.; Wingfield, P. T.; Gronenborn, A. M., *Biochemistry.* **1991**, *30*, 2315-2323.
15. Craig, S.; Schmeissner, U.; Wingfield, P.; Pain, R. H., *Biochemistry.* **1987**, *26*, 3570-3576.
16. Chrunyk, B. A.; Evans, J.; Lilquist, J.; Young, P. R.; Wetzel, R., *J. Cell. Biochem.* **1991**, *15G*, 190:R211.
17. Wetzel, R.; Chrunyk, B. A., in *Biocatalyst Design for Stability and Specificity;* M. Himmel; G. Georgiou, Eds. ACS Publications: Washington, D. C., 1992; In Press.
18. Chrunyk, B. A.; Evans, J.; Lillquist, J.; Young, P.; Wetzel, R., Ms. submitted.
19. Chrunyk, B.; Wetzel, R., Ms. submitted.
20. Simon, P. L.; Fenderson, W. F.; LoCastro, S.; Silvestri, J.; Lillquist, J. S.; Young, P. R.; Bhatnagar, P., *Cytokine* **1989**, *1*, 79.
21. Simon, P. L.; Kumar, V.; Lillquist, J. S.; Bhatnagar, P.; Lee, J. C.; Porter, T.; Green, D.; Sathe, G.; Young, P. R., *J. Biol. Chem.* **1992**, in press.
22. Mott, J. E.; Grant, R. A.; Ho, Y.-S.; Platt, T., *Proc. Natl. Acad. Sci.* **1985**, *82*, 82-92.
23. Haase-Pettingell, C. A.; King, J., *J. Biol. Chem.* **1988**, *263*, 4977-4983.
24. Finn, B. E.; Chen, X.; Jennings, P. A.; Saalau-Bethell, S. M.; Matthews, R., in *Protein Engineering - A Practical Approach;* A. R. Rees; Sternberg, M. J. E.; Wetzel, R., Eds.; 1992; pp. 167-189.
25. Schellman, J. A., *Biopolymers.* **1978**, *17*, 1305.
26. Pace, C. N.; Shirley, B. A.; Thomson, J. A., in *Protein Structure - A Practical Approach;* T. E. Creighton, Ed.; IRL Press: Oxford, England, 1990; pp. 311-330.
27. Mulkerrin, M. G.; Wetzel, R., *Biochem.* **1989**, *28*, 6556-6561.
28. Sturtevant, J. M.; Yu, M.-H.; Haase-Pettingell, C.; King, J., *J. Biol. Chem.* **1989**, *264*, 10693-10698.
29. Epps, D. E.; Yem, A. W.; Fisher, J. F.; McGee, J. E.; Paslay, J. W.; Deibel, M. R., Jr., *J. Biol. Chem.* **1992**, *267*, 3129-3135.
30. Veerapandian, B.; Poulos, T. L.; Gilliland, G. L.; Raag, R.; Svensson, L. A.; Winborne, E. L.; Masui, Y.; Hirai, Y., 1990, Protein Data Bank 4I1B.
31. Chrunyk, B. A.; Matthews, C. R., *Biochemistry.* **1990**, *29*, 2149-54.
32. Wetzel, R.; Perry, L. J.; Baase, W. A.; Becktel, W. J., *Proc. Natl. Acad. Sci. USA.* **1988**, *85*, 401-405.
33. Ptitsyn, O. B.; Pain, R. H.; Semisotnov, G. V.; Zerovnik, E.; Razgulyaev, O. I., *FEBS Letters.* **1990**, *262*, 20-24.
34. Baldwin, R. L., *Chemtracks-Biochemistry and Molecular Biology.* **1991**, *2*, 379-389.
35. Baldwin, R. L., *Proc. Natl. Acad. Sci. USA.* **1986**, *83*, 8069-8072.
36. Wetzel, R.; Perry, L. J.; Veilleux, C, *Biotechnology.* **1991**, *9*, 731-737.

RECEIVED November 11, 1992

Chapter 5

Experimental Investigation of the In Vivo Kinetics of Inclusion Body Formation

Jim Klein and Prasad Dhurjati

Department of Chemical Engineering, University of Delaware, Newark, DE 19716

Double radiolabel pulse-chase experiments were performed to study the kinetics of inclusion body formation *in vivo* using an E. coli strain expressing a mutated form of the *Salmonella typhimurium* CheY protein. CheY was seen to migrate slowly from a soluble form to an insoluble form. This migration strongly influenced the overall partitioning of CheY between the soluble and insoluble fractions. The migration also has significant mechanistic implications. An assumption commonly made about inclusion body formation is that the soluble native protein will not become incorporated into the insoluble fraction. The experimental results suggest that this assumption can be inaccurate.

Overexpression of protein in bacteria often results in a partitioning of the recombinant protein between soluble and insoluble forms and this partitioning has important consequences for the product recovery process. The ability to influence or even predict how a protein will partition would be very beneficial.

Experimental work in this area has been rather limited. The present understanding of inclusion body formation process derives primarily from the work of Jonathan King and coresearchers (*1-5*). They demonstrated that, for the P22 tailspike protein, inclusion bodies are formed from protein folding intermediates. A competition exists for folding intermediates between protein folding which results in native protein and aggregation which results in inclusion bodies. Other researchers studying protein folding and aggregation *in vitro* identified a similar competition between folding and aggregation (*6-8*). The controlled *in vitro* environment facilitated the collection of detailed kinetic information.

The development of an accurate kinetic description of inclusion body formation requires additional *in vivo* studies. It has been proposed that *in vitro* kinetics be applied to *in vivo* reactions (*9*). *In vivo* environments, however, differ significantly from environments which can be created *in vitro*. Chaperones, for example, are known to influence the kinetics of folding and aggregation reactions (*10-13*).

Our study is an investigation of the *in vivo* kinetics of inclusion body formation. Based on work in the literature, it was anticipated that recombinant protein would partition itself into soluble and insoluble forms after translation. This would occur on a time scale of seconds or minutes, typical time scales for protein folding. The partitioning was then expected to remain constant since a common assumption about inclusion body formation is that soluble native recombinant protein does not become incorporated into the insoluble fraction (9, 14). However, this was not observed in our experiments. Our results indicate that a pathway exists for native protein to enter the insoluble fraction by unfolding and aggregating with other unfolded molecules.

Experimental Procedures

Materials. 11.0 mCi/ml, 0.0013 mg/ml L-[^{35}S]methionine and 1.0 mCi/ml, 0.0022 mg/ml L-[^{3}H]leucine were obtained from New England Nuclear. Scintiverse II scintillation fluid from Fisher Scientific was used. Ultra Pure grade urea was obtained from Schwarz/Mann Biotech. 0.2 µM Supor-200 filters were obtained from Gelman Sciences. pH 3-10 Pharmalyte 2D, pH 4.0-6.5 Pharmalyte, bovine pancreas DNase I (90% protein), bovine pancreas Ribonuclease A (approx. 95% purity), ampicillin, electrophoresis grade SDS, Grade I chicken egg white lysozyme, Trizma Base, Trizma HCl and 2-mercaptoethanol were obtained from Sigma.

Bacterial Strains and Culture Conditions. *E. coli* strain JM109 (15) was used in all experiments. Plasmid pME124s15c (16) was obtained from J. Stock, Princeton University. pME124s15c, a pUC12 derivative, codes for a *Salmonella typhimurium* CheY protein with a single point mutation. Plasmid pCT603 (17) codes for the *Clostridium thermocellum* endoglucanase EGD. Protein expression from both plasmids is induced with IPTG. Kinetic information was obtained by running several small cultures in parallel rather than sampling at different time intervals from one large culture. At different time intervals, one of the parallel cultures was stopped. Each such culture yielded partitioning information for a specific chase-time. Parallel cultures were used to avoid changing growth conditions due to volume changes caused by sampling. All cultures had a volume of 10 ml and were grown in 50 ml centrifuge tubes. An agitating incubator at 37°C was used throughout. Media used were: LB (10 g/l bacto-trypton, 5 g/l yeast extract, 10 g/l NaCl, 3.6 mM NaOH), M9 (0.2% glucose, 6.0 g/l Na_2HPO_4, 3.0 g/l KH_2PO_4, 5 g/l NaCl 0, 1.0 g/l NH_4Cl, 0.01 µg/l thiamine , 1.0 mM $MgSO_4$), and M9AA (M9 supplemented with 40 µg/ml of each amino acid except Pro, Leu, and Met). All cultures contained ampicillin at 1 mg/l.

Radiolabeling. The experimental protocol used was a modification of the protocol described by Mosteller *et. al* (18). Inoculum was prepared for startup cultures by mixing together five cultures which had grown overnight in M9 to stationary phase. Inoculum sizes varied between 0.3 and 1.6 ml depending upon the startup culture growth conditions. The sizes were determined such that an $OD_{\lambda=600}$ between 0.8 and 1.0 was obtained after approximately two hours of incubation. Five identical startup cultures were grown consisting of inoculum, fresh media with ampicillin (M9 or LB), and IPTG. Once an $OD_{\lambda=600}$ between 0.8 and 1.0 was reached 100 µCi L-[^{3}H]leucine, 15 µg, was added to each culture. Incubation continued for 15 minutes (17 minutes during the "High Level" experiment) and then unlabeled L-leucine was added to a concentration of 100 µg/ml. After incubation continued five minutes more to permit completion of protein translations already initiated, the five cultures were mixed. 10 ml was removed and harvested. Harvesting was done by mixing a culture with ice cold media and placing it on ice. The remaining 40 ml of culture was filtered using a 0.2 µM Supor-200 filter, rinsed with 37°C chase media (LB or M9 with IPTG, amp, and 100 µg/ml unlabeled L-leucine), and resuspended in 40 ml of 37°C chase media. Different amounts of these resuspended cells were used to inoculate a series of chase cultures grown in parallel. Inoculum sizes were determined to give $OD_{\lambda=600}$ readings near 1.0 at the end of the chase periods. For example a chase culture harvested after one hour consisted of 5 ml of resuspended cells and 5 ml of chase media whereas a chase culture harvested after two hours consisted of 3 ml of resuspended cells and 7 ml of chase media. After all the chase cultures had been harvested, they were pelleted, resuspended in sample buffer (25 mM pH 8.0 Tris), pelleted again and resuspended in 50 µl of sample buffer. Samples were stored at -70°C.

Cells were labeled with [^{35}S]methionine in a similar manner for use as references with the [^{3}H]leucine labeled samples. Five identical startup cultures were grown to an $OD_{\lambda=600}$ between 0.8 and 1.0. Then 50 µCi L-[^{35}S]methionine, 7 µg, was added to each culture. Incubation continued for 20 min. and then the cultures were harvested. The five cultures were mixed, resuspended in sample buffer, pelleted again, resuspended in a total of 250 µl of sample buffer and stored at -70°C.

Sample Preparation. After thawing and mixing to assure homogeneity, equal portions of [^{35}S]methionine cell suspension were added to each sample of [^{3}H]leucine labeled cells. The use of the ^{35}S labeled cells compensated for material lost and variations in sample handling. The CheY ^{3}H/^{35}S ratio in each sample remains the same regardless of: a) any material loss during the sample preparation or electrophoresis steps, b) the amount of sample loaded onto the gels, and c) how much of the spot is cut from the gels. This is true for both soluble and insoluble CheY. The use of a second isotope thus greatly increased the precision of the results obtained.

25 µl of lysozyme buffer (0.01 g/ml lysozyme, 0.45 M EDTA, pH 8.0) were also added to each sample. The samples were then incubated on ice for 30 min. prior to being freeze/thawed twice at -70°C. 5 µl each of 5.0 mg/ml RNase and 5.0 mg/ml DNase were added to the samples which were then incubated at room temperature for five minutes. One drop of Triton-X was mixed into each and incubation was continued at 4°C for another 20 minutes. After incubation, the samples were centrifuged for 10 min. at 32,000 x g and 4°C. Portions of the supernatants were drawn off and placed on ice. These were the "soluble fraction" samples. The pellets were twice resuspended in 0.4 ml of sample buffer to wash off the remaining soluble material before finally being resuspended in 50 µl of sample buffer and placed on ice. These were the "insoluble fraction" samples.

5 µl of pH 3.0-10.0 ampholytes, 5 µl of β-mercaptoethanol and a mass of urea equal to the volume of the sample were added to each of the soluble and insoluble samples. Samples were incubated with occasional agitation at room temperature for a minimum of 1 hour. Excess urea and cell debris were removed by centrifugation at 32,000 x g for 10 minutes.

Electrophoresis. Samples were analyzed by two dimensional gel electrophoresis as described in (*19*) using the Iso-DALT system from Integrated Separation Systems. The first dimension gels, IEF, were cast using an ampholyte mixture of

0.9 ml of pH 3.0-10.0 Pharmalyte and 0.5 ml of pH 4.0-6.5 Pharmalyte. The second dimensions, SDS-PAGE, were run on 10% - 20% gradient gels. Gels were coomassie stained. The spot corresponding to the recombinant protein was physically cut from each gel and the protein was eluted from it using Solvable (DuPont Co., Wilmington, DE).

Treatment of Data. The protein samples, after being eluted from the gel material, were counted for radioactivity using a Beckman LS-100C liquid scintillation counter. All values were greater than 6,000 cpm and most were on the order of 20,000 - 50,000 cpm. Background readings were on the order of 40 cpm. The amounts of ^{35}S and ^3H in each sample were determined correcting for ^{35}S spillover into the ^3H window. $^3H/^{35}S$ ratios for each protein sample were thus obtained, but needed to be adjusted to account for differences in the chase cultures. The samples run on the electrophoresis gels had been prepared using identical amounts of [^{35}S]methionine labeled cell suspension, however, the amounts of [^3H]leucine labeled material varied because different amounts of [^3H]leucine labeled cells had been used to inoculate each chase culture. To account for this the $^3H/^{35}S$ ratios were multiplied by correction factors. For example in one particular experiment, the 0 hour chase, 1 hour chase, and 2 hour chase cultures consisted respectively of 10 ml, 5 ml and 3 ml of [^3H]leucine labeled cell suspension and 0 ml, 5 ml and 7 ml of chase media. The raw $^3H/^{35}S$ ratios for CheY from the insoluble fraction gels were 3.67, 2.33, and 2.89 and from the soluble fraction gels were 3.50, 1.20, and 0.24. Multiplication with the correction factors 10/10, 10/5 and 10/3 gave the values 3.67, 4.66 and 9.65 for the insoluble protein samples and 3.50, 2.41, and 0.79 for the soluble protein samples. These indicate the relative amounts of radiolabeled CheY in the two fractions of each culture at the end of the chase periods. These were normalized against the 0 hour chase values giving 1.00, 1.27, and 2.63 for the insoluble protein samples and 1.00, 0.69, and 0.23 for the soluble protein samples.

Results

Pulse-Chases (CheY). Pulse-chase experiments were performed with the strain JM109 pME124s15c which overexpresses a mutated form of the CheY protein. These experiments were run to check the assumption commonly made that the recombinant protein would partition rapidly into insoluble and soluble forms and that the ratio of insoluble to soluble protein would subsequently remain constant. Each experiment can be characterized in terms of three stages: (a) a start-up period in which cultures were brought to a stage of active growth and recombinant protein production, (b) a pulse period in which cultures were incubated in the presence of a radiolabeled amino acid, and (c) a chase period in which chase cultures were inoculated using washed radiolabeled cells. Chase cultures were run in parallel and harvested after different lengths of chase time. After harvesting, the amounts of labeled soluble and insoluble CheY protein in each chase culture were determined.

Three variations of the pulse-chase experiment were performed to obtain partitioning results at different levels of recombinant protein production. Recombinant protein levels were varied using different IPTG concentrations and media (Recombinant protein expression and inclusion body formation are significantly greater in LB than either M9 or M9AA.) The conditions in the three experiments were as follows: (a) Low Level - The IPTG concentration was maintained at 0.05 mM and M9 medium was used in all three stages of the experiment. (b) Intermediate Level - The IPTG concentration was kept at 1.0 mM and M9 was used in the startup and pulse stages followed by LB in the chase stage.

(c) High Level - IPTG was again kept at 1.0 mM. LB was used in the startup and chase stages. An M9 medium supplemented with amino acids, M9AA, was used in the pulse stage to obtain as high a growth rate as possible under defined conditions. The M9AA medium did not contain the amino acid used as the radiolabel and so sufficient uptake of radiolabel could be obtained. The CheY spots on the 2D gels from the High and Intermediate Level experiments were much more intense than the corresponding spots from the Low Level experiment. Inclusion bodies were clearly visible with a light microscope prior to the pulse stage during the High Level experiment, but not during the other two experiments. Thus, the levels of recombinant protein in the three experiments varied as anticipated.

The results of these three experiments appear in figures 1 and 2. Figure 1 is a plot of the normalized amounts of insoluble radiolabeled CheY versus chase time. Figure 2 contains the corresponding plots of soluble CheY. The protein amounts are normalized with the amount present at the start of the chase stage so that the magnitude of change over the course of the chase is readily determined. In each experiment, the amount of radiolabeled recombinant protein in the insoluble fraction increased continuously by a factor between 2.5 and 5. The amounts in the soluble fraction decreased correspondingly to around 30% of the initial levels. The magnitudes of the changes are significant and occurred over a time scale of hours. These results were unexpected and are difficult to explain based on prevalent models and assumptions of typical protein folding kinetics. Cell growth and product formation kinetics could have been altered by the changes in media and temperature resulting from the washing step and this could have affected the rates of change in the plots. However, the significant result is the fact that there were, indeed, changes of a large magnitude which occurred over a long time period.

Proteolysis (CheY). The rate of proteolysis was measured to determine the extent to which it was responsible for the decreases in the amounts of radiolabeled soluble protein. A pulse-chase experiment was run using 0.05 mM IPTG and M9 throughout. However, unlike the previous experiments where the cell lysates were fractionated, whole cell lysates were run on the gels. Changes in the total amount of radiolabeled CheY were thus measured. A slow decrease in the level of labeled CheY was seen such that 80% of the original radiolabeled CheY remained after four hours of chase. The relative amounts of soluble and insoluble radiolabeled CheY at the start of the chase were determined in a separate experiment run under the same conditions as the proteolysis pulse-chase experiment. It found that approximately 80% of the labeled CheY was soluble at the start of the chase. A similar value was calculated for the CheY partitioning using a material balance as discussed later. If the assumption is made that only soluble CheY is susceptible to proteolysis, approximately of 75% of the original soluble radiolabeled CheY should have remained at the end of the Low Level experiment. The measured result by comparison shows that only approximately 30% of the original soluble material remained. Proteolysis is, therefore, seen to be rather modest and insufficient to account for the decreases in the amounts of radiolabeled CheY seen in the pulse-chase experiments.

Pulse-Chase (EGD). A pulse-chase experiment was performed using a strain which expresses the protein EGD. The results confirm that radiolabel was not being incorporated into newly translated protein during the chase periods The experiment was performed with JM109 pCT603 using an IPTG concentration of 0.05 mM and M9 throughout. The recombinant protein expressed by this strain, endoglucanase EGD, contains a 3 kDa portion which is required for the protein's incorporation into the inclusion body but which is also rapidly clipped off by the cell (17, 20). Figure 3 shows the normalized amounts of soluble and insoluble radiolabeled EGD

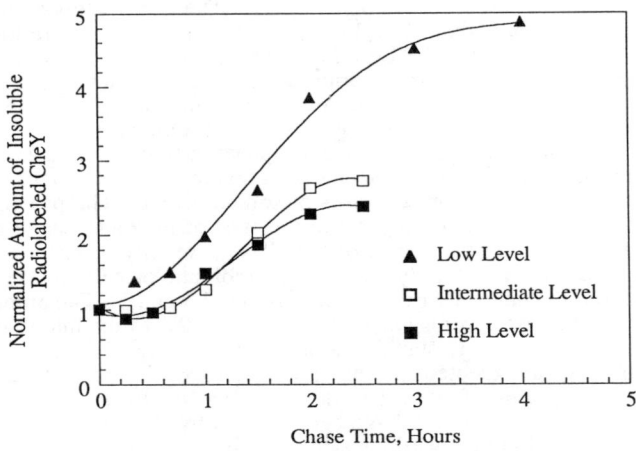

Figure 1. Low, Intermediate and High CheY Expression Pulse-Chases: Insoluble Fraction.

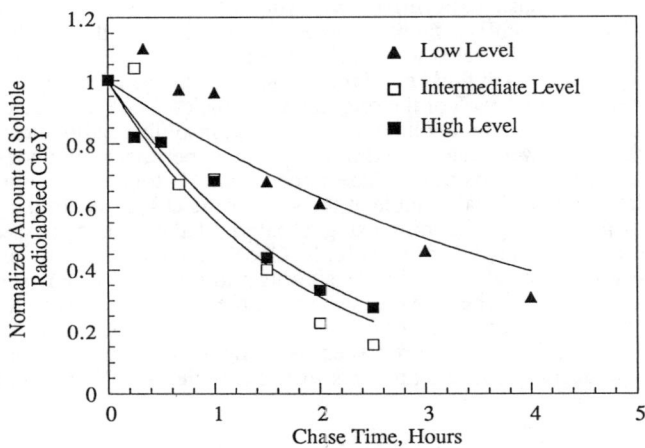

Figure 2. Low, Intermediate and High CheY Expression Pulse-Chases: Soluble Fraction.

plotted against the chase time. The amount of radiolabeled EGD in the insoluble fraction is seen to have increased only slightly before leveling off. Soluble EGD similarly showed no significant change throughout the chase period. The fact that the levels of radiolabeled soluble and insoluble EGD did not change significantly during the chase has significant implications. It confirms that incorporation of radiolabel into newly translated protein was not occurring during the chase periods.

Chloramphenicol Pulse-Chase (CheY). Experimental results were reported in the literature which appeared to contradict the results of the Low, Intermediate and High Level pulse-chase experiments. Krueger *et al.* reported the that when chloramphenicol was added to an *E. coli* strain which formed CheB inclusion bodies, the amounts of soluble and insoluble CheB remained unchanged for at least an hour (*14*). Chloramphenicol is an antibiotic which acts by blocking protein production. An apparent contradiction between the results exists because chloramphenicol would not be expected to affect the behavior of fully translated molecules. A pulse-chase experiment was therefore run in which chloramphenicol was present in the chase media. The IPTG concentration was maintained at 0.05 mM and M9 was used throughout. Figure 4 shows the normalized amounts of soluble and insoluble radiolabeled CheY plotted against chase time. Consistent with the literature report, the amounts of labeled CheY in the two fractions were essentially constant throughout the chase. The difference in the results of the experiments with and without chloramphenicol is surprising. A possible mechanism which accounts for this difference is discussed later.

Discussion

The results of the pulse chase experiments without chloramphenicol indicate that soluble recombinant CheY migrated into the insoluble fraction over a long time period. The crux of the argument for a migration of recombinant protein from the soluble form to the insoluble form is a material balance: the amount in the insoluble fraction increased and this increase was matched by a corresponding decrease in the soluble fraction. Likely alternative interpretations will be considered but are not supported by the experimental evidence. The migration of protein from the soluble to the insoluble fraction plays a significant role in determining the overall partitioning of recombinant protein between the two fractions. A mechanism for the migration is proposed which involves an equilibrium between native and unfolded protein conformations and which is consistent with all the experimental results.

Material Balance. A material balance can be used to show that a movement of protein from the soluble fractions to the insoluble fractions occurred during the pulse-chase experiments without chloramphenicol. The amounts of radiolabeled CheY which entered the insoluble fractions were approximately equal to the amounts which left the soluble fractions. This is equivalent to saying that the total amount of labeled CheY remained constant since all the CheY protein is accounted for in these two fractions. Therefore, since there are no sources of radiolabeled CheY during the chases and none has been lost, the material which has left one fraction must have entered the other.

A check can be made that the amount of labeled CheY which left the soluble fractions equals that which entered the insoluble fraction by verifying that the following equation holds:

$$\text{Sol}(0) - \text{Sol}(t) = \text{IB}(t) - \text{IB}(0) \tag{1}$$

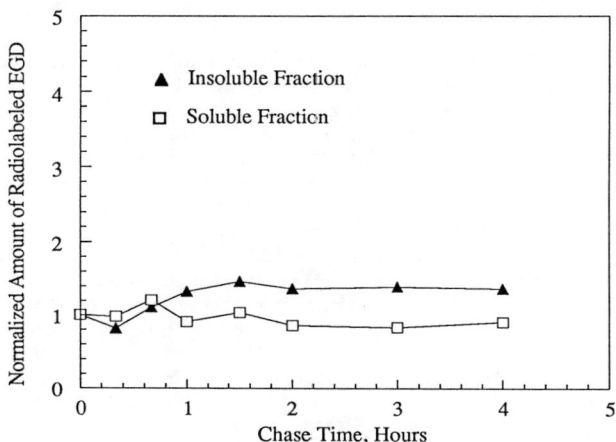

Figure 3. EGD Pulse-Chase: Soluble and Insoluble Fractions.

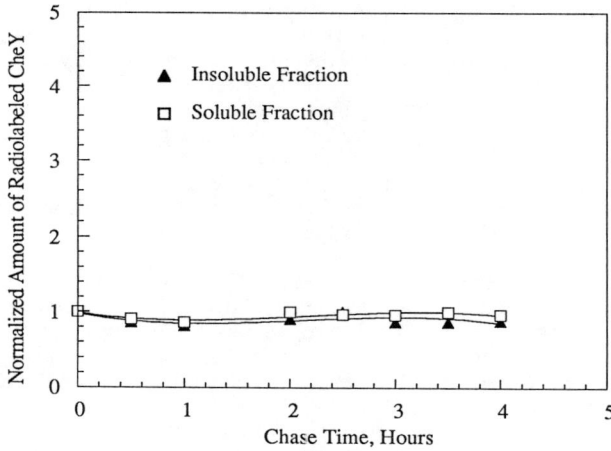

Figure 4. Chloramphenicol Pulse-Chase: Soluble and Insoluble Fractions.

where: Sol(t) and IB(t) are the amounts of soluble and insoluble radiolabeled CheY present after t hours of chase. Rearranged equation 1 is:

$$\frac{IB(0)}{Sol(0)} \frac{\left(\frac{IB(t)}{IB(0)} - 1\right)}{\left(1 - \frac{Sol(t)}{Sol(0)}\right)} = 1 \qquad (2)$$

The experimental data available is sufficient to obtain values for the left-hand side of this equation once values are known for the initial partitioning, IB(0)/Sol(0). These can be determined from equation 2 using a least squares best fit. The initial partitioning values are 0.16, 0.50, and 0.55 for the Low Level, Intermediate Level and High Level experiments respectively. Using these values for the initial partitioning, values for the left hand side of equation 2 are found to be mostly within 10% of 1.0. None of the values deviated from 1.0 by more than 20%. Clearly there is good agreement between the amounts of material which left the soluble fractions and the amounts which entered the insoluble fractions. These results apply to all calculations made using the last three or four chase time data points of the Low, Intermediate and High Level experiments, i.e. the long chase time data points. The changes in the amounts of radiolabeled soluble and insoluble CheY measured at short chase times were on the same order as the precision of the experiments and so these data points cannot be used in the calculations.

Alternative Possibilities. No evidence could be found to support any alternatives to the interpretation that recombinant protein migrated from the soluble form to the insoluble form. Two alternative possibilities seemed most plausible: 1) that radiolabeled CheY continued to be produced during the chase periods, and 2) that the centrifugation step intended to separate the soluble and insoluble fractions was only partially successful at pelleting the insoluble material. It has already been shown that proteolysis of CheY is insufficient to account for the decreases in the amounts of soluble radiolabeled recombinant protein.

The first alternative possibility, that radiolabel continued to be incorporated into newly translated protein during the chases, is in conflict with the results of the EGD pulse-chase experiment. This possibility might have explained the increases in the amounts of insoluble radiolabeled CheY. Sufficient free radiolabeled amino acid for the continued incorporation could have resulted from turnover of radiolabeled protein or incomplete removal of excess radiolabel during the washing step. The amount of radiolabeled EGD in the insoluble fraction did not, however, increase throughout the chase as the CheY had. This may be because the segment of the soluble EGD molecules required for aggregation had been proteolytically clipped off. Whatever the reason, translation of EGD was continuing during the chase and newly translated EGD would have been entering the insoluble fraction. Therefore, continued incorporation of radiolabel into newly translated molecules during the chase was not occurring.

Other measurements confirm this conclusion. Two random spots from the insoluble fraction gels of the Intermediate and High Level experiments were tracked along with the CheY spots. Neither indicated that a significant increase in the amount of radioactivity in the insoluble fraction had occurred. Also measurements were made of the free radioactivity in the media during the chase period of an experiment in which excess radiolabel from the pulse was not removed. No decrease in amount of free radiolabeled amino acid was observed. Measuring the amount of free radioactivity in the media had proven to be an accurate means of following the uptake of radiolabel during pulse periods. These observations serve to confirm the conclusion that the production of radiolabeled CheY ceased at the end of the pulse period.

Experimental confirmation was obtained which eliminated the other alternative possibility, that the centrifugation used in these experiments was insufficient to achieve complete recovery of the insoluble recombinant protein. If complete recovery of insoluble material was not being achieved, small particles of insoluble material may have remained in the soluble fractions until they had grown sufficiently to be recovered in the insoluble fraction. The observed changes in the amounts of soluble and insoluble protein might then have resulted from particle growth during the chase period. This possibility was discounted with the following experiment: A culture was grown and radiolabeled such that when harvested it resembled the 0 hour chase culture of the Low Level experiment. The culture was split into several identical parts which were centrifuged at different speeds ranging from 6,000 to 15,000 rpm. The amounts of radiolabeled recombinant protein in the supernatants, i.e. the 'soluble fractions', were determined using 2D gels as before. They were found to be quite consistent, within 5% of each other. Similar results were obtained when the experiment was repeated using a culture grown under conditions similar to the 1 hour chase culture of the High Level experiment. The insoluble material is seen to be easily recovered since the amounts of radiolabeled recombinant protein recovered were independent of the centrifuge speeds used. A trend of increasing recoveries with increasing centrifuge speeds would have revealed that recovery was difficult. Reports in the literature confirm that centrifugation forces no greater than the 5,000 x g present at 6,000 rpm and far less than the 32,000 x g present at 15,000 rpm can achieve virtually complete inclusion body recovery (21, 22).

Protein Partitioning. The partitioning of the radiolabeled recombinant protein at each point in the chase can be determined:

$$\frac{IB(t)}{Sol(t)} = \frac{IB(0)}{Sol(0)} \frac{\left(\frac{IB(t)}{IB(0)}\right)}{\left(\frac{Sol(t)}{Sol(0)}\right)} \tag{3}$$

The results appear in figure 5 in which the ratio of insoluble radiolabeled CheY to soluble radiolabeled CheY is plotted against chase time. It is apparent that the migration from the soluble form to the insoluble form plays a significant part in determining the overall partitioning of the recombinant protein. The initial partitioning of protein may not strongly favor the formation of the insoluble form. Eventually, though, most of the recombinant protein ends up as insoluble protein. This behavior has important implications for the preferential recovery of protein in either the soluble or insoluble form.

Implications for the Mechanism of Inclusion Body Formation. The fact that the results of the pulse-chase experiments with and without chloramphenicol are so different is surprising and complicates interpretation of the results in terms of a mechanistic model. The prevalent model of inclusion body formation holds that newly translated protein is susceptible to aggregation only until it folds. If this were true for CheY the slow incorporation of protein into the insoluble fractions would be due to a particularly long lived folding intermediate. The stability of this intermediate would be exceptional. Another possible explanation of the results is

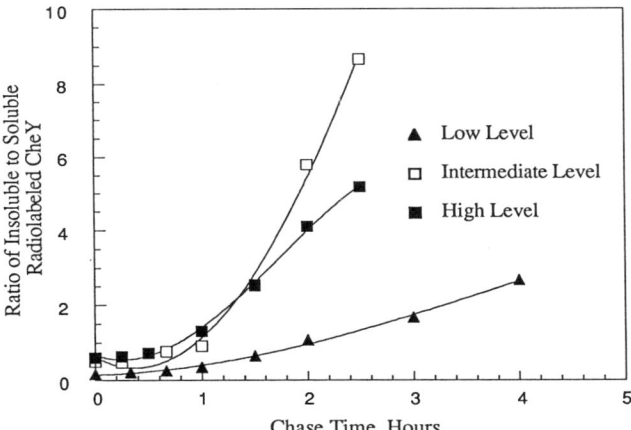

Figure 5. Low, Intermediate and High CheY Expression Pulse-Chases: Ratios of Insoluble to Soluble Radiolabeled CheY.

that the mutated CheY protein is "sticky". Namely that the soluble protein is either susceptible to aggregation in its native state or that it never folds completely. Neither of these explanation is consistent with the results of the chloramphenicol experiment. In both cases, the migration from the soluble to the insoluble fraction should continue during the chase, at least initially, since the concentration of the aggregating species would remain near what it had been during the pulse. Further, Stock *et. al.* (*16*) demonstrated that this particular CheY mutant retained its activity indicating that the protein does fold completely.

A model for inclusion body formation which is consistent with all the experimental data can be arrived at by eliminating the assumption that the folding reaction leading to native protein is irreversible. The reverse reaction would provide a pathway for native protein to become incorporated into the inclusion body by unfolding and aggregating with other unfolded molecules. This interpretation would explain why the incorporation of soluble protein into the insoluble fraction was so slow. It further explains why the migration from the soluble fraction to the insoluble fraction is dependent upon whether or not translation is ongoing. When translation is occurring, as in the Low, Intermediate and High Level experiments, the concentration of unfolded protein should be relatively high since it is supplied with newly translated unfolded molecules. However, when no translation is occurring as in the chloramphenicol experiments reported here and in Krueger *et al.*(*14*), the concentration of unfolded protein would be relatively low. Since it is likely that aggregation is a higher order reaction than folding, the folding reaction would be favored over aggregation at low concentrations of unfolded protein. The same interpretation can be applied to *in vitro* protein solutions. The amount of aggregation observed is often not significant because the concentration of unfolded protein is low. *In vitro* systems have only the reverse folding reaction as a source of unfolded protein and typical protein folding equilibrium constants would not yield high concentrations of unfolded protein. However, there have been reports of aggregation *in vitro* and it has been observed that the tendency to aggregate *in vitro* correlated with the tendency of the proteins to forms inclusion bodies *in vivo* (*14*).

Conclusion

Pulse-chase experiments were run to check the hypothesis that recombinant protein partitions rapidly into insoluble and soluble forms and that the ratio of insoluble to soluble protein subsequently remains constant. The hypothesis was not supported by the experimental results. Large changes in the amounts of soluble and insoluble radiolabeled CheY occurred throughout the chase periods. These changes resulted from a migration of the protein from the soluble form to the insoluble form. However, no changes occurred in the amounts of soluble and insoluble radiolabeled CheY when chloramphenicol was present in the chase media. Both of these experimental results are consistent with a model for inclusion body formation which excludes the commonly made assumption that the soluble native form of the protein does not become incorporated into the insoluble fraction.

Acknowledgments

The authors are very grateful to D. Wetlaufer for his expert advice and assistance. We are also grateful to J. Stock at Princeton University for providing the plasmid pME124s15c. This work was funded in part by a grant from the National Science Foundation PYI Award ECE-8552492.

References

1. Goldberg, D.; King, J. *Biochem.* **1982**, 79, 3403-3407.
2. Goldberg, D.; Smith, D. H.; King, J. *Proc. Natl. Acad. Sci.* **1983**, 80, 7060-7064.
3. Yu, M.; King, J. *Proc. Natl. Acad. Sci.* **1984**, 81, 6584-6588.
4. Hasse-Pettingell, C. A.; King, J. *J. Biol. Chem.* **1988**, 263, 4977-4983.
5. Mitraki, A.; King, J. *Bio/Tech.* **1989**, 7, 690-697.
6. Zettlmeissl, G.; Rudolph, R.; Jaenicke, R. *Biochem.* **1979**, 18, 5567-5571.
7. Cleland, J. L.; Wang, D. I. C. *Biochem.* **1990**, 29, 11072-11078.
8. Goldberg, M. E.; Rudolph, R.; Jaenicke, R. *Biochem.* **1991**, 30, 2792-2797.
9. Kiefhaber, T.; Rudolph, R.; Kohler, H.; and Buchner, J. *Bio/Tech.* **1991**, 9, 825-829.
10. Goloubinoff, P.; Christeller, J. T.; Gatenby, A. A.; Lorimer, G. H. *Nature* **1989**, 342, 884-888.
11. Viitanen, P. V.; Lubben, T. H.; Reed, J.; Goloubinoff, P.; O'Keefe, D.; Lorimer, G. H. *Biochem.* **1990**, 29, 5665-5671.
12. Buchner, J.; Schmidt, M.; Fuchs, M.; Jaenicke, R.; Rudolph, R.; Schmid, R. X.; Kiefhaber, T. *Biochem.* **1991**, 30, 1586-1591.
13. Blum, P.; Velligan, M.; Lin, N.; Matim, A. *Bio/Tech.* **1992**, 10, 301-304.
14. Krueger, J.; Stock, A. M.; Schutt, C. E.; Stock, J. B. In *Protein Folding;* Gierasch, L. and King, J., Eds.; American Association for the Advancement of Science: Washington D.C., 1990; 136-142.
15. Yanisch-Perron, C.; Vieira, J.; Messing, J. *Gene* **1985**, 33, 103-119.
16. Stock, A. M.; Mottonen, J. M.; Stock, J. B.; Schutt, C. *Nature* **1989**, 337, 745-749.
17. Tokatlidis, K.; Dhurjati, P.; Millet, J.; Beguin, P.; Aubert, J. *FEBS* **1991**, 282, 205-208.
18. Mosteller, R. D.; Goldstein, R.V.; Nishimoto, K. R. *J. Biol. Chem.* **1980**, 255, 2524-2532.
19. Dunbar, B. S. *Two-Dimensional Electrophoresis and Immunological Techniques*; Plenum Press: New York, NY, 1987.
20. Tokatlidis, K.; Salamitou, S.; Beguin, P.; Dhurjati, P.; and Aubert, J. *FEBS* **1991**, 291, 185-188.
21. Taylor, G.; Hoare, M.; Gray, D. R.; Marston, F. A. O. *Bio/Tech.* **1986**, 4, 553-557.
22. Middelberg, A. P. J.; Bogle, I. D. *Biotech. Prog.* **1990**, 6, 255-261.

RECEIVED November 11, 1992

Chapter 6

Molecular Chaperones and Their Role in Protein Assembly

Saskia M. van der Vies[1], Anthony A. Gatenby[2], Paul V. Viitanen[2], and George H. Lorimer[2]

[1]Département de Biochimie Médicale, Centre Médical Universitaire, 1 Rue Michel-Servet, CH−1211 Geneva 4, Switzerland
[2]Central Research and Development, E. I. du Pont de Nemours and Company, Experimental Station, Wilmington, DE 19880−0402

It has recently become clear that a specialized group of proteins termed molecular chaperones has evolved to assist protein assembly in the cell, a process that was originally considered a spontaneous event. Molecular chaperones function by preventing the formation of biologically inactive structures by binding non-covalently to exposed protein surfaces that have the tendency to interact incorrectly with other components in the cell. Molecular chaperones are required for a number of fundamental processes such as protein synthesis, protein translocation, DNA replication and recovery from stresses like heat. Their existence has implications for biotechnology.

To understand how information present in the amino acid sequence is utilized to produce biologically functional proteins is one of the fundamental questions in biology. The early observations made by Anfinsen and colleagues (*1*) that denatured purified ribonuclease refolds spontaneously in the absence of any other proteins into an active enzyme, formed the basis for the "self assembly hypothesis". This hypothesis states that all the information required to assemble a polypeptide chain into a biological three-dimensional structure is contained within the amino acid sequence. The successful assembly of many other more complex structures from their denatured components *in vitro*, such as tobacco mosaic virus and the bacterial ribosomes, led to the assumption that protein assembly *in vivo* should also be a spontaneous event. However, in many cases the rates of protein assembly *in vitro* and *in vivo* differ. The reassembly of proteins *in vitro* occurs often with low yield particularly at high protein concentrations and physiological temperatures (reviewed in *2*). Results of studies in a number of different experimental systems have recently led to the realization that protein assembly *in vivo* is more complex than was originally thought, and requires the involvement of other proteins that have collectively been termed molecular chaperones (*3*).

Molecular chaperones are currently defined as a class of unrelated families of cellular proteins that mediate the correct assembly of other polypeptides, but are not themselves components of the final functional structures (*4*). It has been proposed that molecular chaperones assist protein assembly by promoting productive assembly pathways, thereby preventing molecules from forming biologically inactive structures, such as those found in inclusion bodies and/or aggregates. The term "assembly" used here covers both the folding of nascent polypeptides as well as any additional oligomerisations into larger protein structures. The formation of inactive structures is considered a result of incorrect interactions of hydrophobic and/or charged surfaces that are normally involved in domain interactions. The role of molecular chaperones is to prevent these incorrect interactions. It is argued that there is a need for molecular chaperones because the intrinsic properties of proteins assure that incorrect interactions are possible.

The general concept of molecular chaperones (reviewed in *5*) originated as a result of studies on the biogenesis of the chloroplast enzyme ribulose bisphosphate carboxylase-oxygenase, or rubisco, which fixes carbon dioxide in photosynthesis (reviewed in *6*). However this concept also has implications for other area's of research. In biotechnology, for instance, problems are often encountered when recombinant proteins are produced in heterologues hosts. The possibility that human chaperone diseases exist whereby some proteins fail to assemble correctly because of mutations in certain chaperones, may have medical implications.

The novel concept of molecular chaperones has stimulated research in several laboratories resulting in a rapidly growing chaperone literature. Here we describe the role of molecular chaperones and discuss their influence on protein assembly. Several other reviews have been published recently (*7-10*).

The Role of Molecular Chaperones *In Vivo*

Molecular chaperones perform an essential role and function during many cellular processes where changes in protein-protein interaction and consequently the exposure of hydrophobic and/or charged protein surfaces occur. These surfaces are normally sheltered in domain interactions, and are essential for sustaining the native protein structure. However, when transiently exposed, these surfaces have the potential to interact with other cellular components which may result in the formation of biologically inactive structures. Molecular chaperones prevent these incorrect interactions by binding non-covalently to the transiently exposed protein surfaces. Some processes that have been suggested to require molecular chaperones are discussed below, while others have been reviewed elsewhere (*7*).

Protein synthesis is a vectorial process whereby the genetic information present in the RNA is translated sequentially into a polypeptide chain. Thus all the information required to assemble the protein, which is present in the amino acid sequence, may not be available at the same time. If the rate of protein assembly is faster than the rate of protein synthesis, the amino terminal region may commence folding before the complete RNA has been translated. The rate of protein refolding *in vitro* is frequently completed in seconds (*2*), while the synthesis of an average protein in yeast takes about 2 minutes (*11*). Therefore, *in vivo* the amino terminus may become engaged in incorrect interactions with itself or other components in the cell producing inactive structures. Chaperones may be required to prevent these incorrect interactions. For example, the chaperonins, one family of molecular chaperones (Table I), bind to newly synthesized polypeptides in cell extracts of *Escherichia coli* (*12*), while chaperones of the heat shock 70 (hsp70) family bind transiently to polypeptides during their synthesis *in vivo* (13).

Proteins that are transported across bacterial and eukaryotic membranes traverse in an unfolded or partially folded conformation. It has been suggested that molecular chaperones maintain newly synthesized precursor polypeptides in a partially folded conformation in the cytosol prior to their translocation (*4*). When the protein has been transferred to the cellular location where it functions (proteins may have to traverse more than one membrane), correct assembly occurs. This process has also been proposed to require the assistance of molecular chaperones. Ample evidence suggests that members of the hsp70-BiP and chaperonin families of molecular chaperones are required during protein transport (reviewed in *14-16*).

Many kinds of stress such as heating, chilling, wounding and infection either elicit or enhance the expression of specific genes. All organisms examined respond to heat by inducing the synthesis of a group of proteins called heat-shock proteins (reviewed in *17*). Extremes of heat cause protein denaturation and often the formation of insoluble aggregates. It has been suggested that one role of molecular chaperones is to unscramble these aggregates (*18*). Many, but not all, molecular chaperones are stress proteins that also function in the absence of stress. There is no evidence so far that chaperones can unscramble insoluble aggregates and allow correct reassembly of the polypeptide chains. However, it has been reported that the *E.coli* DnaK chaperone mediates the reappearance of enzyme activity from heat denatured RNA polymerase *in vitro* (*19*).

The degradation of proteins involves the exposure of protein surfaces that contain substrate recognition sites for the proteolytic machinery. These sites are often present on interactive surfaces that are normally buried in the interior of the folded protein molecule. Molecular chaperones may assist in presenting the polypeptide in a conformation suitable for degradation. The observation that the binding of the *E. coli* DnaK protein, a chaperone belonging to the hsp70 family (Table I), to a mutant form of alkaline phosphatase is required for the ATP-dependent degradation of the enzyme is supportive of this idea (*20*). While studies with isolated proteins *in vitro* have shown that polypeptides bound to the *E.coli* chaperonin are susceptible to proteolytic degradation (*21-23*).

The Molecular Chaperone Class

The number of currently recognized chaperones is growing rapidly as their involvement in an increasing number of cellular processes is appreciated. It is now evident that chaperones are found in every cell compartment where protein assembly occurs. The molecular chaperones class presented in Table I is defined functionally

Table I. Proteins Regarded as Molecular Chaperones

Class	Members	Functions
Nucleoplasmin	Nucleoplasmin Protein XLNO-38 Protein Ch-NO38 Nucleoplasmin S	Assembly of nucleosomes and ribonucleoprotein particles in eukaryotes
Chaperonins	Chaperonin 60 Chaperonin 10	Folding of nascent polypeptides and transported proteins in bacteria, plastids, mitochondria and the cytosol
Heat shock proteins 70	Hsp68, 72, 73; DnaK;BiP; clathrin uncoated ATPase; grp75, 78, 80; hsc 70, KAR2; SSA1-4; SSB1; SSC1; SSD1	Assembly of newly synthesized and transported proteins in pro- and eukaryotes.
Heat-shock protein 90	Hsp 83, 87; HtpG	Masking of steroid receptors
Signal recognition particle	Several homologues in the cytosol of pro-and eukaryotes	Arrest of translation and prevention of incorrect folding of precursor molecules prior to translocation
Prosequences	Prosubtilisin, pro-alpha lytic protease	Protease assembly
Ubiquitinated proteins	Precursor ribosomal proteins	Ribosome assembly
Trigger factor		Protein transport in bacteria
Sec B protein		Protein transport in bacteria
papD protein		Pilus assembly in bacteria

while the different families within this class are defined on the basis of sequence similarity.

Nucleoplasmin is considered the archetypical molecular chaperone. Laskey et al (24) were the first to use this term, to describe the function of a soluble nuclear protein that mediates the assembly of nucleosome cores. It was Ellis who extended the term molecular chaperones to describe a wider range of proteins (3).

Some chaperones (Table I) are covalently attached to the molecules whose assembly they mediate, such as the pro-sequences of the pro-subtilisin (25) and pro-alpha-lytic protease (26) and the aminoterminal ubiquitin sequence present on some small ribosomal proteins in eukaryotes (27). In all three cases, the presence of the pro-sequences improves the correct assembly of functional structures, but are subsequently removed. The advantage of these cotranslational chaperones is presumably that they don't have to detect their substrate among the enormous variety of other protein molecules.

It can be argued that the presequences found on precursor proteins, which are transported across membranes should also be regardes as molecular chaperones, because they prevent the rest of the polypeptide chain to fold into a transport-incompetent conformation. However, in several cases it has been found that the chaperone function of these presequences is not sufficient to allow efficient transport of polypeptide chains without the assistance of other separate chaperones; e.g. the signal recognition particle in the case of cotranslational protein transport (28) and the trigger factor, the secB protein and the bacterial chaperonin 60 in the case of posttranslational protein transport (12, 29, 30). However, these observations raise the interesting possibility that the correct assembly of some proteins requires the assistance of more than one chaperone (see later). Clearly the most interesting question is by what mechanism the molecular chaperones function; what does each chaperone recognize in a range of functionally unrelated proteins and what is the underlying molecular mechanism by which they mediate protein assembly. Here we shall focus on the chaperonins, the structurally most complex of the molecular chaperones and discuss their effect on protein assembly.

The Chaperonin Family

Chaperonins constitute a family of sequence-related molecular chaperones that are found in all bacteria, mitochondria and plastids examined (4). The recent discovery of a protein from the eukaryotic cytosol that shows sequence-similarity with the known chaperonins may indicate that chaperonins are probably present in all compartments of the cell (31 and reviewed in 32).

There are two types of chaperonins that are sequence-related to each other (33). The larger type is called chaperonin 60 (cpn60) since its subunit Mr is about 60,000 while the smaller type is called chaperonin 10 (cpn10) with a subunit Mr of about 10,000. The purified chaperonin 60 from E.coli (originally known as the groEL protein), chloroplasts (originally known as the rubisco subunit binding protein) and yeast mitochondria (also known as hsp 60) all have a distinct structure involving 14 subunits arranged in two stacked rings containing 7 subunits each (known as the "double donut", 34-36). Interestingly the cytosolic chaperonin from the archaebacterium Sulfolobus occurs as a complex of two stacked nine-membered rings (37), while the mammalian mitochondrial chaperonin 60 (38-41) has been purified as a single ring of 7 identical subunits. The bacterial, mitochondria and cytosolic chaperonin 60 are homo-oligomers in contrast to the chloroplast chaperonin which contains equal amounts of sequence-related subunits, α and β (*33, 42*). The biological significance of these different arrangements is unknown but deserves attention. The chaperonin 10 (originally known as the groES protein in *E.coli*) occurs as an oligomer of 7 identical subunits arranged in a single ring (*43*) and has recently been identified in mammalian mitochondria (*44, 45*) and spinach chloroplasts (*46*). Interestingly the chloroplast chaperonin, unlike the mitochondrial and bacterial cpn 10, appears as a binary protein with a Mr of 24,000 consisting of two cpn10-like domains fused together. The purified chaperonin 60 possesses weak ATPase activity which is effectively inhibited upon binding of the chaperonin 10. This binding itself requires ADP or ATP (*43, 47*).

The *E.coli* Chaperonins. The chaperonins from *E.coli* were the first to be studied in detail and their involvement in protein assembly *in vivo* and *in vitro* is discussed herein. Several reviews and papers have been published describing the chaperonins from other species (*4, 6, 36-42, 48-50*).

Protein Assembly *In Vivo* is Mediated by Chaperonins. The *E.coli* chaperonins were originally identified because mutations in their genes prevented the replication of several bacteriophages (reviewed in *51*). The chaperonins 60 and 10 interact functionally *in vivo* (*52*) and are essential for the formation of the bacteriophage lambda "preconnector", a protein structure on which the phage head proteins assemble (*53*). The chaperonins are essential for cell viability at all temperatures (*54*) and the genes are present on one operon (*52*) whose expression is induced upon heat shock (*55*). Additional studies have revealed that the two types of chaperonins are involved in a number of processes in uninfected bacterial cells. For example, genetic evidence suggests a role in DNA replication (*56, 57*), cell division (*58*), protein secretion (*59, 60*) and oligomeric protein assembly (*61, 62*). Most of the cellular processes that are mediated by chaperonin 60 also require chaperonin 10. However, chaperonin 60 alone is sufficient for maintaining polypeptides in conformations that are suitable for translocation across membranes *in vitro*. It appears that cpn60 inhibits the folding of the precursor and holds it in a conformation that interacts correctly with the transport machinery (*12*).

E.coli Chaperonins Influence Protein Assembly *In Vitro*. The observation that the correct reassembly of chemically denatured purified rubisco requires the function of the two *E.coli* chaperonins and MgATP provided the first direct evidence in support of the proposed assisted self-assembly mechanism (*63, 64*). The interaction of the *E.coli* chaperonins with a number of other purified proteins has since been documented: for example, pre-β-lactamase (*65*), rhodanese (*66,*), dihydrofolate reductase (*21, 22*), lactate dehydrogenase (*67*), citrate synthase (*68*), α-glucosidase (*69*), an antibody F_{ab} fragment (*70*) and a variety of thermophilic enzymes (*71*). The *E.coli* chaperonins appear to influence the reassembly by binding non-covalently to a range of unrelated proteins. From these *in* *vitro* studies some generalizations can be deduced which are summarized below (reviewed in *72*).

A schematic representation of the refolding of proteins *in vitro* and the influence of the *E.coli* chaperonins is presented below, where N = native; I = folding intermediate; U = unfolded; cpn60 = chaperonin 60 and cpn10 = chaperonin 10, the * indicates a changed conformation of I.

$$U \rightleftharpoons I \rightleftharpoons (cpn60)_{14}\text{-}I^* \underset{(cpn10)_7}{\overset{MgATP \quad K^+}{\rightleftharpoons}} N$$

$$\updownarrow N \qquad \updownarrow aggr$$

Most purified proteins that are treated with high concentrations of chaotropes such as urea or guanidinium hydrochloride, loose their biological activity and adapt an unstructured conformation. Dilution of the chaotrope results in a rapid formation of folding intermediates, some of which are stable and have been characterized (reviewed in *73*). The folding intermediate of the bacterial enzyme rubisco is stable at low temperatures (4°C) and physiological conditions of ionic strength and pH (*23*). At a given temperature and up to a well-defined concentration (referred to as the critical aggregation concentration, C.A.C.) the folding intermediate is stable and will fold to the native state in the absence of the chaperonins. At concentrations higher than the C.A.C., aggregation occurs until the concentration of the folding intermediate is reduced to the C.A.C. after which the residual molecules fold into an active conformation. However, in the presence of the chaperonin 60, the stable folding intermediate forms a binary complex in a reaction that is kinetically indistinguishable from its refolding to the native state suggesting that the chaperonin recognizes structural features present in the folding intermediate. Under appropriate conditions, unfolded rubisco (*23, 47*), rhodanese (*65, 74*), citrate synthase (*68*) and α-glucosidase (*69*) aggregate upon dilution of the chaotrope. However, when cpn60 is present during dilution, aggregation is prevented because the folding intermediates are trapped by the chaperonin. Thus, depending on the experimental conditions, the cpn60 either prevents aggregation or folding to the native state by binding to folding intermediates.

Characterization of the rubisco folding intermediate has revealed structural properties similar to the so-called "molten-globule" (23), an intermediate formed during the refolding of many proteins *in vitro* (75, 76). The observation that the dihydrofolate reductase folding intermediate is prevented from binding to the cpn60 in the presence of an excess of casein, a protein with properties of a partially folded protein, suggests a "molten-globule"-like conformation for the substrate polypeptide (21). Dihydrofolate reductase, rhodanese and rubisco polypeptides bound to cpn60 are in a partially folded conformation as judged by their sensitivity to proteolytic degradation (21-23). Analysis of the spectral properties of the binary complex of cpn60 with either dihydrofolate reductase, rhodanese or α-glucosidase may also suggest that the chaperonin stabilizes a structure that resembles a "molten-globule" (21, 69). However, the precise structural motif recognized by cpn60 is unknown. Most substrate proteins analyzed contain a variety of secondary structural elements in their folded states and, thus, provide little clues for the recognition site. Two dimensional NMR analysis has revealed that a protein fragment corresponding to the aminoterminus of native rhodanese adopts an α-helical conformation while bound to the chaperonin 60 (77). However, it has recently been shown that an antibody Fab fragment, a protein completely devoid of α-helices in the native state, interacts with the chaperonin 60 during refolding *in vitro* in a manner similar to that observed for all proteins examined (70). Thus, the recognition site may not be based on a defined secondary structure but rather on the accessibility of the interactive side chains that constitute hydrophobic surfaces.

Polypeptides bound to cpn60 are released by MgATP and potassium ions. Release may also result in generation a biologicallly active protein depending on the nature of the polypeptide. The binding of ATP and/or non-hydrolysable ATP-analogous to cpn60 alone has been shown to induce large conformational changes measured by changes in protease sensitivity (78). Such conformational changes are sufficient to reduce the affinity of cpn60 for the substrate polypeptide. If the polypeptide is released under conditions where it refolds spontaneously, it progresses to the native state. Examples of such proteins are β-lactamase, dihydrofolate reductase and α-glucosidase. However, under conditions where aggregation is the predominant process the released polypeptide irreversibly aggregates, as in the case of rubisco, rhodanese and citrate synthase. This aggregation is prevented only when the polypeptide is released in the presence of the chaperonin 10 and MgATP and folding to the native state occurs. These observations indicate that the polypeptide released from cpn60 in the presence of MgATP and cpn10 has a different conformation than those polypeptides released with MgATP alone (79). Under physiological conditions where protein concentrations and temperature are higher than those required to refold most proteins *in vitro*, both cpn60 and cpn10 are probably required for correct protein assembly.

Evidence is emerging that some members of the different chaperone families may function cooperatively to assist protein assembly. The transient interaction of newly imported polypeptides with the hsp70 chaperone in the mitochondrial matrix is necessary for complete translocation but is not sufficient for correct assembly of the polypeptide chain, which requires sequential transfer to chaperonin 60 (*80*). A similar conclusion was reached for the *in vitro* refolding of rhodanese (*81*). While the dnaK chaperone is capable of binding a rhodanese folding intermediate, transfer to the chaperonin 60 is necessary for correct folding. Additional proteins, originally identified because of their requirement in DNA replication (reviewed in *82*), have been shown to influence the refolding of rhodanese *in vitro*. The dnaJ protein functions cooperatively with the dnaK chaperone in binding the rhodanese folding intermediate, while the grpE protein enhances the efficiency of transfer to cpn60 (81). Whether these sequential events occur *in vivo* remains to be determined.

Implications for Biotechnology

There is much commercial interest in producing large amounts of biologically active recombinant proteins in heterologous systems. However, in many cases problems such as formation of insoluble inclusion bodies and/or inactive aggregates are encountered. The formation of such aggregates is probably the result of incorrect interactions of folding intermediates that do not participate along productive folding pathways (*83*). It has been reported that increased levels of the *E.coli* chaperonins increase the amount of correctly assembled foreign rubisco enzyme when its subunits are produced from cloned genes present on a plasmid (*62*). The accumulation of correctly folded procollagenase *in vivo* increases 10 fold upon overexpression of either the *E.coli* chaperonins or the dnaK chaperone (*84*). It appears that by increasing the amount of chaperones by heat shock, production from a multicopy plasmid, or simply by the presence of unfolded proteins (*85*), the folding capacity of the cell can be enhanced leading to the increased production correctly assembled protein. Although the above reports are encouraging, there are still many foreign proteins whose assembly is not enhanced upon increased amount of chaperone proteins (*84*). Thus it appears that simple overexpression of chaperones is not a universal method for increasing the amount of biologically active recombinant proteins and every case may have to be evaluated individually. We recommend that such evaluations include the determination whether a foreign protein requires chaperones for its assembly in the cell in which it occurs naturally as it may be necessary to co-express those chaperones in the heterologous host.

Literature Cited.

1. Anfinsen, C.B. *Science* **1973**, *181*, 223.
2. Jaenicke, R. *Prog. Biophys. Mol. Biol.* **1987**, *49*, 117.
3. Ellis, R.J. *Nature* **1987**, *328*, 378.
4. Hemmingsen, S.M.; Woolford,C.; van der Vies; S.M.; Tilly, K; Dennis, D.T.; Georgopoulos, C.P.; Hendrix, R.W.; Ellis, R.J. *Nature* **1988**, *333*, 330.
5. Ellis, R.J. *Semin. Cell Biol.* **1990**, *1*, 1.
6. Ellis, R.J. *Science* **1990**, *250*, 954.

7. Ellis, R.J.; van der Vies, S.M. *Annu. Rev. Biochem.* **1991,** *60,* 321.
8. Rothman, J.E. *Cell* **1989,** *59,* 591.
9. Gething M.-J.; Sambrook, J. *Nature* **1992,** *355,* 33.
10. Hartl, F.-U.; Martin, J.; Neupert, W *Annu. Rev. Biophys. Biomol. Struct.* **1992,** *21,* 293.
11. Petersen, N.S.; McLaughlin, C.S. *J. Mol. Biol.* **1973,** *81,* 33.
12. Bochkareva, E.S.; Lissin, N.M.; Gorshovich, A.S. *Nature* **1988,** *336,* 254.
13. Beckman, R.P.; Mizzen, L.A.; Welch, W.J. *Science* **1990,** *248,* 850.
14. Hartl, F.-U.; Neupert, W. *Science* **1990,** *247,* 930.
15. Langer, T.; Neupert, W. *Curr. Topics Microbiol. Immunol.* **1991,** *167,* 3.
16. Gething, M.-J.; Sambrook, J. *Semin. Cell Biol.* **1990,** *1,* 65.
17. Lindquist, S; Graig, E.A. *Annu. Rev. Genet.* **1988,** *22,* 631.
18. Pelham, H.R.B. *Cell* **1986,** *46,* 959.
19. Skowkyra, D.;Georgopoulos, C.; Zylicz, M. *Cell* **1990,** *62,* 939.
20. Sherman, M.Y.; Goldberg, A.L *EMBO J.* **1992,** *11,* 71.
21. Martin, J.; Langer, T.; Boteva, R.; Schramel, A.; Horwich, A.L.;Hartl, F.-U. *Nature* **1991,** *352,* 36.
22. Viitanen, P.V.; Donaldson, G.K.; Lorimer, G.H.; Lubben, T.H.; Gatenby, A.A. *Biochemistry* **1991,** *30,* 9716.
23. van der Vies, S.M.; Viitanen, P.V.; Gatenby, A.A.; Lorimer, G.H.; Jaenicke, R. *Biochemistry,* **1992,** *31,* 3635.
24. Dingwall, C.; Laskey, R.A. *Semin. Cell Biol.* **1990,** *1,* 11.
25. Zhu, X.; Ohta, Y.; Jordan, F.; Inouye, M. *Nature* **1989,** *339,*483.
26. Silen, J.L.; Agard, D.A. *Nature* **1989,** *341,* 462.
27. Finley, D.; Bartel, B.; Varsharsky, A. *Nature* **1989,** *338,* 394.
28. Walter, P.; Lingappa, V.R. *Annu. Rev. Cell Biol.* **1986,** *2,* 449.
29. Lecker, S.; Lill, R.; Ziegelhofer, T.; Bassford, P.J.; Kumamoto, C.A.; Wickner, W. *EMBO J.* **1989,** *8,* 2703.
30. Weiss, J.B.; Ray, P.H.; Bassford, P.J. *Proc. Natl. Acad. Sci. USA* **1988,** *85,* 8977.
31. Grimm, R.; Spetk, V.; Gatenby, A.A.; Schafer, E. *FEBS lett.* **1991,** *286,* 155.
32 Ellis, R.J. *Nature* **1992,** *358,* 191.
33. Martel, R.; Cloney, L.P.; Pelcher, L.; Hemmingsen, S.M. *Gene* **1990,** *94,* 181.
34. Hendrix, R.W. *J.Mol.Biol.* **1979,** *129,* 375.
35. Pushkin, A.V.; Tsuprun, V.L.; Solovjeva, N.A., Shubin, V.V.; Evstigneeva, Z.G.; Kretovich, W.L. *Biochim. Biophys. Acta* **1982,** *704,* 379.
36. Hallberg, R. L. *Semin. Cell Biol.* **1990,** *1,* 37.
37. Trent, J.D.; Nimmesgern, E.; Wall, J.S.; Hartl, F.-U.; Horwich, A.L. *Nature* **1991,** *354,* 490.
38. Jindal, S.; Dudani, A.K.; Singh, B.; Harley, C.B.; Gupta, R.S. *Mol. Cell. Biol.* **1989,** *9,* 2279.

39. Picketts, D.J.; Mayanil, C.S.K.; Gupta, R.S. *J. Biol. Chem.* **1989**, *264*, 12001.
40. Viitanen, P.V.; Lorimer, G.H.; Seetharam, R.; Gupta, R.S.; Oppenheim, J.; Thomas, J.O.; Cowan, N.J. *J. Biol. Chem.* **1992**, *267*, 695.
41. Miller, S.G.; Leclerc, R.F.; Erdos, G.W. *J. Mol. Biol.* **1990**, *214*, 407.
42. Hemmingsen, S.M.; Ellis, R.J. *Plant Physiol.* **1986**, *80*, 269.
43. Chandrasekhar, G.N.; Tilly, K.; Woolford, C.; Hendrix, R.; Georgopoulos, C. *J. Mol. Biol.* **1986**, *261*, 12414.
44. Lubben, T.H.; Gatenby, A.A.; Donaldson, G.K.; Lorimer, G.H.; Viitanen, P.V. *Proc. Natl. Acad. Sci. USA* **1990**, *87*, 7683.
45. Hartman, D.J.; Hoogenraad, N.J.; Condron, R.; Hoj, P.B. *Proc. Natl. Acad. Sci. USA* **1992**, *89*, 3394.
46. Bertsch, U.; Soll, J.; Seetharam, R.; Viitanen, P.V. *Proc. Natl. Acad. Sci. USA* **1992**, *89*, 8696.
47. Viitanen, P.V.; Lubben, T.H.; Reed, J.; Goloubinoff, P.; O'Keefe, D.P.; Lorimer, G.H. *Biochemistry* **1990**, *29*, 5665.
48. Lubben, T.H.; Donaldson, G.K.; Viitanen, P.V.; Gatenby, A.A. *Plant Cell* **1989**, *1*, 1223.
49. Osterman, J.; Horwich, A.L.; Neupert, W.; Hartl, F.-U. *Nature* **1989**, *341*, 125.
50. Hemmingsen, S.M. *Semin. Cell Biol.* **1990**, *1*, 47.
51. Zeilstra-Ryalls, J.; Fayet, O.; Georgopoulos, C. *Annu. Rev. Microbiol.* **1991**, *45*, 301.
52. Kochan, J.; Murialdo, H. *Virology* **1983**, *131*, 100.
53. Fayet, O.; Ziegelhofer, T.; Georgopoulos, C. *J. Bacteriol.* **1989**, *171*, 1379.
54. Grossmann, A.D.; Strauss, D.B., Walter, W.A., Gross, C.A. *Genes Dev.* **1987**, *1*, 179.
55. Tilly, K.; Georgopoulos, C. *J.Bact.* **1982**, *149*, 1082.
56. Fayet, O.; Louarn, J.-M.; Georgopoulos, C. *Mol. Gen. Genet.* **1986**, *202*, 435.
57. Jenkins, A.J.; March, J.B.; Oliver, I. R.; Masters, M. *Mol. Gen. Genet.* **1986**, *202*, 446.
58. Miki, T.; Orita, T.; Furuno, M.; Horiuchi, T. *J. Mol. Biol.* **1988**, *201*, 327.
59. Philips, G.J.; Shilvary, T.J. *Nature* **1990**, *344*, 882.
60. Kusukawa, N.; Yura, T.; Ueguchi, C.; Akiyama, Y.; Ito, K. *EMBO J.* **1989**, *8*, 3517.
61. Van Dyk, T.K.; Gatenby, A.A.; LaRossa, R.A. *Nature* **1989**, *342*, 451.
62. Goloubinoff, P.; Gatenby, A.A.; Lorimer, G.H. *Nature* **1989**, *337*, 44.
63. Goloubinoff, P.; Christeller, J.T.; Gatenby, A.A.; Lorimer, G.H. *Nature* **1989**, *342*, 884.
64. Ellis, R.J.; Hemmingsen, S.M. *TIBS* **1989**, *14*, 8.
65. Laminet, A.A.; Ziegelhoffer, T.; Georgopoulos, C.; Plucktun, A. *EMBO J.* **1990**, *9*, 2315.

66. Mendoza, J.A.; Rogers, E.; Lorimer, G.H.; Horowitz, P.M. *J. Biol.Chem.* **1991,** *266,* 13044.
67. Badcoe, I.G.; Smith, C.J.; Wood, S.; Halsall, D.J.; Holbrook, J.J.; Lund, P.; Clarcke, A.R. *Biochemistry* **1991,** *30,* 9195.
68. Buchner, J.; Schmidt, M.; Fuchs, M.; Jaenicke, R.; Rudolph, R.; Schmid, F.X.; Kiefhaber, T. *Biochemistry* **1991,** *30,* 1586.
69. Holl-Neugebauer, B.; Rudolph, R.; Schmidt, M.; Buchner, J. *Biochemistry* **1991,** *30,* 11609.
70. Schmidt, M.; Buchner, J. *J. Biol. Chem.* **1992,** *267,* 16829.
71. Taguchi, H.; Konishi, J.; Ishii, N.; Yoshida, M. *J. Mol. Biol.* **1991,** *266,* 22411.
72. Gatenby, A.A.; Donaldson, G.K.; Baneyx, F.; Lorimer, G.H.; Viitanen, P.V.; van der Vies, S.M. in *Biocatalysis Design for Stability and Specificity*; Himmel, E.M.; Georgiou, G.;American Chemical Society: Washington, 1992, in press
73. Kim, P.S.; Baldwin, R.L *Annu. Rev. Biochem.* **1982,** *51,* 459.
74. Mendoza, J.A.; Lorimer, G.H.; Horowitz, P.M. *J. Biol. Chem.* **1992,** *267,* 17631.
75. Ptitsyn, O.B. *J.Protein Chem.* **1987,** *6,* 273.
76. Kuwajima, K. *Proteins* **1989,** *6,* 87.
77. Landry, S.J.; Gierash, L.M. *Biochemistry* **1991,** *30,* 7359.
78. Baneyx, F.; Gatenby, A.A. *J. Biol. Chem.* **1992,** *267,* 11637.
79. Viitanen, P.V.; Gatenby, A.A.; Lorimer, G.H. *Prot. Sci.* **1992,** *1,* 363.
80. Manning-Krieg, U.C.; Scherer, P.E.; Schatz, G. *EMBO J.* **1991,** *10,* 3273.
81. Langer, T.;Chi, L.; Echols, H.; Flanagan, J.; Hayer, M.K.; Hartl, F. U. *Nature* **1992,** *356,* 683.
82. Georgopoulos, C. *TIBS* **1992,** *17,* 295.
83. Mitraki , A.; King, J. *Bio/Technology,* **1989,** *7,* 1141.
84. Lee, S.C.; Olins, P.O. *J.Biol. Chem.* **1992,** *267,* 2849.
85. Parsell, D.A.; Sauer, R.T. *Genes Dev.* **1989,** *3,* 1226.

RECEIVED November 11, 1992

Chapter 7

Role of Molecular Chaperones in Transport of Proteins into the Mammalian Endoplasmic Reticulum

Solubilization-Achieving Proteins, Folding-Accelerating Proteins, and Translocation-Mediating Proteins

H. Wiech and R. Zimmermann

Zentrum Biochemie, Abteilung Biochemie II der Georg-August Universität Göttingen, Gosslerstr. 12d, D–3400 Göttingen, Germany

> In signal peptide dependent transport of proteins across the membrane of the endoplasmic reticulum (ER) one can distinguish between transport mechanisms which involve ribonucleoparticles and those which employ molecular chaperones. Both mechanisms appear to converge at the membrane of the ER, specifically, at the level of the ATP-dependent translocase. The function of members of the Hsp70- and Hsp90-protein families in maintaining the transport-competent conformation of precursor proteins and in assisting in protein folding are discussed in a unifying model. According to their different activities molecular chaperones are divided into three groups. An enzyme catalyzing a rate determining step in the folding process is termed a *folding accelerating protein* (FOAP). A cytosolic factor keeping a precursor protein in a water soluble and non-aggregated state is characterized as *solubilization achieving protein* (SOAP). If the molecular chaperone is directly involved in maintaining the transport-competent conformation of a certain class of precursor proteins it is referred to as a *translocation mediating protein* (TRAMP).

The actual understanding of the development and maintenance of specific cellular compartments represents a central problem in modern cell biology. All proteins are synthesized in the cytosol (excluding the few proteins which are synthesized by mitochondrial and chloroplast ribosomes). Due to the subcellular compartmentalization the sites of synthesis and functional location are often separated by at least one biological membrane. Thus, non-cytosolic proteins must be directed to a variety of different subcellular locations where they fulfill their specific functions. Therefore, the emergence of a nascent polypeptide chain from a ribosome signals the beginning of a complex process that leads to a correctly folded, localized, processed and assembled protein product. Consequently, folding of proteins must be rigorously controlled, both temporally and spatially. Hence, the amino acid sequence of the nascent chain encodes not only information necessary to determine its native three-dimensional structure but also information which is required for the cellular machinery to control the timing of folding. Furthermore, mechanisms have to exist which ensure the specific transport of proteins across membranes and the assembly of proteins into membranes (for review see *1, 2, 3*).

In the case of protein transport into the mammalian endoplasmic reticulum (ER) most of the precursor proteins possess a transient N-terminal extension, called signal peptide (*4*). The biophysical properties of the signal peptide influence the remaining part of the protein with respect to adopting a native like conformation and targeting to the ER. Therefore, the emergence of the signal peptide from the ribosome is used as a recognition signal for proteinaceous components involved in the timing of folding as well as in the translocation process.

It appears that there are two cytosolic systems which can contribute to the fidelity of the respective precursor proteins (Figure 1). One mechanism involves two ribonucleoparticles, the ribosome (RIB) and the signal recognition particle (SRP; *5*), and their respective receptors at the membrane surface, ribosome receptor (RR; *6, 7, 8*) and docking protein (DP; *9, 10*) (Figure 1, pathway A). SRP seems to be able to support precursor protein translocation with respect to membrane specificity (together with its receptor, the docking protein) and, in collaboration with the ribosome, it seems to be able to keep the precursors in a state where the signal peptide is exposed and where the precursor stays water soluble as well as in an "unfolded" state. In other words, mammalian SRP has a variety of functions which are brought about by the inhibition of elongation of the nascent precursor polypeptide after SRP has bound to the signal peptide as it emerged from the ribosome (*11*). Hence, in this process the rate of protein synthesis is slowed down and energy in form of GTP is necessary to release the nascent chain from SRP and to complete the reaction (*12*). In the other mechanism, ribonucleoparticles are not involved. Therefore, this mechanism is termed ribonucleoparticle-independent transport (Figure 1, pathway B; *13*). Here the interactions of a fully synthesized precursor protein with the *cis*-acting molecular chaperone Hsc70 and at least one so far unknown component (MoCh) keep the precursor protein in a transport competent state (*14, 15*). In this transport competent state, molecular chaperones help the precursors to stay soluble as well as loosely folded and keep their signal peptides exposed (*13, 16*). In addition, the precursors which can make use of this system seem to require structural features which allow them to stay in this form on their own, at least for a certain time. In this mechanism the rate of folding and/or aggregation is slowed down in an ATP-dependent fashion.

Both mechanisms converge at the level of the membrane (Figure 1, step C) where they use the same translocation machinery (SR, *17*; NSM, *18*; ASM, *19*; TRAM, *20*; Tase, *21*). In both cases the translocation of the precursor protein is dependent on ATP and the signal peptide is cleaved off by signal peptidase (Figure 1, step D; SPase, *22*). On the *trans*-side of the membrane molecular chaperones such as BiP (*23, 24*) appear to be involved in assisting proteins in adopting their final three-dimensional structure (Figure 1, step E).

Our aim is to understand the role of molecular chaperones, such as members of the Hsp70 protein family, in folding of secretory proteins in the cytosol and the lumen of the endoplasmic reticulum as well as to identify all the various molecular chaperones involved. Furthermore, we try to elucidate the nature of the transport competent state. Therefore, we study protein transport and protein folding in *in vitro* reactions.

Protein Transport

The Assay. We have mainly been interested in the ribonucleoparticle-independent pathway which is predominantly used by precursor proteins with less than 75 amino acids residues (*13, 25, 26*). To analyze the role of molecular chaperones in this process we employ the following assay for translocation of proteins *in vitro*. The [^{35}S]-methionine labeled precursor protein M13 procoat protein (which does not depend on translocase for membrane insertion!) is synthezised in a cell free transcription and translation system, derived from *E.coli* (Figure 2, panel I). This procoat protein is further enriched by removal of the ribosomes and ammonium sulfate precipitation. The

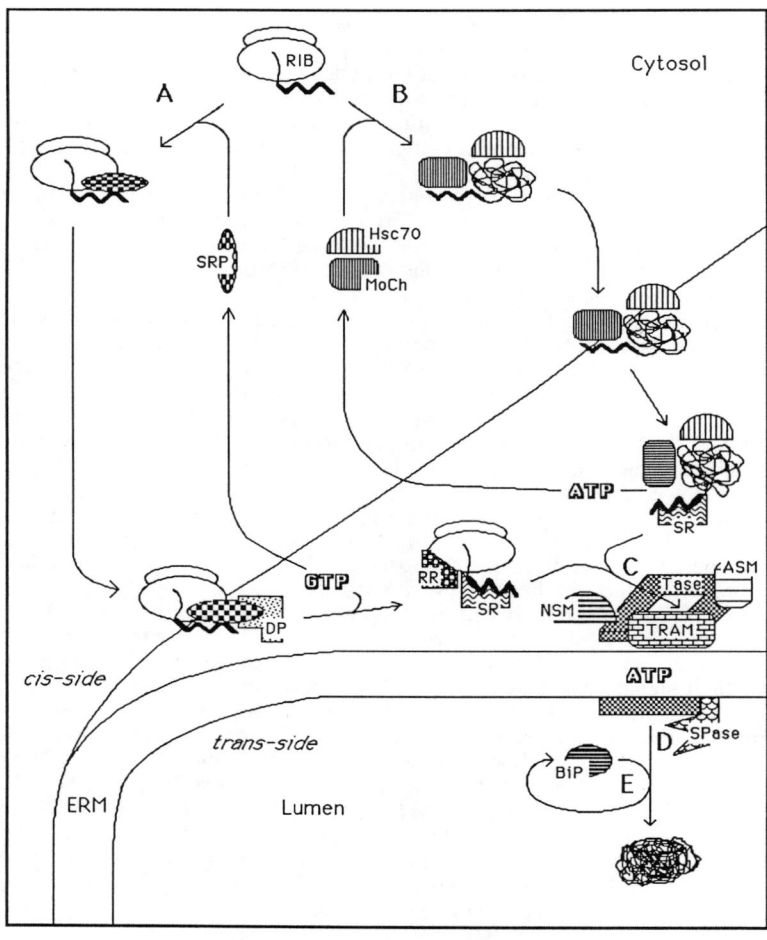

Figure 1. Model for Ribonucleoparticle Dependent and Ribonucleoparticle Independent Transport of Proteins into the Mammalian Endoplasmic Reticulum
RIB   :ribosome
SRP   :signal recognition particle
Hsc70   :stress protein of the Hsp70 protein family
MoCh   :molecular chaperone (i.e. NEM-sensitive component = NSC)
DP   :docking protein
RR   :ribosome receptor
SR   :signal peptide receptor
NSM   :NEM-sensitive membrane protein
ASM   :Azido-ATP-sensitive membrane protein
Tase   :hypothetical translocase
TRAM   :translocation chain-associating membrane protein
SPase   :signal peptidase
BiP   :heavy chain binding protein
ERM   :membrane of the endoplasmic reticulum

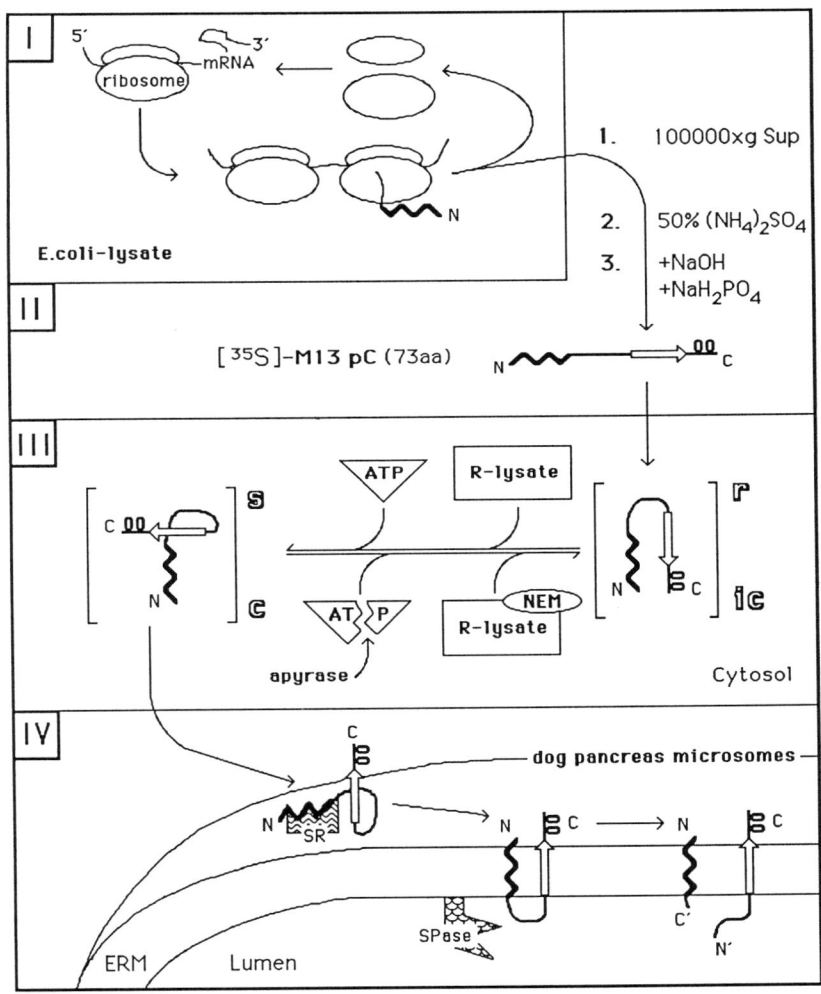

Figure 2. Schematic Illustration of the *in vitro* Assay to Analyze the Activity of *cis*-Acting Molecular Chaperones in Protein Transport into the ER

s	:sensitive	r	:resistant
c	:competent	ic	:incompetent
R-lysate	:rabbit reticulocyte lysate	SPase	:signal peptidase
NEM	:N-ethylmaleimide	Sup	:supernatant
ATP	:adenosine-5'-triphosphate	ERM	:ER membrane
SR	:signal peptide receptor		

precursor protein is then denatured at alkaline pH and renatured by the addition of sodium dihydrogen phosphate (Figure 2, panel II). The supplementation of the transport reaction with dog pancreas microsomes does not result in translocation (Figure 2, panels III and IV). Therefore, the structure of the procoat protein is called transport incompetent (Figure 2, panel III; indicated as "ic"). The addition of rabbit reticulocyte lysate (R-lysate) and MgATP, however, changes the properties of the precursor protein in such a way that a transport reaction can take place (Figure 2, panel IV). In contrast to the transport incompetent state, the transport competent procoat protein (Figure 2, panel III, indicated as "c") is less resistant (i.e. sensitive, indicated as "s") to protease digestion. The conversion from the resistant (indicated as "r") to the sensitive precursor state ("s") is dependent on the hydrolysis of ATP and a protein which can be inactivated by N-ethylmaleimide (NEM). Increasing amounts of reticulocyte lysate stimulate the transport reaction in the presence of hydrolyzable ATP as can be seen by the processing of the precursor protein to the mature form. Following addition of high amounts of proteinase K, only the mature form is protected by the microsomal membranes and consequently has reached its assembled state (Figure 2, panel IV). In the absence of ATP and in the presence of apyrase, or if NEM-inactivated reticulocyte lysate is added, the insertion of the procoat protein into the membrane does not occur due to its transport incompetent state.

Hsc70 is Involved. Supplementing of the reaction mixture with purified Hsc70 stimulates the transport reaction synergistically if limiting amounts of reticulocyte lysate are present (15). Under these conditions Hsc70 (NEM-insensitive) and at least a second component of the lysate (NEM-sensitive, termed NSC) are necessary to keep the precursor in a transport competent state. Addition of purified Hsp90, however, cannot substitute for Hsc70 (27). This reflects a high degree of substrate specificity for the molecular chaperones involved in the translocation process.

Progress Report. Fractionation of the reticulocyte lysate by ammonium sulfate precipitation revealed that the transport stimulating activity was enriched 30-times in the 10-66 % ammonium sulfate fraction (Figure 3, compare processing and sequestration of lane 3 with lanes 1,2 and 4). The stimulation of transport was ATP-dependent (Figure 3, lane 5) and NEM-sensitive (Figure 3, lane 6). Addition of purified Hsc70 further stimulated the reaction indicating that at least two components, Hsc70 and NSC, were necessary to allow the transport reaction (data not shown).

Therefore, the assay was modified for the characterization of NSC. If the amount of isolated Hsc70 in the reaction was kept constant, it was possible to analyze the activity of the fractionated reticulocyte lysate. The fractionation of the transport stimulating activity is shown in Figure 4. After gel filtration chromatography of the 10-66 % ammonium sulfate fraction of the reticulocyte lysate, a stimulation of the transport reaction was detected only in fractions with a molecular mass of approximately 200 kDa +/- 40 kDa (Figure 4, lanes 13 to 15; compare with the starting material (NRL) in lane 2). In this fraction the activity was enriched 400-times. Using native PAGE we were able to show that binding of procoat protein to this 200 kDa-complex in the presence of isolated Hsc70 was ATP-dependent and NEM-sensitive (data not shown). When the 200 kDa fraction was inactivated by NEM treatment, the binding of procoat protein to a low molecular mass protein (between 15-50 kDa) was detected. At present, further biochemical approaches to characterize the complex have been unsuccessful.

Protein Folding

Hsp90 Plays a Role. In collaboration with Ursula Jakob and Johannes Buchner we studied mitochondrial citrate synthase as a model protein for analyzing the kinetics of renaturation of denatured protein. We have observed that Hsp90 (in contrast to Hsc70) is a molecular chaperone stimulating protein folding by preventing non-native proteins

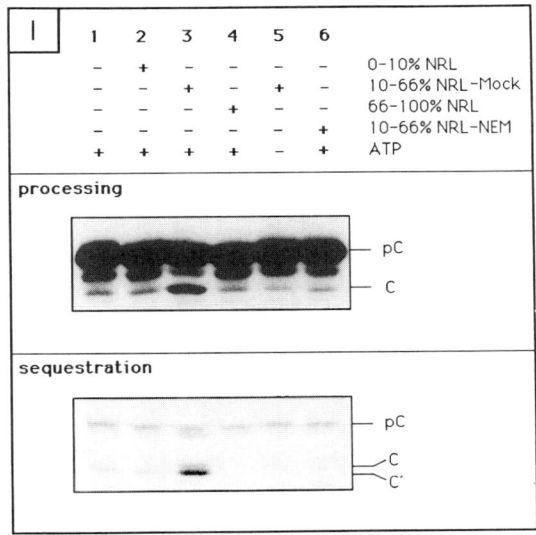

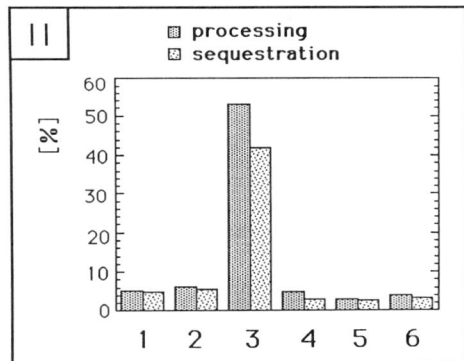

Figure 3. Analysis of the Assembly of Procoat Protein
The assay for *cis*-acting molecular chaperones was performed as described previously (27). The respective aliquots were supplemented in the presence of ATP with no cytosolic factors (lane 1), with the 0-10 % (i.e. % saturation) ammonium sulfate fraction (lane 2), with the 10-66 % ammonium sulfate fraction (lane 3, Mock treatment), with the 66-100 % ammonium sulfate fraction (lane 4), with the 10-66 % ammonium sulfate fraction (lane 6, after NEM treatment) or in the absence of ATP (after apyrase treatment 80 units/ml) with the 10-66 % ammonium sulfate fraction (lane 5, Mock treatment). After addition of dog pancreas microsomes (5 A_{280}/ml) the samples were incubated for 15 minutes at 37 °C and analyzed as described (27) for processing and sequestration. Finally the samples were subjected to an urea-SDS-PAGE followed by fluorography (I) and densitometry (II).
NRL-Mock: The 10-66 % ammonium sulfate fraction was incubated for 30 minutes at 25 °C in the presence of 10 mM NEM and 50 mM DTT.
NRL-NEM: The 10-66 % ammonium sulfate fraction was incubated for 20 minutes at 25 °C in the presence of 10 mM NEM, the reaction was stopped by addition of 50 mM DTT and a subsequent incubation step was performed for 10 minutes at 25 °C.

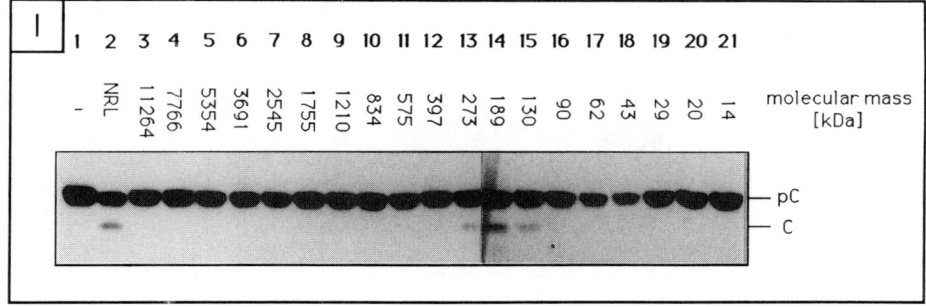

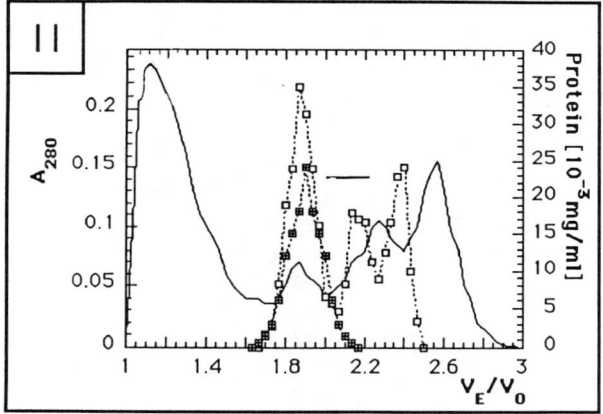

Figure 4. Analysis of the 10-66 % Ammonium Sulfate Fraction of the Rabbit Reticulocyte Lysate by Gel Filtration Chromatography on Superose 6

The rabbit reticulocyte lysate was fractionated with ammonium sulfate and the 10-66 % (i.e. % saturation) ammonium sulfate fraction was than applied to a Superose 6 column. The fractions were analyzed with respect to their total protein content (A_{280}) as well as their Hsc70- (□) and Hsp90- (⊞) content (protein, µg/ml) (II). Furthermore, the fractions were tested in the assay for *cis*-acting molecular chaperones as described in Figure 3 with the modification that all samples were supplemented with 100 µg/ml of isolated Hsc70 (I). In lane 1 no cytosolic fraction and in lane 2 the 10-66% ammonium sulfate fraction was added. The bar in II indicates the active fractions of assay I.

from unproductive intermolecular interactions (*28*). The results are consistent with a model in which one Hsp90 dimer binds one or two molecules of the non-native "substrate protein" (for detailed description see Figure 5 and below). Since nucleoside triphosphates do not have any effect on the function of Hsp90, we speculate that the release of the substrate protein is driven by folding of the non-native protein. Thus, Hsp90 and most likely Grp94, its endoplasmic reticulum counterpart, represent a class of molecular chaperones which exhibit a GroEL/ES-like effect on protein folding but which function with a completely different molecular mechanism.

A Model for Protein Folding *in vitro*

The classic *in vitro* protein refolding experiments carried out with isolated ribonuclease (*29*) led to the dogma that both the structure of the folded state and the mechanism of folding are encoded in the amino acid sequence (defined as primary structure). Consequently, under suitable conditions, folding can be envisaged as a spontaneous process (see Figure 5, panel II, pathway starting at A, continuing over B to C). This hypothesis was supported by the identification of folding transition states and the characterization of folding pathways. The experimental conditions used, however, were often non-physiological and reversible unfolding could not be achieved for all proteins (for review see *30, 31, 32, 33*).

Although a detailed description of the folding pathway is still missing, it has been established for small monomeric proteins that folding occurs through the succession of a finite number of intermediate conformational states. In an *in vitro* model system, native proteins can be converted to a random coil state by modifying external parameters (i.e. temperature or pressure) and solvation conditions (i.e. pH-value, destabilizing salts). After dilution of the unfolded protein into native buffer conditions (Figure 5, panel II, reaction A) the refolding process ("non-vectorial folding") can start (Figure 5, panel II, reaction B).

In vitro refolding (Figure 5) can be initiated by the collapse of hydrophobic regions of the polypeptide chain in the interior of the protein molecule. This process is accompanied by the rapid formation (in the milli-second time scale) of stable secondary structures (*34, 35, 36*). These secondary structure elements form the frame for specific tertiary contacts between parts of the polypeptide chain. The progressing folding process seems to follow a limited number of pathways with distinct intermediates (Figure 5, panel II, endproduct of reaction B). These intermediates (for "molten globule" see *37, 38*; for "compact intermediates" see *32)* possess a significant portion of secondary structure, show a compact form, but have no defined tertiary structure and they expose more hydrophobic surfaces than native folded molecules. The formation of the intermediates is thought to occur in a cooperative fashion and is associated with a large change in enthalpy and heat capacity. It seems that these intermediates are in fast equilibrium with the completely unfolded state (Figure 5, panel II, reaction B) and are only slowly converting to the native state (Figure 5, panel II, reaction C).

Therefore, the rate determining steps often occur at a late time point in the folding process, shortly before the protein reaches its final native conformation. Furthermore, the rate determining steps involve the reorganization of tertiary contacts. These steps are often limited by the reshuffling of incorrectly formed disulfide bonds (*39*) and/or by proline isomerization (*40, 41*). In addition, rate-limiting steps in the folding of large proteins may comprise domain pairing (*42*). Finally, the rate of reconstitution of oligomeric proteins is often determined by slow association processes (*43*).

The analysis of the folding of proteins may be complicated by the formation of aggregates (Figure 5, panel I, reactions B2 and C2; i.e. the products of non-productive association and/or precipitation of polypeptides chains) in competition with correct folding reactions (*44, 45*).

The driving force for folding is the tendency to achieve minimum potential energy. Surprisingly, the stabilization energy of a native protein is relatively low ($\Delta G=-40kJ/mol$

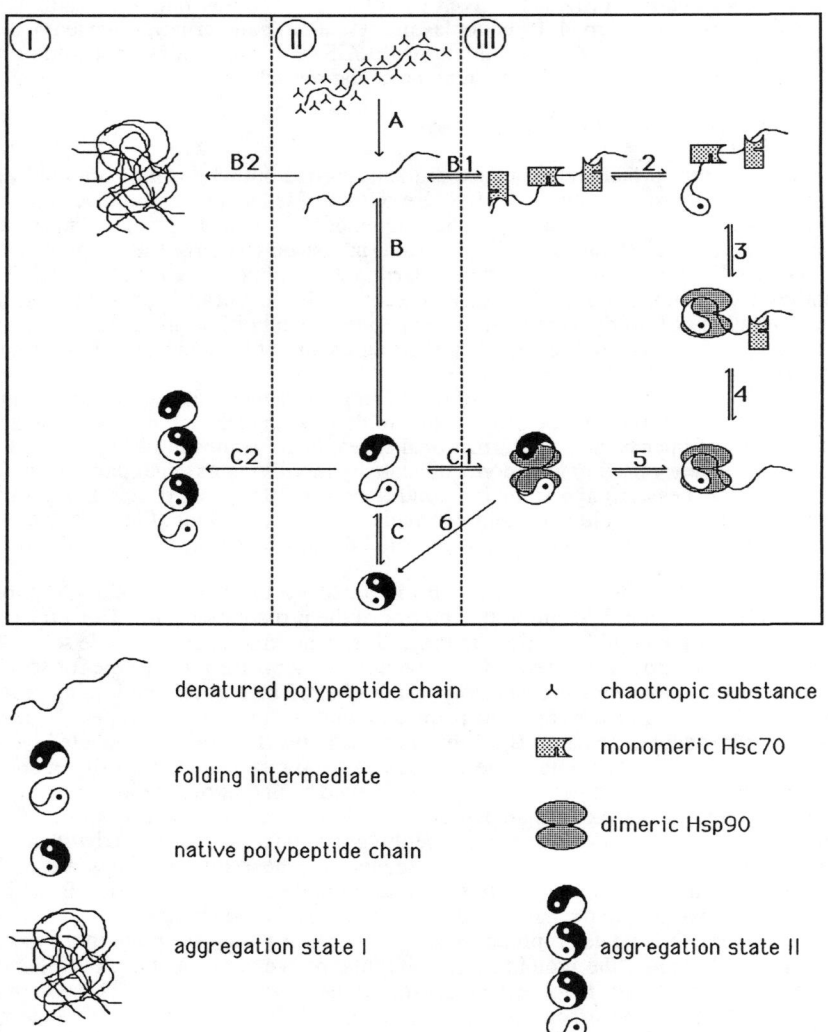

Figure 5. Hypothetical Model for Protein Folding *in vitro*

on the average) and this marginal stability is the result of a delicate balance between opposing forces. This guarantees a high degree of flexibility which is a prerequisite for enzymatic function and supports transient unfolding of a protein following the input of small amounts of energy.

SOAPS and FOAPS. Two mechanims seem to be involved in facilitating correct protein folding. One is the catalysis of slow steps on the folding pathway and thus the reduction of the time of exposure of hydrophobic surfaces in folding intermediates (Figure 5, reactions C1 and 6). Examples for *folding accelerating proteins* (termed FOAPs) are the protein disulfide isomerase (PDI; *46,47*) and the peptidylproline-*cis-trans*-isomerase (PPI; *48, 49*). The second process involves molecular chaperones that bind to folding intermediates and thus prevent unproductive aggregation reactions (Figure 5, panel III, reactions B1 and C1). In this case, the transient binding of a *solubilization achieving protein* (defined as SOAP) to unfolded or partially folded polypeptide chains may allow the substrate protein to reach the right conformation while bound to the surface of the molecular chaperone ("non-vectorial folding"). SOAPs thereby suppress competing unspecific self-aggregation reactions which are the consequence of interactions between different parts of the polypeptide due to the exposure of interactive surfaces of the folding intermediates. No evidence was found that SOAPs actively guide correct folding and assembly, i.e. act by selecting the correct folding pathway.

Folding of proteins represents an isomerization reaction (Figure 5, panel II; 1. order reation) and, therefore, is independent on the protein concentration. On the other hand, aggregation is a reaction which is strongly dependent on the amount of refolding protein (*44*; illustrated in Figure 5, panel I; n. order reaction). At increasing concentrations of the denatured polypeptide, the rate of folding stays constant (Figure 5, panel II, reations B and C) whereas the rate of aggregation increases and can be much faster than the folding process (Figure 5, reations B2 and C2). Consequently, the recovery of native protein during *in vitro* protein folding is determined by a kinetic competition between the folding reaction and the aggregation (41). Reactions that sequester the quickly formed folding intermediates through transient binding (Figure 5, panel III, reactions B1 and C1) compete with their aggregation (Figure 5, panel I, B2 and C2). The stabilization of folding intermediates by way of interaction with molecular chaperones (FOAPs or SOAPs) allows the rate determining steps in the folding process to proceed (Figure 5, panel III, reaction 6). Thus, the use of molecular chaperones in *in vitro* refolding experiments increases the yield of active and native protein.

Citrate Synthase as a Model. The use of mitochondrial citrate synthase as a model protein for analyzing the kinetics of protein renaturation revealed that there is a striking difference between the rate of aggregation which posesses a half time of 15 seconds and the half time for the reactivation process which is 5 minutes. Therefore, aggregation is preceding the rate determining steps in protein folding (in Figure 5, panel II, and in the case of citrate synthase the rate determining step may be the dimerization reaction C). If a protein has reached its correct native conformation Hsp90 (or GroEL/ES) are not able to interact with or to unfold the protein substrate (*50, 28*). Consequently SOAPs are only active at early steps in protein folding where crucial intermediates with the potential to aggregate are prominent.

Preliminary results gained by analyzing the activity of isolated Hsc70 and/or Hsp90 together with other cytosolic fractions in the refolding process of citrate synthase led us to assume a sequential action of Hsc70, Hsp90 and unidentified components of the cytosol (Figure 5, panel III, reactions B1 or C2 and 2, 3, 4, 5, 6). According to this view, Hsc70 interacts with an early intermediate which may be characterized by containing only secondary structure elements (Figure 5, panel III, reaction B1). This reaction does not result in the formation of active dimeric citrate synthase, probably because the binding of Hsc70 is transient. Therefore, the irreversible aggregation reaction is dominant (in Figure 5 competition between B1 and 2 with B2).

In contrast, Hsp90 binds to a folding intermediate which may at least partially assume a "molten globule" conformation (Figure 5, panel III, reaction C1) and chaperones the reactivation of the substrate protein (in Figure 5 reaction 6).

Surprisingly, if Hsc70 and Hsp90 are both present in the renaturation assay, no refolding of citrate synthase occurs (in Figure 5 the reactions B1, 2, and 3 take place and result in a stable intermediate of citrate synthase bound to Hsc70 and Hsp90). Formation of active citrate synthase was only detected after supplementation with a cytosolic fraction. This observation suggests that an unknown component in the cytosol may release Hsc70 from the stable intermediate (in Figure 5 the reaction 4). As illustrated in Figure 5, the reactions 5 and 6 can now take place and result in the formation of active citrate synthase.

Summarizing, the *in vitro* refolding of citrate synthase was useful in analyzing the characteristics of Hsc70 and Hsp90 as molecular chaperones and in delineating the role of unknown cytosolic components which modulate their activity. In addition, the concerted action of all components will eventually align the *in vitro* with the *in vivo* situation. Therefore, new insights into the mechanism of folding of proteins will be gained.

A Model for Protein Folding *in vivo*

A nascent polypeptide chain upon emerging from the ribosome and making contact with the solvent cytosol starts to fold and acquires a different conformation relative to its primary structure (Figure 6, panel II, reaction A). This may be a stepwise process accompanying the growth of the polypeptide chain from the amino- to the carboxy-terminus ("vectorial folding"). In contrast to the time necessary for synthesis for a typical protein (which is in the range of minutes) the partially folded state can be achieved rapidly (in Figure 6, panel II, reaction B and C; in the range of milliseconds to seconds) and the slower rate-determing steps occur late in the folding process (Figure 6, panel II, reaction D). Monomeric single domain proteins in particular, may reach their native structure without any assistance and complete the process in a biologically reasonable time (Figure 6, panel II, pathway A, B, C, D). The correct folding of a multi domain protein, however, can be either independent or dependent of the action of other proteins which, due to a temporary interaction with the translated protein, assist or catalyse the folding process (Figure 6, panel III, reactions B1, B2, C1, C2, D1, 3, 4, 5, 6). *Folding accelerating proteins* (FOAPs), *solubilization achieving proteins* (SOAPs) or *translocation mediating proteins* (termed TRAMPs) are located either in a soluble form in the cytosol or may be bound to the ribosome, thereby enhancing the possibility of binding to the emerging protein.

Often, it is possible to isolate partially folded polypeptides as a complex with specific cellular proteins especially members of the heat shock protein families (*51, 52, 53, 54, 55, 56*). Some of the molecular chaperones involved in *in vivo* protein folding seem to have a broad specifity whereas others assist only in the assembly of defined macromolecules.

By analogy to the model describing protein folding *in vitro*, there may also be an *in vivo* sequential interaction of Hsc70, Hsp90 and as yet unidentified cytosolic components with the partially folded substrate protein in order to efficiently reach a native structure. Depending on the mechanism, the folding protein may interact with SOAPs and/or FOAPs at different times during the elongation or after termination of protein synthesis (compare in Figure 6 the reactions B1, B2 with the reations C1, C2 or with reaction D1). The stabilization of the resulting conformation -due to the binding of molecular chaperones to interactive surfaces- would allow enough time for the

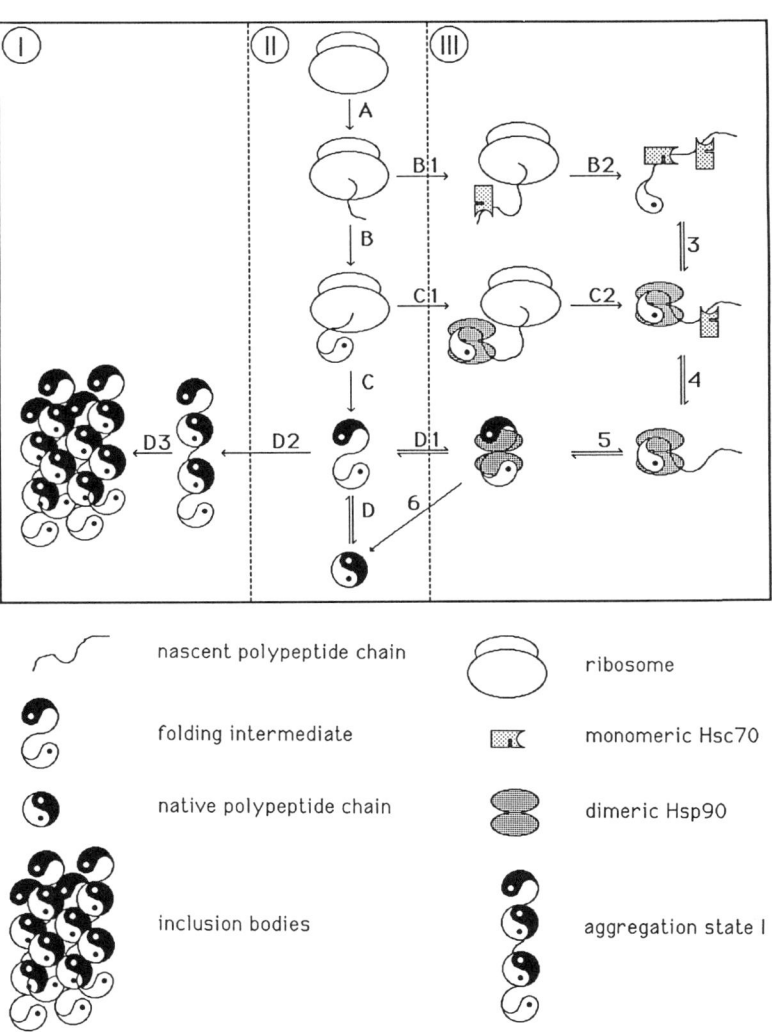

Figure 6. Hypothetical Model for Protein Folding *in vivo*

completion of the rate determining step in the folding process and thus result in the correctly assembled protein.

Overexpresssion of recombinant proteins in prokaryotic and eukaryotic hosts has shown that there is a correlation between aggregation *in vivo* and the rate of expression of a protein. By analogy to the *in vitro* situation, one can speculate that under these conditions aggregation is also competing with correct folding and association. High concentrations of nascent or unfolded polypeptide chains favour aggregation (Figure 6, panel I, reaction D2 and D3; 57). The end products *in vivo* represent dense insoluble protein particles ("inclusion bodies") which are stablized through unproductive interactions (Figure 6, panel I, reaction D3; 58).

To investigate protein folding *in vivo*, assays have been developed that are not dependent on obtaining large quantities of partially folded proteins which are sufficiently pure for physicochemical measurements. They include the formation of disulfide bonds, the use of conformation specific antibodies, the protease sensitivity of proteins and sucrose density gradient centrifugation. These techniques have been useful for probing the tertiary and quarternary structure of radiolabelled proteins (for review see 3). They also make possible the study of the acquisition of the final native structure of different proteins and should, in the future, help to analyze the folding pathways in more detail.

A Model for Transport Competence

Proteins can be classified into two major groups on the basis of their functional location in the cell, i.e. cytosolic proteins and non-cytosolic proteins. The folding pathway of various transition intermediates belonging to these classes is schematically illustrated in Figure 7. According to our view, three parameters influence the properties of a protein in the cytosol. The first parameter is the Gibbs free energy (represented in Figure 7 by the y-axis) which can be used as a measure of the evolution of the folding process. The tendency to achieve a minimum of potential energy is connected with the adoption of a native conformation. Second, a folding intermediate is only a substrate for molecular chaperones for a defined time interval (represented in Figure 7 by the x-axis). This relaxation time indicates how long a protein will stay in a certain conformation or complex before it reaches the next state. Third, depending on its conformation, a folding intermediate may possess characteristics (termed transport competence) which may or may not allow the transport across a membrane *per se* . In some cases the interaction with a molecular chaperone is required (in Figure 7 represented by the z-axis).

In the early phase of elongation, the polypeptide emerging from the ribosome possesses primary structural features (Figure 7, p^{1o}). As elongation proceeds, the nascent polypeptide chain adopts a secondary structure as a result of contacts with the hydrophilic surroundings (Figure 7, p^{2o}). When the N-terminal amino acids of the protein act as a recognition signal for targeting to another compartment (i.e for a non-cytosolic protein) the first point of decision on which folding pathway will be used occurs (in Figure 7 illustrated by the interaction of p^{2o} with SRP).

Cytosolic Proteins. For a cytosolic protein, the rapid formation of folding intermediates precedes the rate determining step on the folding pathway (for details see Figure 6, in Figure 7 the reaction from (p^{2o}) to $p^{(MG)}$). These intermediates can either follow an unproductive pathway which leads to an enzymatically inactive native like form of the protein (for details see also Figure 6, in Figure 7 the reaction from $p^{(MG)}$ to (p^{3o}) and $(p^{3o})^{\dagger}$) or, after a defined time interval reach the native state *per se* (Figure 7, the reaction from $p^{(MG)}$ to $[p^{3o}]$). If the efficiency of the folding process and the acquisition of the native state (i.e. for the assembly of oligomeric proteins or where the catalysis of rate determining steps is necessary) requires transient binding to a molecular chaperone, the crucial folding step is accelerated via the binding to a FOAP and

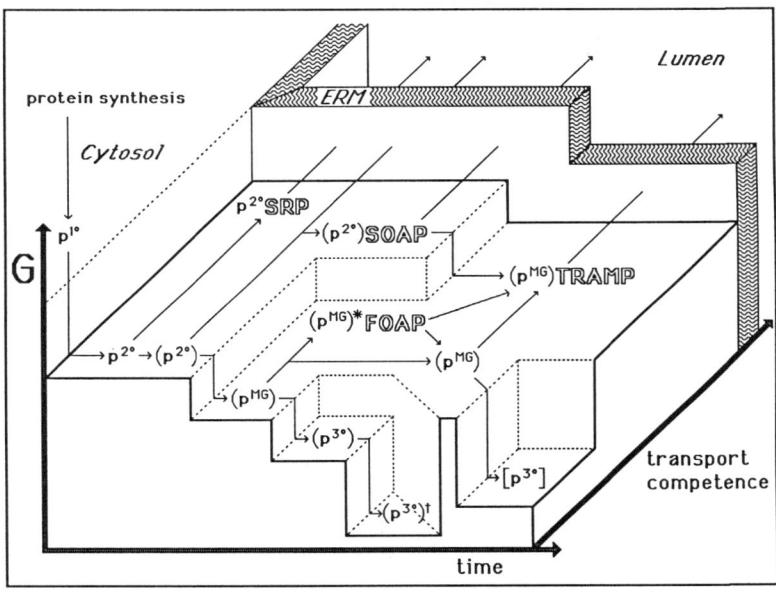

Figure 7. Hypothetical Diagram for Transport Competence
G   :Gibbs free energy
time   :relaxation time
transport competence :folding state of the protein

p^{1o} :just elongated polypeptide chain with primary structure
p^{2o} :nascent polypeptide chain with secondary structure
(p^{2o}) :fully synthesized protein with secondary structure
(p^{MG}) :"molten globule"-intermediate
(p^{3o}) :protein with nearly native tertiary structure
$(p^{3o})^{\dagger}$:inactive, aggregated protein
$[p^{3o}]$:native, active protein with tertiary structure
$(p^{MG})^{*}$:isomeric form of "molten globule"-intermediate
SRP   :signal recognition particle
SOAP   :*solubilisation achieving protein*
FOAP   :*folding accelerating protein*
TRAMP   :*translocation mediating protein*
ERM   :membrane of the endoplasmic reticulum

subsequent release of the native protein (Figure 7, reaction from $p^{(MG)}$ through $p^{(MG)}*FOAP$ and than $p^{(MG)}$ to $[p^{3o}]$). Alternatively or in addition, the transient interaction of a substrate protein with a SOAP may be important to avoid the aggregation of the protein (for reasons of clarity not illustrated in Figure 7). This binding step could be particularly crucial in the case of multidomain or oligomeric proteins.

Non-Cytosolic Proteins. As discussed earlier, for proteins which have to cross or become inserted into a membrane, there is an inverse correlation between the stability of the folded state of a precursor protein and the efficiency of its transport across a membrane (59, 16). In agreement with this observation it was shown that substrate analogs which stabilize the native conformation of an enzyme inhibit the transport of a precursor protein across biological membranes (60, 61, 13).

SRP. The formation of sterically unfavourable structures that interfere with translocation can be avoided by a tight coupling of transport and protein biosynthesis. This is the case for secretory proteins which use the ribonucleoparticle-dependent pathway (Figure 7, reaction from p^{2o} to $p^{2o}SRP$). As previously discussed, the ribosome and SRP preserve the *status nascendi* of a protein. In other words, as soon as the signal peptide emerges from the ribosome the association of SRP with the signal peptide results in a partial inhibition or a slowing down of the rate of elongation.

Signal Peptide. Other precursor proteins which are unable to interact with this system may, spontaneously, adopt a membrane transport competent form after completion of biosynthesis (Figure 7, (p^{2o}) binds to the membrane of the ER and can be translocated). This may be simply achieved by the presence of an additional amino acid sequence (the signal sequence) which, through unfavourable interactions with the remaining polypeptide chain, may prevent the formation of a stable three dimensional structure (62, 63). This retardation in folding could be sufficient to block or at least slow down the formation of a transport incompetent structure.

SOAP. On the other hand, the translocation of preproteins can depend on the temporary binding to a *cis*-acting molecular chaperone, i.e. a SOAP, (Figure 7, reaction from (p^{2o}) to (p^{2o})SOAP). The bound precursor protein may be released spontaneously which results in the formation of a transport competent state (Figure 7, the complex (p^{2o})SOAP dissociates and (p^{2o}) translocates across the membrane). Preproteins which are not able to interact during a defined time interval with a SOAP, show transient low solubility and thus tend to aggregate presumably because parts of their hydrophobic surface (especially true for membrane proteins) remain exposed to the cytosol. As a result of attempting to achieve a state of minimum potential energy without the assistance of a molecular chaperone they end up in a dead end state. On the other hand, preproteins which are denatured by chaotropic substances and are unfolded can be transported efficiently after dilution of the chaotropic substance. Under these conditions the dependence on certain cytosolic factors can be circumvented (64, 65, 66, 19).

TRAMP. Additionally, the transport competent state can result from interactions with more specific (in comparison to a SOAP) molecular chaperones, i.e. TRAMPs, (Figure 7, the reaction from (p^{2o})SOAP to $p^{(MG)}$TRAMP or the pathway from $p^{(MG)}$ to $p^{(MG)}$TRAMP). The association or dissociation (or both steps) may depend on the input of energy in the form of ATP.

The nature of the interaction between a protein substrate and a molecular chaperone could be the binding to unfolded exposed stretches of the polypeptide backbone and should persist longer than is required for protein synthesis (e.g. if binding occurs cotranslationally), whereas the dissociation must occur at a useful rate to keep pace with synthesis and translocation; otherwise, it interferes with transport as well as folding. The release of the polypeptide chain may occur through a stepwise exposure of segments of the polypeptide chain to folding conditions and may be the consequence of sequence specificity. The distinctive affinities of the components of the complex for the different peptide segments may ensure that especially critical sequences (i.e. the signal peptide) will be bound for a relatively long time interval (i.e. low k_S-value). In contrast, binding of molecular chaperones to other (cryptic) parts of the polypeptide chain may be characterized by a high k_S-value. Therefore, such segments would be stabilized in their actual conformation only for a short time interval.

Translocation. On the membrane of its destination the translocation competent conformation of a preprotein may need further help from the activity of a TRAMP. These molecular chaperones, depending on their affinities for peptide segments, dissociate and release the preprotein into the membrane translocase which initiates translocation. The translocation competent conformation of the precursor protein is thought to be a molten globule state (see above).

Considering the thermodynamic aspect, the difference in stability between this flexible intermediate and the native state is small, consequently a small input of energy (e.g. provided through binding of a molecular chaperone) is sufficient either to arrest folding in that state or to convert a folded protein back to a molten globular conformation. The difference in Gibbs free energy between the native and unfolded state or molten globule state of the protein is small and therefore the hydrolysis of one ATP is sufficient to drive such an unfolding reaction. We imagine that a possible mechanism could involve tight binding of a precursor protein to a complex between SOAP/TRAMP and ATP. However, the mechanism of molecular coupling of ATP-hydrolysis and unfolding is not at all known at present.

Kinetically, the unfolding of a partially folded intermediate is rapid, refolding to the native state, however, is slow. Therefore, a preprotein can easily be further unfolded as soon as it leaves the cytosol and enters the translocase in the membrane. Furthermore, unfavourable refolding immediately after dissociation from the molecular chaperone is blocked since the kinetics of refolding to the native state are comparatively slow.

The same principles as illustrated in the model for transport competence with respect to the ER (Figure 7) may be valid in the case of transport of non-cytosolic proteins to other intracellular compartments (e.g. mitochondria and chloroplasts). In this respect SOAPs (e.g. Hsc70) represent molecular chaperones with a more general function concerning the maintenance of transport competence (in the case of non-cytosolic proteins). SOAPs are involved in various cellular processes and they are members of conservative and homogeneous protein families. In contrast, TRAMPs (e.g. the NEM-sensitive component) interact with specific recognition segments of the preproteins. Consequently, these components together with the receptors on the membrane increase -in a cascade like way- the specificity and efficiency of the whole translocation process. This protein group is made up by heterogeneous members.

On the *trans*-side of the ER membrane similar reactions like the ones described for the cytosol (*cis*-side) may take place. Only in this case the starting point is not the ribosome (as indicated in Figure 7) but the translocase which releases the "quasi-nascent" polypeptide chain into the lumen. Then, due to an interaction with a *trans*-acting SOAP (i.e. BiP and Grp94) and/or a FOAP (e.g. protein disulfide isomerase) the mature protein may reach its final assembled state. A similar situation exists in mitochondria (i.e. mtHsp70 and Hsp60; *67*) and in chloroplasts in the assembly of imported proteins.

Acknowledgments

We wish to thank Johannes Buchner and Ursula Jakob at the University of Regensburg, Germany, for a fruitful and most stimulating collaboration. This work was supported by the "Deutsche Forschungsgemeinschaft" and by the "Fonds der Chemischen Industrie".

Literature Cited

(1) Wiech, H.; Stuart, R.; Zimmermann, R. *Seminars in Cell Biol.* **1990**, 1, 55-63.
(2) Wiech, H.; Klappa, P.; Zimmermann, R. *FEBS Lett.* **1991**, 285, 182-188.
(3) Gething, M.J.; Sambrock, J. *Nature* **1992**,355, 33-45.
(4) von Heijne, G. *Nucleic Acids Res.* **1986**, 14, 4683-4690.
(5) Siegel, V.; Walter, P. *Trends Biochem. Sci.* **1988**, 13, 314-316.
(6) Tazawa, S.; Unuma, M.; Tandokoro, N.; Asano, Y.; Ohsumi, T.; Ichimura, T.; Sugano, H. *J. Biochem.* **1991**, 109, 88-98.
(7) Kellaris, K. V.; Bowen, S.; Gilmore, R. *J. Cell Biol.* **1991**, 114, 21-33.
(8) Nunnari, J.M.; Zimmerman, D.L.; Ogg, S.C.; Walter, P. *Nature* **1991**, 352, 638-640.
(9) Hortsch, M.; Labeit, S.; Meyer, D.I. *Nucleic Acids Res.* **1988**, 16, 361-362.
(10) Tajima, S.; Lauffer, L.; Rath, V.L.; Walter, P. *J. Cell Biol.* **1986**, 103, 1167-1178.
(11) Ainger, K.J.; Meyer, D.I. *EMBO J.* **1986**, 5, 951-955.
(12) Connolly, T.; Gilmore, R. *Cell* **1989**, 57, 599-610.
(13) Schlenstedt, G.; Gudmundsson, G.H.; Boman, H.G.; Zimmermann, R. *J. Biol. Chem.* **1990**, 265, 13960-13968.
(14) Wiech, H.; Sagstetter, M.; Müller, G.; Zimmermann, R. *EMBO J.* **1987**, 6, 1011-1016.
(15) Zimmermann, R.; Sagstetter, M.; Lewis, M.J.; Pelham, H.R.B. *EMBO J.* **1988**, 7, 2875-2880.
(16) Müller, G.; Zimmermann, R. *EMBO J.* **1988**, 7, 639-648.
(17) Robinson, A.; Kaderbhai, M.A.; Austen, B.A. *Biochem. J.* **1987**, 242, 767-777.
(18) Zimmermannn, R.; Sagstetter, M.; Schlenstedt, G. *Biochimie* **1990**, 72, 95-101.
(19) Klappa, P.; Mayinger, P.; Pipkorn, R.; Zimmermann, M.; Zimmermann, R. *EMBO J.* **1991**, 10, 2795-2803.
(20) Görlich, D.; Hartmann, E.; Prehn, S.; Rapoport, T.A. *Nature* **1992**, 357, 47-52.
(21) Simon, S.M.; Blobel, G. *Cell* **1991**, 65, 371-380.
(22) Shelness, G.S.; Blobel, G. *J. Biol. Chem.* **1990**, 265, 9512-9519.
(23) Vogel, J.P.; Misra, L.M.; Rose, M.D. *J. Cell Biol.* **1990,** 110, 1885-1895.
(24) Sanders, S.L.; Whitfield, K.M.; Vogel, J.P.; Rose, M.D.; Schekman, R.W. *Cell* **1992**, 69, 353-365.
(25) Müller, G.; Zimmermann, R. *EMBO J.* **1987**, 6, 2099-2107.
(26) Schlenstedt, G.; Zimmermann, R. *EMBO J.* **1987**, 6, 699-703.
(27) Wiech, H.; Buchner, J.; Zimmermann, M.; Zimmermann, R.; Jakob, U. submitted for publication.
(28) Wiech, H.; Buchner, J.; Zimmermann, R.; Jakob, U. *Nature* **1992**, 358, 169-170.
(29) Anfinsen, C.B. *Science* **1973**, 181, 223-230.
(30) Jaenicke, R. *Prog. Biophys. Molec. Biol.* **1987**, 49, 117-237.
(31) Baldwin, R.L. *Trends biochem. Sci.* **1989**, 14, 291-294.
(32) Creighton, T.E. *Biochem. J.* **1990**, 270, 1-16.
(33) Jaenicke, R. *Biochemistry* **1991**, 30, 3147-3161.
(34) Udgoanker, J.B.; Baldwin, R.L. *Naure* **1988**, 335, 694-699.
(35) Roder, H.; Elöve, G.A.; Englander, S.W. *Nature* **1988**, 355, 700-704.
(36) Udgoankar, J.B.; Baldwin, R.L. *Proc. Natl. Acad. Sci. USA* **1990**, 87, 8197-8201.

(37) Kuwajima, K. *Proteins struct. funct. Genet* **1989**, 6, 87-103.
(38) Ptitsyn, O.B. *J. Prot. Chem.* **1991**, 6, 272-293.
(39) Creighton, T.E. *Prog. Biophys. Mol. Biol.* **1978**, 33, 231-297.
(40) Brandts, J.F.; Halvorson, H.R.; Brennan, M. *Biochemistry* **1975**, 14, 4953-4963.
(41) Kiefhaber, T.; Rudolph, R.; Kohler, H.-H.; Buchner, J. *Biotechnology* **1991**, 9, 825-829.
(42) Teschner, W.; Rudolph, R.; Garel, J.R. *Biochemistry* **1987**, 26, 2791-2796.
(43) Jaenicke, R.; Rudolph, R. *Meth. Enzymol.* **1986**, 131, 218-250.
(44) Zettlmeißl, G.; Rudolph, R.; Jaenicke, R. *Biochemistry* **1979**, 18, 5567-5571.
(45) Zetina, C.R.; Goldberg, M.E. *J. Mol. Biol.* **1980,** 137, 401-414.
(46) Freedman, R.B. *Trends Biochem. Sci.* **1984**, 9, 438-441.
(47) Freedman, R.B.; Bulleid, N.J.; Hawkins, H.C.; Paver, J.L. *Biochem. Soc. Symp.* **1989**, 55, 167-192.
(48) Fischer, G.; Bang, H.; Mech, C. *Biomed. Biochim. Acta* **1984**, 43, 1101-1111.
(49) Lang, K.; Schmid, F.X.; Fischer, G. *Nature* **1987**, 329, 268-270.
(50) Buchner, J.; Schmidt, M.; Fuchs, M.; Jaenicke, R.; Rudolph, R.; Schmid, F.X.; Kiefhaber, T. *Biochemistry* **1991**, 30, 1586-1591.
(51) Haas, I.G.; Wabl, M. *Nature* **1983**, 306, 387-389.
(52) Gething, M.J.; McCammon, K.; Sambrook, J. *Cell* **1986**, 46, 939-950.
(53) Bole, D.G.; Hendershot, L.M.; Kearney, J.F. *J. Cell. Biol.* **1986**, 102, 1558-1566.
(54) Bochkareva, E.S.; Lissin, N.M.; Girshovich, A.S. *Nature* **1988**, 336, 254-257.
(55) Lubben, T.H.; Donaldson, G.K.; Viitanen, P.V.; Gatenby, A.A. *Pl. Cell* **1989**, 1, 1223-1230.
(56) Lecker, S.; Lill, R.; Ziegelhoffer; T., Georgopoulos, C.; Bassford jr., P.J.; Kumamoto, C.A.; Wickner, W. *EMBO J.* **1989**, 8, 2703-2709.
(57) Rudolph, R. In *Modern Methods in Protein-, Nucleic Acid Analysis*, Tschesche, H. Ed.; Walter de Gruyter, Berlin, New York, **1990**, pp 149-171.
(58) Marston, F.A.O. *Biochem. J.* **1986**, 240, 1-12.
(59) Randall, L.L.; Hardy, S.J.S. *Cell* **1986**, 46, 921-928.
(60) Eilers, M.; Schatz, G. *Nature* **1986**, 322, 228-232.
(61) Rassow, J.; Guiard, B.; Wienhues, U.; Herzog, V.; Hartl, F.-U.; Neupert, W. *J. Cell Biol.* **1989**, 109, 1421-1428.
(62) Park, S.; Lui, G.; Topping, T.B.; Cover, W.H.; Randall, L.L. *Science* **1988**, 239, 1033-1035.
(63) Laminet, A.A.; Plückthun, A. *EMBO J.* **1989**, 8, 1469-1477.
(64) Crooke, E.; Wickner, W. *Proc. Natl. Acad. Sci. USA* **1987**, 84, 5216-5220.
(65) Chirico, W.J.; Waters, G.M.; Blobel, G. *Nature* **1988**, 332, 805-810.
(66) Sanz, P.; Meyer, D.I. *EMBO J.* **1988**, 7, 3553-3557.
(67) Koll, H.; Guiard, B.; Rassow, J.; Ostermann, J.; Horwich, A.L.; Neupert, W.; Hartl., F.-U. *Cell* **1992**, 68, 1163-1175.

RECEIVED October 26, 1992

Chapter 8

Folding of Recombinant Human Insulin-Like Growth Factor-1 in Yeast

Bhabatosh Chaudhuri and Christine Stephan

Department of Biotechnology, K−681.106, Ciba-Geigy Ltd., Basel, CH−4002 Switzerland

The role of propeptides, molecular chaperones and folding enzymes in the *in vivo* folding of human insulin-like growth factor-1 (IGF-1), expressed in the yeast *Saccharomyces cerevisiae*, has been studied. In order to acquire its native conformation intramolecular disulfide-bridged IGF-1 needs to gain entry into the yeast secretory pathway. Secretion is possible using the alpha-factor leader (αF_L), a polypeptide which consists of a 19-amino acid signal sequence and a 66-amino acid proregion (proαF_L). It has been observed that classical signal sequences alone do not allow translocation of nascent IGF-1 across the membrane of the endoplasmic reticulum (ER). Translocation is permitted when the proαF_L is covalently-linked to a signal sequence, when the stress-70 protein BiP (the yeast *KAR2* gene product) is over-expressed, and when the activity of the endogenous *cis* → *trans* prolyl isomerase in a yeast strain is partially annulled. Constructs which use the proαF_L to maintain a conformation necessary for translocation are the most efficient in permitting secretion of IGF-1. However, it appears that after allowing translocation, the proregion, which is processed later in the Golgi by the Kex2 endoprotease, has a deleterious effect on the folding of IGF-1. Early removal of the proregion, by co-expression of a mutant Kex2 protein which is retained in the ER, prevents formation of inactive disulfide-bonded IGF-1 dimers, the preponderant species in the secreted product.

IGF-1 is one of the two structurally-related human insulin-like growth factors (IGFs). IGF-1 bears not only a high degree of homology to the primary amino-acid sequence of human insulin, but it also exhibits physiological effects similar to insulin in the living cell (*1*). Proinsulin, the single-chain insulin precursor, contains three polypeptide domains, the A, B and C chains. The amino-terminal B chain and the carboxyl-terminal A chain are connected by a C peptide. When mature insulin forms, the C peptide is processed, but the two chains B and A stay inter-linked through three disulfide bridges. IGF-1, a single-chain 70-amino acid polypeptide, can be divided into structural domains similar to proinsulin. The amino acid residues in the B and A domains of IGF-1 are more than 50% identical to the B and A chains of proinsulin, and the position of the three disulfide bonds in insulin and IGF-1 are preserved. The three-dimensional structure of IGF-1 can, therefore, be modelled reasonably on the basis of the known structure of insulin (2). The model has been confirmed by the structure of IGF-1 obtained in solution using 2-D NMR spectroscopy (3).

In spite of these similarities to insulin, IGF-1 has some unique features which may have some bearing on its folding in a eucaryotic cell. Unlike proinsulin, the C peptide in single-chain IGF-1 remains uncleaved. There is also a D peptide, an 8-amino acid carboxyl-terminal extension, which is not found in proinsulin (1). The analysis of the human cDNA suggests that IGF-1 is formed as a precursor, which contains an additional peptide of 35-amino acid residues at its carboxyl terminus (4). In a mammalian cell, the role of the C and D peptides, and of the carboxyl-terminal proregion in the folding and secretion of IGF-1 is unknown. It appears that the folded structure which the molecule acquires during secretion undergoes further modification. Although unusually high concentrations (i.e. about 1000 times more than that of insulin concentrations present in the plasma) of the two IGFs are secreted from human cells, yet most of the IGFs (at least 80% of the total IGF-1) circulate, as inactive entities, tightly bound to carrier proteins (5). It is likely that IGF-1 is biologically active only when released from the carrier protein–IGF-1 complex.

Recombinant IGF-1 has been expressed in a variety of different organisms (1). In order to obtain active IGF-1, we have used the secretory pathway of the unicellular eukaryote, the yeast *Saccharomyces cerevisiae*. In this report, an attempt to secrete the correctly folded molecule is described.

Experimental Procedures

Materials. All enzymes used for DNA manipulations were obtained from Boehringer Mannheim. Reagents needed for sodium dodecyl sulfate-polyacrylamide gel electrophoresis (SDS-PAGE), immunoblotting and enzyme-linked immunoabsorbant assay (ELISA) were from Bio-Rad (Richmond, CA), excepting poly(vinylidene) difluoride (PVDF) membranes and goat serum which were bought from Millipore (Bedford, MA) and Gibco (Basel, Switzerland), respectively. Nitrocellulose filters used for visualising proteins secreted from yeast colonies, were available from Schleicher and Schuell (Feldbach, Switzerland). Centricon-3 concentrators were purchased from Amicon (Beverly, MA). Yeast extract and Bactopeptone, required for preparing IGF-1 expression medium, were obtained from Difco Laboratories (Detroit, MI) and casamino acids from Sigma (St Louis, MO). Anti-(rabbit immunoglobulins) conjugated to alkaline phosphatase antibodies (the secondary antibody) were purchased from Tago (Burlingame, CA). Rabbit polyclonal IGF1-antiserum was generated using active, monomeric IGF-1 purified by high-performance liquid chromatography (HPLC), and was obtained from K.Einsle (Ciba-Geigy, Basel). The polyclonal anti-Kar2 antibodies were raised in rabbits against an 11-amino acid peptide from the yeast immunoglobulin heavy-chain binding protein (BiP) by using a published procedure (6). The antibodies were provided by A.Hinnen (Ciba-Geigy, Basel). The *Escherichia Coli* signal peptidase was a gift from W.Wickner (UCLA, CA).

Strains and Transformations. All newly constructed plasmids were transformed in *E.coli* HB101. Yeast transformations (7) were performed in *S.cerevisiae* strains AB110 (α, *his4-580, leu2, ura3-52, pep4-3,* [*cir°*]) and SE104, a [*cir°*] derivative of GRF18 (α, *leu2-3, leu2-112, ura3Δ5*), where the functional copies of the proteinases A and B, and the carboxypeptidases S and Y have been removed by gene replacement. The chromosomal copy of the cytoplasmic cyclophilin gene (c-cyclophilin, *CYP1*) (8) in the yeast strain SE104 was disrupted by transformation with the linearized plasmid pUC19/*cyp1::LEU2* where a flushed 2.9kb *BglII* fragment containing the functional copy of the LEU2 gene has been inserted at the unique *NcoI* site (also flushed with T4 polymerase) of the *CYP1* gene (8).

Cloning of BiP and Cytoplasmic Cyclophilin Genes from Yeast. The yeast genes encoding the binding protein, BiP (*KAR2*) (6,9) and c-cyclophilin (*CYP1*) (8), were cloned by using the polymerase chain reaction (PCR) (10) and oligodeoxyribonucleotide primers (Microsynth, Zürich, Switzerland). The coding region along with the transcription terminator of the *KAR2* gene was isolated using oligomers 5' TACAGCTGGATCCTACCATGTTTTTCAACAGACTAAGCGCTGGC 3' (a), and 5' CTCGAGCCTTTCAACTCTCTC 3' (b). The complete *CYP1* gene, which includes the promoter, the coding region and the terminator, was obtained by using the oligomers 5' ATATGAATTCTAGAACCTTTCATCATC 3' (a), and 5' ATTAAAGCTTGATTGAAATTAAAACAA 3' (b). In each case, (a) and (b) represent the 5' ends of the sense strand and the 3' ends of the anti-sense strand of the published *KAR2* and *CYP1* sequences (6,8). Genomic DNA from the wild type yeast strain S288C was amplified in 30 cycles of PCR. A *BamHI-XhoI* fragment containing the ~2495 base pair (bp) coding region of the *KAR2* gene, which includes ~2180 bp of the coding sequence and ~315bp of the *KAR2* terminator, was subcloned in pUC20 and was completely sequenced (11). Four point mutations in the coding region which altered the following amino acids of the yeast BiP polypeptide, Tyr^{544} to Cys^{544}, Phe^{1347} to Leu^{1347}, Asn^{2133} to Asp^{2133} and Phe^{2313} to Leu^{2313} were identified. We do not know whether there is any major difference in functionality between the BiP which we have expressed from a PCR clone and the wild type. Preliminary results indicate that this mutant form of yeast BiP undergoes translocation and is retained in the ER (Latham, S.; Chaudhuri, B., unpublished data). The restriction map of the cloned *EcoRI-HindIII* 1385bp *CYP1* gene, subcloned in pUC19, is identical to the one already published (8). The PCR product has been partially sequenced (11). No differences have been found between the clone obtained from PCR and the published sequence.

Plasmids.

Plasmids bearing IGF-1 Expression Cassettes which contain different Secretion Signals. The normal expression cassette used for the secretion of IGF-1 (7) consists of a 400bp glyceraldehyde-3-phosphate dehydrogenase promoter (*GAPDH*p) and the coding sequence for the 85-amino acids of the *S.cerevisiae* alpha-factor leader (αF_L). The cassette ends with the tetrapeptide Leu-Asp-Lys-Arg which provides the Kex2 endoprotease cleavage site and it is followed by the coding sequence for mature IGF-1 and the alpha-factor terminator (αF_T) (*7*). The 255bp αF_L DNA sequence was replaced by the 57bp yeast invertase signal sequence (Inv_{SS}), the 51bp yeast acid-phosphatase signal sequence ($PHO5_{SS}$), the 57bp alpha-factor signal sequence (αF_{SS}), and the hybrid leaders, the 255bp Inv_{SS}proαF_L and the 249bp $PHO5_{SS}$proαF_L (where proαF_L denotes the proregion of the αF_L) by standard techniques of DNA manipulation (*12*). Detailed descriptions of these constructions have been reported earlier (*13*). *Bam*HI fragments of the different expression cassettes were subcloned in the *E.coli-S.cerevisiae* shuttle vector pDP34B (*7,14*), containing the complete 2-micron yeast plasmid sequence, and the yeast genomic *URA3* and the *dLEU2* sequences as selectable markers. The plasmids encoding the IGF-1 expression cassettes with the αF_L and the αF_{SS} as secretion signals have been named pBC23 and pBC27, respectively.

BiP Expression Cassette under the Control of the *CYC1* Promoter for Co-expression with IGF-1. A ~1100bp *Sal*I-*Bam*HI fragment encoding the yeast iso-cytochrome b gene (*CYC1*) promoter was isolated from the plasmid pLG669-7 (*15*) which contains a *CYC1-lacZ* gene fusion. The *Sal*I-*Bam*HI promoter fragment and ~2495bp *Bam*HI-*Sac*I encoding the *KAR2* gene and its transcription terminator (*6*) were subcloned in *Sal*I/*Sac*I digested pUC20. The isolated ~3600bp *Sal*I-*Sac*I fragment was purified from the plasmid pUC20/*CYC1*p-*KAR2*, the 5' and the 3' ends were flushed with T4 polymerase and the fragment was ligated in the flushed and dephosphorylated *Bgl*II site of pBC23 to yield a plasmid containing both the IGF-1 and BiP expression cassettes. This plasmid was named pBC28 and was used to express IGF-1.

Cytoplasmic Cyclophilin Expression Cassette for Co-expression with IGF-1. A 1385bp *Bgl*II fragment encoding the *CYP1* gene (*8*), obtained as a PCR product, was subcloned in the unique *Bgl*II site of pBC23 to yield the plasmid pBC29. Clones harboring plasmids which encode both the IGF-1 and c-cyclophilin expression cassettes were chosen for further expression of IGF-1.

Plasmids bearing Kex2p variants. At first, the expression cassette for the soluble form of the Kex2p (sKex2p) (*16*) was obtained by deleting from the *KEX2* gene (*17*) the DNA encoding the C-terminal 200-amino acids. The gene construct for sKex2pHDEL was made by adding linkers, encoding the His-Asp-Glu-Leu (HDEL) (*18*) peptide sequence and two stop codons, to the 3' end of the gene encoding sKex2p. The construction has already been described (*14*). The expression cassettes containing the genes encoding for wild type (wt) Kex2p and the mutant sKex2pHDEL were isolated as *Bam*HI fragments and subcloned in the unique *Bgl*II site of pBC23, to yield plasmids pBC24 and pBC26, respectively.

Expression of IGF-1. The plasmids were transformed in the yeast strains AB110 and SE104, both free of the 2-micron yeast plasmid but containing a mutant non-functional allele of the *URA3* gene. Three transformants from each transformation were grown for 72h in an IGF-1 expression medium which lacks uracil (*7*). This allowed plasmid-bearing cells to be maintained during growth. IGF-1 expressed by these cells was measured both quantitatively and qualitatively.

Quantitative Determination of IGF-1 in the Yeast Culture Media. Correctly folded, active, monomeric IGF-1 was quantified by reversed-phase HPLC and the total amount of secreted IGF-1–like molecules was estimated by ELISA. The details of these procedures have been described elsewhere (*7*).

Immunoblot Analysis. Samples for SDS-PAGE followed by Western blot analysis were performed according to previously published protocols (*7*).

Colony blot analysis. Yeast transformants were picked and streaked as circular zones on plates containing a solid IGF-1 expression medium. After incubating the plates for 24h at 30°C, the cells were blotted on to nitrocellulose filters (*13*). The secreted proteins remaining fixed to the filters were detected by a method identical to the protocol used for Western blot analysis (*7*).

Results and Discussion

Signal Sequences, Unfolding of Nascent Proteins and Secretion from Yeast. For targeting newly translated polypeptides to the secretory pathway a transient N-terminal extension, consisting of a hydrophobic core of amino acids, is required (*19*). This extra stretch, usually 15 to 30-amino acids long and termed signal peptides, mediate translocation of nascent polypeptides into the endoplasmic reticulum (ER). Translocation represents the first step in the eukaryotic secretory pathway (*20*).

It has been suggested that recognition of signal peptides by a cytosolic ribonucleoprotein complex, the signal recognition particle (SRP), is an essential facet of the translocation process (*21*). Nonetheless, it has remained intriguing how a disparate set of signal peptides, sharing only a hydrophobic core, can be at all recognized by the cell. Randall and Hardy have tried to explain this phenomenon by proposing that translocation takes place because the peptides which act as signal sequences enforce a delay in the folding of a nascently synthesized polypeptide (*22*). There is evidence that in yeast, the cytosolic heat shock proteins hsp70 (the 70 kilodalton heat shock proteins) and the ER-resident BiP (the *KAR2* gene product) may be also involved in translocation competence of proteins, perhaps by assisting the relaxation of any newly acquired tertiary structure (*23-25*). It is believed that for translocation to occur the nascent polypeptide must be in an unfolded state.

A variety of signal peptides have been used to direct the secretion of foreign proteins from yeast. Only rarely have the signal peptides, intrinsic to a particular protein, been able to direct the processing and secretion of a mature protein from yeast. In most cases, secretion of foreign proteins have been achieved from *S.cerevisiae* by fusing signal sequences of yeast secretory proteins to the coding region of different heterologous polypeptides (*26*).

Secretion of IGF-1 in yeast. Probably the most commonly used yeast secretion signal is the prepro sequence of the prepro-α-factor, the precursor of a 13-residue peptide pheromone secreted from haploid α mating type cells (27). Secretion of IGF-1 can also be directed by the prepro sequence, usually referred to as the alpha-factor leader (αF_L). The 85-amino acid αF_L sequence is an unusually long polypeptide which can target proteins to the secretory pathway. Besides possessing a pre- or signal sequence, the αF_L also contains a proregion (the proαF_L).

In a homologous situation where the αF_L permits the secretion of α-factor to allow mating of α cells with the opposite mating type **a** cells, the signal sequence is cleaved during translocation. In the lumen of the ER a precursor intermediate proαF_L-α-factor is formed (28) to which N-linked core sugars are attached (29). The proαF_L contains three consensus sequences (27) for N-glycosylation (Asn-Xaa-Ser/Thr) sites (29). Signal sequence cleavage and glycosylation of the proαF_L has also been observed when the heterologous proαF_L–IGF-1 fusion protein enters the secretory pathway (7).

For the maturation of α-factor, the glycosylated proαF_L–α-factor undergoes processing either in the late Golgi or in the secretory vesicles by the product of the *KEX2* gene which specifically recognizes pairs of exposed dibasic amino acids, Lys-Arg (30). A similar Kex2p-mediated removal of the proαF_L occurs in the maturation of IGF-1 (7,14). It is not known if the proregion is absolutely necessary for the secretion of the α-factor (31). It is possible that, at least, in the secretion of heterologous proteins, the proαF_L is redundant.

Classical Signal Sequences Do Not Allow Translocation of IGF-1 into the ER. Four yeast secretion signals, the 19-amino acid invertase signal sequence (Inv_{SS}) encoded by the *SUC2* gene (32), the 17-amino acid-phosphatase signal sequence ($PHO5_{SS}$) encoded by the *PHO5* gene (33), and the pre (αF_{SS}) and prepro (αF_L) sequences of prepro-α-factor encoded by the *MFα* gene (27), have been used to attempt the secretion of IGF-1 from *S.cerevisiae*. Expression cassettes, under the control of the same *GAPDH* promoter but encoding different secretion signals, have been used for the expression of IGF-1. It has been observed that only the αF_L sequence permits secretion. Yeast strains harboring the plasmids which encode the Inv_{SS}, the $PHO5_{SS}$ and the αF_{SS} do not secrete IGF-1 into the culture medium. Comparison of the levels of IGF-1–specific mRNA, in the total RNA obtained from cells harboring signal sequence constructs, does not show any discernible differences from strains where the αF_L is used as a secretion signal. This result excludes the possibility that drastically reduced mRNA levels could be a reason for the absence of secretion of IGF-1 when only signal sequences are employed (13).

When IGF-1 is expressed fused to signal sequences, IGF-1–like molecules accumulate inside the cell which can be detected on immune-blots with anti–IGF-1 polyclonal antibodies. Homogenisation of spheroplasted cells permits separation of crude membranes. Membrane protection experiments reveal that the intracellular

IGF-1–like species are not membrane-encompassed and, therefore, are probably in the soluble cytosolic fraction. Further analysis on polyacrylamide gels shows that the molecules have a slower mobility than authentic monomeric IGF-1. They can undergo cleavage by *E.coli* signal peptidase (*34*) to yield the mature IGF-1 molecules (*13*).

These observations imply that classical signal sequences do not permit the translocation of IGF-1 into the ER. The signal sequence attached IGF-1 molecules accumulate in the cytoplasm. It is likely that during synthesis these polypeptides fold too quickly into a compact conformation which prevents translocation (*7,22*). Since an unfolded conformation is believed to be a prerequisite (*23*) for the first step in secretion, the translocation event fails to take place. In this context, experiments investigating whether any of the known molecular chaperones or folding enzymes (*35*) play a role in the process of translocation of IGF-1 would be interesting (*7,36*).

Role of BiP in Folding/Unfolding of Proteins. The immunoglobulin heavy-chain binding protein (BiP) is a member of the 70kDa heat-shock protein family (*35*). In mammals, BiP is a 78kDa protein which is retained in the ER. BiP is generally induced in cells which undergo stress. The yeast homolog of BiP, encoded by the *KAR2* gene, is essential for karyogamy (*6*). The *KAR2* gene product is one of the few proteins in yeast which contains, at the C-terminus, the *S.cerevisiae* ER-retention signal, His-Asp-Glu-Leu (HDEL) (*18*). It has been proposed that for proteins destined for secretion, the *ERD2* gene product in yeast (*37*) recognizes the C-terminal HDEL tetrapeptide in an intermediate compartment between the ER and the Golgi. After recognition, the HDEL containing polypeptide is recycled back into the ER, thus allowing retention of the protein in the ER through the Erd2p receptor-mediated pathway. Other proteins are probably also involved in this process of recognition and recycling, but they remain to be identified.

It has been shown that malfolding of mutant proteins and incorrectly glycosylated proteins causes an elevated level of intracellular BiP (*35*). BiP is thought to act as a quality control agent (*38*), permanently binding to malfolded and unglycosylated proteins and preventing these malfolded structures from leaving the ER. Proteins bound to BiP are finally destroyed in the ER via a novel degradation pathway (*38*).

Not only does BiP have a role in modulating protein folding in the lumen of the ER, but it is also thought that BiP is involved in the translocation of precursors across the ER membrane. Experiments in which levels of BiP were down-regulated, by expressing a temperature sensitive yeast *kar2* mutant, have shown that secretory proteins accumulate as precursors in the cytosol at the nonpermissive temperature (*24,25*). It has been suggested that BiP binds to nascent proteins whilst they are still translocating, preventing the first part of the polypeptide from folding, so that once it has been sufficiently translocated the protein can fold correctly. Thus, the binding of BiP probably causes newly synthesized proteins to maintain an unfolded conformation which would then allow translocation. BiP can be released from bound proteins by ATP hydrolysis. Pelham proposes that, through a cyclic system of ATP hydrolysis, the protein involved is able to fold in steps every time BiP dissociates and then associates with it (*18*). Therefore, BiP acts by stabilizing nascent proteins until they adopt their correct conformation.

Over-expression of BiP Unfolds Nascent IGF-1 to Allow Translocation/ Secretion. We have expressed the yeast BiP homolog (the *KAR2* gene product) (*6,9*) under the control of the comparatively weak *CYC1* promoter (*15*). The promoter is repressed in the presence of glucose. The IGF-1 expression cassettes containing the αF_{SS} and the αF_L were subcloned in pDP34B (*14*) to yield the plasmids pBC27 and pBC23, respectively. The BiP expression cassette was subcloned in pBC27 to yield pBC28. The plasmids pBC23, pBC27 and pBC28 were transformed in the yeast strains AB110 and SE104. Transformants from both yeast strains were analyzed on a colony blot filter assay which can detect secreted IGF-1 from yeast cells blotted onto nitrocellulose filters. Proteins adhering to the filter were detected using anti–IGF-1 polyclonal antibodies (*7*). Secretion of IGF-1 is seen in transformants bearing plasmids pBC23 and pBC28 but not in the transformants bearing pBC27 (Figure 1).

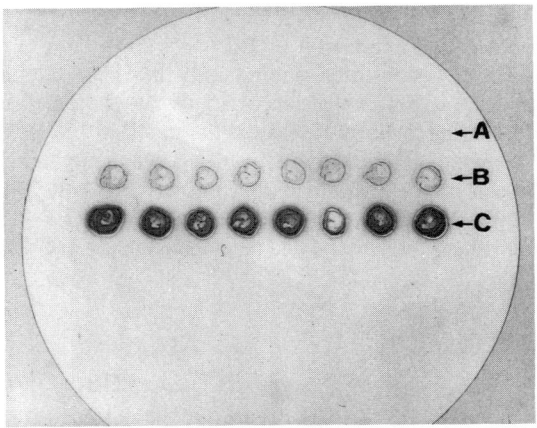

Figure 1. IGF-1 secretion from yeast strains over-expressing *KAR2*, as detected on a colony assay. Eight transformants from the strain SE104 bearing the plasmids (**A**) pBC27 (αF_{SS}–IGF-1), (**B**) pBC28 (αF_{SS}–IGF-1 and *KAR2*) and (**C**) pBC23 (αF_L–IGF-1) were streaked out on an agar-based IGF-1 expression medium (*7*). Secreted IGF-1 blotted on to nitrocellulose filters were incubated with the polyclonal IGF-1 anti-serum. The nitrocellulose-bound antigen-antibody complex was visualized using the alkaline phosphatase-conjugated secondary antibody and the Bio-Rad immunoassay kit.

Transformants in SE104 were grown in liquid cultures in the IGF-1–expression medium with glucose as the sole carbon source (*7*). The glucose concentrations are exhausted at about 14h of growth allowing de-repression of the *CYC1* promoter. Supernatants of cells grown for 72h were analyzed by HPLC. Identical titers of IGF-1 were obtained in the strains AB110 and SE104 (Table I).

Table I. Over-expression of BiP Aids in the Translocation of Nascent IGF-1 in Yeast

Plasmid	Expression Cassette(s)	Signal or Leader Sequence	HPLC titers in µg/ml [a]
pBC23	IGF-1	αF_L	12
pBC27	IGF-1	αF_{SS}	0[b]
pBC28	IGF-1 & BiP	αF_{SS}	3

[a] The titers are an average of values obtained from three different transformants.
[b] The supernatants were concentrated 10-fold with Centricon-3 filters.

The HPLC data confirm that expression of BiP on a multi-copy plasmid does allow secretion of IGF-1 from a construct bearing only a signal sequence. It has been observed that none of the IGF-1 expression plasmids allow more than four copies per cell, when plasmid-bearing cells are selected in uracil. It is possible that IGF-1 expression plasmids encoding *KAR2* have even a lower copy number. Intracellular levels of BiP expressed in yeast strains harboring plasmids pBC27 and pBC28 have been compared on Western blots (Figure 2). Levels of BiP in yeast transformants bearing pBC28 are distinctly higher and can be detected using anti-Kar2 antibodies.

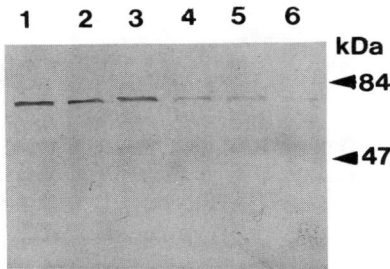

Figure 2. Western blot analyses of intracellular BiP, using 10% SDS-PAGE. 100 OD_{600} cells were mechanically lysed using 0.3 g glass beads (0.5 mm diameter) with 200 µl of sample buffer (4% SDS, 0.1 M Tris pH6.8, 4 mM EDTA) and 1.5µl of the cell lysate was loaded in each slot of the polyacrylamide gel. After blotting with poly (vinylidene) difluoride membrane, the proteins were incubated with anti-Kar2 antibodies and were visualized as in Figure 1. Lanes 1-3, transformants of SE104 harboring pBC28 (αF_{SS}–IGF-1 and *KAR2*); lanes 4-6, SE104 harboring pBC27 (αF_{SS}–IGF-1). The 47 and 110 kilodalton (kDa) bands used as markers belong to the pre-stained high-range standard proteins (Bio-Rad).

These results imply that early association of BiP with IGF-1 precursors can afford a conformation which allows translocation. The constitutive levels of BiP, expressed from a chromosomal copy, are not enough to maintain the newly synthesised IGF-1 in an unfolded state. However, when cells produce more BiP from an extra-chromosomal expression cassette, translocation-competence is somewhat facilitated resulting in secretion of mature IGF-1 into the culture supernatant. In this context, it ought to be mentioned that the expression of BiP under the control of a strong constitutive promoter (viz. *GAPDH*p), which normally permits higher levels of transcription than the inducible *CYC1* promoter, causes strains to grow very poorly, suggesting that overexpression of BiP with the *GAPDH* promoter may be deleterious for the yeast cell. Moreover, the construction fails to provide any secretion of IGF-1. Analysis of the plasmids isolated from these strains show that they have undergone rearrangement. These results indicate that an optimal expression of the stress protein BiP could transfigure into a positive effect on the translocation of the nascent IGF-1 when signal sequences are used for translocation. Until now, we have failed to show so emphatically such a clear effect of the overexpression of BiP on the secretion of IGF-1, using the αF_L or other hybrid leader sequences as secretion signals (Stephan, C., and Chaudhuri, B., unpublished data).

Immunophilins and Protein Folding. The cellular action of the immunosuppresive drugs Cyclosporin A and FK506 are thought to be mediated via the cytoplasmic receptor proteins cyclophilin and FKBP (FK506 binding protein), respectively (*39*). The cyclophilins and FKBPs from different species are highly conserved, indicating that they may have an important role to play in the cell. Interestingly, cyclophilin and FKBP share no apparent homology. Yet they have been grouped together because they both catalyze the *cis* → *trans* isomerisation of Xaa-Pro bonds present in short synthetic peptides, implying that they have the capability to rotate a *cis* Xaa-Pro bond in the peptide substrate to the *trans* geometric conformation. Recent studies have shown that in addition to cyclophilin and FKBP, there are other cellular proteins which exhibit this isomerase activity (*39,40*). Proteins binding to small immunosuppressant molecules (either cyclic peptides or macrolides) and possessing peptidylprolyl isomerase activity, have been classified as belonging to a new family of proteins, termed immunophilins (*39*).

In vitro experiments have shown that cyclophilin catalyzes the slow isomerisation of Xaa-Pro bonds in proteins, too (*41*). This catalysis accelerates the *in vitro* refolding of proline-containing polypeptides. There is only indirect evidence that cyclophilin may have an important function in the folding of proteins *in vivo*. Mutations in the *Drosophila melanogaster* cyclophilin homolog ninaA, an eye-specific membrane protein, seem to affect the folding and intracellular trafficking of rhodopsin (*42*). It has also been noticed that when chicken embryo fibroblasts are treated with cyclosporin A the folding of the triple helix of type I collagen appears to be delayed (*35*).

S.cerevisiae has at least three cyclophilin-related genes (*8,43*), one cytoplasmic and the other two bearing signal sequences. It is not clear whether the signal sequence-bearing cyclophilins really secrete from yeast. Recently, other cyclophilin genes encoding signal sequences have been cloned in higher eukaryotes and there are data which strongly indicate that cyclophilins in the secretory pathway participate in intracellular protein transport (*44,45*).

Using RNase T1 as a model system, purified cyclophilin, obtained from the intracellular proteins of yeast, has been shown to be an *in vitro* catalyst of protein folding (*46*). It has been speculated that cytoplasmic cyclophilin may also be involved in the unfolding/refolding of newly synthesized proteins *in vivo*, particularly of those which have to traverse cellular membranes (*47*).

Removal of Cytoplasmic Cyclophilin Activity Has an Effect on the Folding of IGF-1 in the Cytoplasm. Secretion of IGF-1 in yeast seems to be impossible using signal sequences alone. Untranslocated IGF-1 molecules, with uncleaved signal sequences, are associated with a soluble fraction after lysis of yeast spheroplasts (*13*). From these observations, it is tempting to think that a *cis* → *trans* isomerisation of one or more of the five Xaa-Pro bonds present in IGF-1 may be responsible for the rapid folding of the IGF-1 precursor in the cytoplasm into a conformation which prevents translocation across the membrane of the rough ER. Although prolyl isomerisation is a slow process *in vitro*, it may not be slow enough *in vivo* for IGF-1 to prevent a premature folding into a conformation which is translocation incompetent.

C-cyclophilin (the *CYP1* gene product) is an abundant protein in yeast, corresponding to about 0.4% of the total cellular mRNA (*8*). We have studied the effect of removing the functional copy of the *CYP1* gene from the genome of the yeast strain SE104. *In vivo* translocation of IGF-1 was monitored by assaying for secretion of IGF-1 from the strains SE104 and SE104*cyp1* on colony blots.

Surprisingly, we find that in the strain SE104*cyp1*, where the cyclophilin activity has been abolished (Chaudhuri, B.; Stephan, C., unpublished data), secretion of IGF-1 is observed from a strain harboring the IGF-1 expression plasmid containing the αF_{SS} (Figure 3). Secretion cannot be detected in the same strain when c-cyclophilin is expressed again, this time from a plasmid which also bears the IGF-1 expression cassette (pBC29).

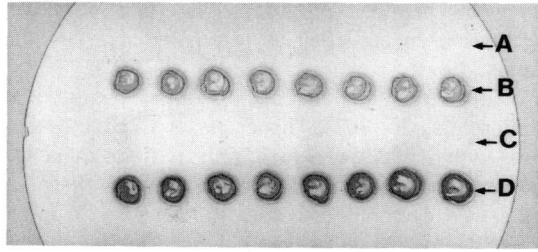

Figure 3. Influence of c-cyclophilin (*CYP1* gene product) on the secretion of IGF-1, as detected on a colony assay (see Figure 1). Yeast transformants harboring the expression plasmids pBC23 (αF_L–IGF-1), pBC27 (αF_{SS}–IGF-1) and pBC29 (αF_{SS}–IGF-1 and *CYP1*) were analyzed. Eight transformants from each of the three transformations were used. (**A**), pBC27::SE104; (**B**), pBC27::SE104*cyp1*; (**C**), pBC29::SE104*cyp1*; (**D**), pBC23::SE104.

It is obvious that in strains which do not express the *CYP1* gene, nascent IGF-1 undergoes translocation, and after signal sequence cleavage, at least part of the molecules traverse the secretion pathway. The correctly folded IGF-1 molecules in the culture supernatant have been identified by HPLC. Using a signal sequence, secretion of IGF-1 from strains lacking *CYP1* is low compared to the secretion from strains transformed with the αF_L-encoding expression plasmids. This indicates that although translocation is allowed, its efficiency is poor in the absence of c-cyclophilin (Table II).

Table II. The Rotamase Cyclophilin Prevents Translocation of Nascent IGF-1 in Yeast

Plasmids	Expression Cassettes	Signal/Leader Sequence for IGF-1 Expression	Yeast Strain	HPLC Titers in µg/ml [a]
pBC23	IGF-1	αF_L	SE104	16
pBC23	IGF-1	αF_L	SE104*cyp1*	13
pBC27	IGF-1	αF_{SS}	SE104	0 [b]
pBC27	IGF-1	αF_{SS}	SE104*cyp1*	4
pBC29	IGF-1 & Cyp1p	αF_{SS}	SE104*cyp1*	0[b]

[a] The titers are an average of values obtained from three different transformations.
[b] The supernatants were concentrated 10-fold with Centricon-3 filters.

These results strongly suggest that the αF_{SS}–IGF-1 precursors, which form in the cytoplasm, do acquire a folded/misfolded structure catalyzed by the *cis* → *trans* prolyl isomerase. Probably multiple prolines are involved in this process. As a result of premature folding/misfolding, translocation is blocked. In the absence of the isomerase, the newly synthesized IGF-1 molecules can probably undergo co-translational translocation because one or more of the Xaa-Pro bonds is maintained in a configuration which influences the entry of IGF-1 into the secretory pathway. It can be very well imagined that deletion of the gene encoding the cytoplasmic FKBP (*48,49*) together with the *CYP1* gene, in a yeast strain, would have a more pronounced effect on the translocation of nascent IGF-1.

Covalent Attachment of the Proregion of the αF_L to a Signal Sequence Allows Translocation of IGF-1. Of the four yeast secretion signals used to direct IGF-1 to the secretory pathway, the αF_L is the only sequence which permits secretion (*13*). The αF_L consists of a 19-amino acid signal sequence (αF_{SS}) and a 66-amino acid proαF_L. The function of the proαF_L in the secretion of the mating pheromone α-factor is unclear. A construction where the proαF$_L$ has been deleted permits translocation of α-factor across yeast microsomal membranes *in vitro*, though poorly (*31*). In the context of the secretion of foreign proteins, the three heterologous proteins, aminoglycoside phosphotransferase encoded by Tn903 that directs kanamycin resistance, human granulocyte-macrophage colony stimulating factor and interleukin-1β, have been reported to be secreted from yeast only by using the αFss (*50*).

The three signal sequences αF$_{SS}$, Inv$_{SS}$ and PHO5$_{SS}$ do not allow translocation of IGF-1 into the yeast ER, which would enable the secretion of IGF-1 into the yeast culture medium (*13*). Since the presence of the proαF$_L$ in the αF$_L$ facilitates secretion, we were eager to find out whether the proαF$_L$ might also help in promoting secretion of IGF-1 when fused to the Inv$_{SS}$ and the PHO5$_{SS}$. Interestingly, fusion of the proαF$_L$ to the signal sequences does allow secretion, indicating that membrane translocation of IGF-1 is possible, *in vivo* (*13*). Transformants from the yeast strains AB110 and SE104 were grown in liquid cultures and the secreted IGF-1, using these novel hybrid leader peptides (ie. Inv$_{SS}$proαF$_L$ and PHO5$_{SS}$proαF$_L$), was measured by HPLC (Table III).

Table III. The Proregion of the αF$_L$ Allows Translocation/Secretion of IGF-1 [a]

Leader Sequence Used	Yeast Strain	HPLC Titers in μg/ml [b]
αF$_L$	AB110	10
αF$_L$	SE104	16
Inv$_{SS}$proαF$_L$	AB110	12
Inv$_{SS}$proαF$_L$	SE104	19
PHO5$_{SS}$proαF$_L$	AB110	4
PHO5$_{SS}$proαF$_L$	SE104	6

[a] The IGF-1 expression cassettes were identical excepting for the leader sequence.
[b] The titers are an average of values obtained from three different transformations.

It appears that signal peptides alone are not sufficient to allow translocation of IGF-1 across the membrane of the ER. For translocation, an unfolded conformation of the nascent protein is required. It is likely that the proαF$_L$ has a role in unfolding the nascent IGF-1, keeping the polypeptide in a conformation conducive to translocation (*13*). We found that the proαF$_L$ need not be the only sequence which permits translocation of IGF-1. In order to structurally define a proregion, primarily acting as a translocatory agent, we have screened a random library, consisting of 150 to 250bp of yeast genomic DNA fragments. It has been possible to isolate alternate sequences which fulfill the role of the proαF$_L$ in allowing translocation of IGF-1 (*51*).

The Role of the Proregion of the αF_L in the Folding of IGF-1. There are a variety of secretory proteins which after signal sequence cleavage form intracellular precursors, called pro-proteins (52). These intermediates contain peptide residues, known as propeptides, not found in the mature, secreted protein. After export of the pro-proteins from the ER, the proregion is proteolytically processed from the precursor, usually in the *trans* Golgi membrane or in the secretory vesicles (53). Cleavage often occurs at pairs of dibasic amino acid residues.

Until recently, the possible cellular function of propeptide sequences was mainly a matter of conjecture. Evidence has now been obtained indicating that proregions can act as molecular chaperones (54). It has been shown both *in vitro* and *in vivo* that the proregions of two prokaryotic zymogens aid in the folding of the corresponding active proteases, subtilisin and α-lytic protease (54). *In vitro* data indicate that the causative agent for the proper folding of yeast carboxypeptidase Y (CPY) is also its proregion, present in the precursor form of CPY (55).

Our results indicate that the proαF_L may have an unfolding action on IGF-1 before translocation is allowed. Upon entry into the ER, nascent IGF-1 is free to acquire its native structure. We have inquired whether the proαF_L sequence is necessary, beyond the translocation step, in the folding of IGF-1. In order to investigate the role of the proαF_L after signal sequence mediated translocation has occurred, it is imperative that the proαF_L be removed in the lumen of the ER rather than in the Golgi. For proteins which have entered the secretory pathway folding is thought to take place in the ER (56). The *KEX2* gene product, Kex2p, removes the proαF_L from the precursor protein proαF_L-IGF-1 only after IGF-1 has acquired its tertiary structure.

The majority of the IGF-1-like molecules secreted from yeast are not the active monomers but inactive disulfide-linked dimers (57,58). It has been proposed that specific charged interactions between two IGF-1 molecules during the process of folding in the cell may cause the formation of dimers (58). Another reason for dimerization could be that IGF-1 folds aberrantly in the ER owing to the presence of the extraneous proαF_L.

ProαF_L causes IGF-1 to form disulfide-linked dimers. To study the folding of the proαF_L-IGF-1 fusion protein in the ER we have co-expressed a novel mutant Kex2 endoprotease which is functional in the ER instead of in the Golgi (14). The wtKex2p is a Ca^{++} dependent serine protease spanning the membrane of the Golgi (59). This Golgi-bound enzyme is known not to secrete. A soluble form of the enzyme (sKex2p), lacking the C-terminal 200-amino acids which includes the membrane-spanning domain, secretes into the culture medium in considerable amounts (16). The *S.cerevisiae* ER-retention signal (18) has been attached to the C-terminus of sKex2p yielding an enzyme (sKex2pHDEL) which is partially retained inside the cell (Latham, S.; Chaudhuri, B., unpublished data). It has been shown that mutations in the proαF_L cause ER-accumulation of proαF_L-IGF-1 precursors. The novel mutant enzyme sKex2pHDEL functions in the ER by removing the proαF_L from the ER-accumulated pool of precursor proteins. After processing, mature IGF-1 is released (14).

We have co-expressed sKex2pHDEL with the αF_L–IGF-1 fusion protein in the yeast strain AB110. The total IGF-1–like molecules in the yeast culture supernatant was measured by ELISA. This assay estimates the sum of the different species of IGF-1, the correctly folded monomers, the positional isomers where two of the disulfide bonds in IGF-1 are interchanged and the mixture of molecules which are disulfide-linked dimers or multimers (7). HPLC gives an accurate measure of the correctly folded monomeric molecules alone. The results (Table IV) indicate that there is an appreciable decrease in the amount of molecules which are not the correctly-folded monomer. HPLC titers subtracted from the ELISA values portray the amount of the inactive molecules formed.

Table IV. The Presence of the ProαF_L during Folding of IGF-1 Influences the Formation of Inactive Molecules

Plasmids	Expression Cassettes	HPLC Titers [a] in μg/ml	ELISA Values [a] in μg/ml	Inactive IGF-1 in μg/ml
pBC24	IGF-1 & wtKex2p	10	78	68
pBC26	IGF-1 & sKex2pHDEL	9	24	15

[a] The values are an average obtained from three different transformations.

Furthermore, Western blot analysis reveals that dimers/multimers, which are disulfide-linked (i.e. aggregates which can be reduced by dithiothreitol), are remarkably reduced when the ER-retained, modified Kex2p is expressed in an yeast cell (Figure 4). This implies that an early processing of the proαF_L in the ER prevents intermolecular disulfide bond formation in molecules secreted from yeast. This has been confirmed by a variety of experiments which include pulse-chase kinetics (Chaudhuri, B.; Latham, S. E.; Stephan, C., unpublished data). These results also emphasize the fact that proαF_L does have a role in the *in vivo* folding of the intramolecular disulfide-linked IGF-1 molecules. In the native conformation of IGF-1 the 6 cysteines involved in the 3 *intra*molecular -S-S- bonds are masked (58). It is likely that during the folding of the proαF_L–IGF-1 polypeptide at least two of the cysteines in IGF-1 are somehow uncovered. This must take place if *inter*molecular -S-S- bonds are to form. The unmasking of cysteines could either take place because of misfolding or because proαF_L slows down the rapid turnover of a folding intermediate where a cysteine residue is momentarily exposed (60).

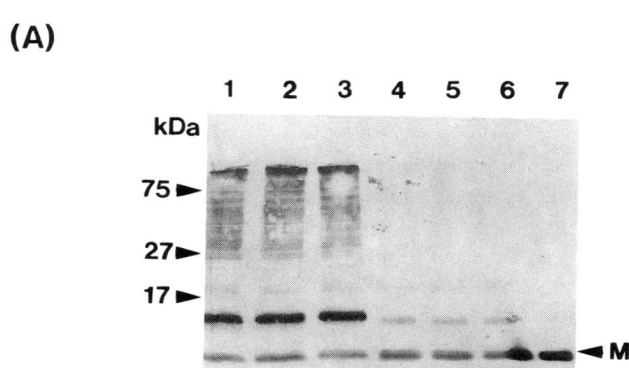

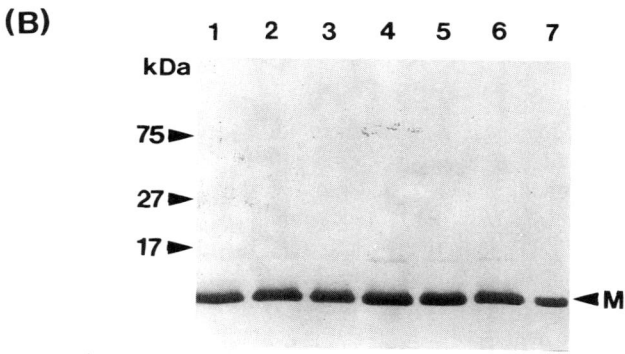

Figure 4. A comparison of IGF-1-like proteins secreted from AB110 transformants, harboring the IGF-1 expression plasmids pBC24 (encoding IGF-1 and wtKex2p) and pBC26 (encoding IGF-1 and sKex2pHDEL), on a Western blot. Three transformants from each of the two transformations were used for analyses. 2ml of cells (OD_{600}/ml=30) were harvested from 72h yeast cultures. The supernatants were concentrated 4-fold using Centricon-3 filters. 10μl of cell supernatants were electrophoresed on a 15% SDS-polyacrylamide gel. Proteins fractionated on the gel were blotted as in Figure 2 and was incubated with IGF-1 anti-serum. Proteins were visualized as in Figure 1. (A) Lanes 1-3: pBC24; lanes 4-6: pBC26; lane 7: 150 ng of HPLC-purified IGF-1 monomer (M). (B) same as in (A), excepting that all samples were treated with the reducing agent dithiothreitol. The 17, 27 and 75 kDa bands used as markers belong to the pre-stained low-range standard proteins (Bio-Rad).

Conclusions

It has been shown that signal sequences alone do not allow membrane-translocation of IGF-1 *in vivo*, probably because nascent IGF-1 folds too quickly in the cytoplasm into a translocation-incompetent conformation. The molecular chaperone BiP helps in translocation, perhaps by enforcing a relaxed conformation of the newly synthesized molecules.

Removal of the endogenous cyclophilin activity in the cell also permits translocation, possibly by reducing the rate of isomerisation of *cis* Xaa-Pro bonds to the *trans* configuration. The *cis* → *trans* prolyl isomerisation is known to have an impact on the kinetics of protein-folding, *in vitro*. In the case of IGF-1, reduced rates of isomerisation probably keeps nascent IGF-1 in an unfolded conformation compatible to translocation.

The proregion of the prepro-α-factor can function as an efficient translocatory agent in the secretion of IGF-1. However, after performing its chaperone-like function (i.e. preventing formation of inappropriate structures which disallow translocation), the proregion has an effect on the folding pathway of IGF-1. The proregion participates in the formation of inactive disulfide-linked dimers and multimers. It is seen that dimer formation is dramatically reduced when the proregion is removed in the lumen of the ER with the help of the mutant, ER-retained Kex2p endoprotease.

Acknowledgements

We express our indebtedness to W.Wickner for a generous gift of the *E.Coli* signal peptidase. We would like to thank L.Guarente for providing us with the plasmid pLG669-7, U.Gausmann and M.Riederer for the *KAR2* gene, J.Heim and H.-J.Treichler for the yeast strain SE104, G.Roemmle and H.-B. Jenny for performing the ELISA and the HPLC, respectively, A.Strauss for confirming some of the HPLC results and S.Latham for technical assistance. We are grateful to A.Hinnen for his continued support. A special thanks is due to H.-F. Vahlensieck and N. Dastoor for help in preparing the manuscript.

Literature Cited

1. Humbel, R. E. *Eur. J. Biochem.* **1990**, *190*, pp. 445-462.
2. Blundell, T.; Wood, S. *Annu. Rev. Biochem.* **1982**, *51*, pp. 123-154.
3. Cooke, R. M.; Harvey, T. S.;Campbell, I. D. Biochemistry **1991**, *30*, pp. 5484-5491.
4. Jansen, M.; van Schaik, F.M.A.; Ricker, A.T.; Bullock, B.; Woods, D.E.; Gabbay, K.H.; Nussbaum, A.L.; Sussenbach, J.S.;van den Brande, J.L. *Nature (London)* **1983**, *306*, pp. 609-611.
5. Zapf, J; Hauri, C; Waldvogel, M; Froesch, E.R. *J. Clin. Invest.* **1986**, *77*, pp. 1768-1775.
6. Rose, M. D.; Misra, L. M.; Vogel, J. P. *Cell* **1989**, *57*, pp. 1211-1221.
7. Steube, K.; Chaudhuri, B.; Märki, W.; Merryweather, J. P.; Heim, J. *Eur .J .Biochem.* **1991**, *198*, pp. 651-657.

8. Haendler, B.; Keller, R.; Hiestand, P. C.; Kocher, H. P.; Wegmann, G.; Movva, N. R. *Gene (Amsterdam)* **1989**, *83*, pp. 39-46.
9. Normington, K.; Kohno, K.; Kozutsumi, Y.; Gething, M.-J.; Sambrook, J. *Cell* **1989**, *57*, pp. 1223-1236.
10. Mullis, K.; Faloona, F.; Scharf, S.; Saiki, R.; Horn, G.; Erlich, H. *Cold Spring Harbor Symp. on Quant. Biol.* **1986**, *51*, pp. 263-273.
11. Sanger, F.; Nicklen, S.; Coulson, A.R. *Proc. Natl. Acad. Sci. USA* **1977**, *74*, pp. 5463-5467.
12. *Molecular Cloning: a Laboratory Manual*; Sambrook, J; Fritsch, E.F.; Maniatis, T., Eds.; Cold Spring Harbor Laboratory: Cold Spring Harbor, NY, **1989**; Vol. 1-3.
13. Chaudhuri, B.; Steube, K.; Stephan, C. *Eur. J. Biochem.* **1992**, *206*, pp. 793-800.
14. Chaudhuri, B.; Latham, S. E.; Helliwell, S. B.; Seeboth, P. *Biochem. Biophys. Res. Commun.* **1992**, *183*, pp. 212-219.
15. Guarente, L.; Ptashne, M. *Proc. Natl. Acad. Sci. USA* **1981**, *78*, pp. 2199-2203.
16. Seeboth, P. G.; Heim, J. *Appl.Microbiol.Biotechnol.* **1991**, *35*, pp. 771-776.
17. Mizuno, K; Nakamura, T; Ohshima, T; Tanaka, S.; Matsuo,H. *Biochem. Biophys. Res. Commun.* **1988**, *156*, 246-254.
18. Pelham, H. R. B. *Annu. Rev. Cell Biol.* **1989**, *5*, pp. 1-23.
19. Von Heijne, G. *J. Mol. Biol.* **1985**, *184*, pp. 99-105.
20. Rothman, J. E.; Orci, L. *Nature (London)* **1992**, *355*, pp. 409-415.
21. Walter, P.; Lingappa, V. R. *Annu. Rev. Cell Biol.* **1986**, *2*, pp. 499-516.
22. Randall, L. L.; Hardy, S. J. S. *Science* **1989**, *243*, pp. 1156-1159.
23. Deshaies, R. J.; Koch, B. D.; Werner-Washburne, M.; Craig, E. A.; Schekman, R. *Nature (London)* **1988**, *332*, pp. 800-805.
24. Vogel, J. P.; Misra, L. M.; Rose, M. D. *J. Cell Biol.* **1990**, *110*, pp. 1885-1895.
25. Nguen, T. H.; Law, D. T. S.; Williams, D. B. *Proc. Natl. Acad. Sci .USA* **1991**, *88*, pp. 1565-1569.
26. Hirsch, H.H.; Suarez-Rendueles, P.; Wolf, D. *In Molecular and Cell Biology of Yeast*; Walton, E.F.; Yarronton, G.T., Eds.; Blackie, Glasgow, London, **1989**; pp. 135-200.
27. Kurjan, J.; Herskowitz, I. *Cell* **1982**, *30*, pp. 933-943.
28. Waters, M.G.; Evans, E.A.; Blobel, G. *J. Biol. Chem.* **1988**, *263*, pp. 6209-6214.
29. Kornfeld, R.; Kornfeld, S. *Annu. Rev. Biochem.* **1985**, *54*, pp. 631-664.
30. Fuller, R. S.; Sterne, R. E.; Thorner, J. *Annu. Rev .Physiol.* **1988**, *50*, pp. 345-362.
31. Rothblatt, J. A.; Webb, J. R.; Ammerer, G.; Meyer, D. I. *EMBO J.* **1987**, *6*, pp. 3455-3463.
32. Taussig, R.; Carlson, M. *Nucleic Acids Res.* **1983**, *11*, pp. 1943-1954.
33. Meyhack, B.; Bajwa, W.; Rudolph, H.; Hinnen, A. *EMBO J.* **1982**, *1*, pp. 675-680.
34. Wolfe, P.B.; Wickner, W.; Goodman, J.M. *J. Biol. Chem.* **1983**, *258*, pp. 12073-12080.
35. Gething, M.-J.; Sambrook, J. *Nature (London)* **1992**, *355*, pp. 33-45.
36. Hober, S.; Forsberg, G.; Palm, G.; Hartmanis, M.; Nilsson, B. *Biochemistry* **1992**, pp. 1749-1756.

37. Semenza, J. C.; Hardwick, K. G.; Dean, N.; Pelham, H. R. B. *Cell* **1990**, *61*, pp. 1349-1357.
38. Hurtley, S. M.; Helenius, A. *Annu. Rev. Cell Biol.* **1989**, *5*, pp. 277-307.
39. Schreiber, S. L. *Science* **1991**, *251*, pp. 283-287.
40. McKeon, F. *Cell* **1991**, *66*, pp. 823-826.
41. Fischer, G.; Schmid, F. X.; *Biochemistry*, **1990**, *29*, pp. 2205-2213.
42. Stammes, M. A.; Shieh, B.-H.; Chuman, L.; Harris, G. L.; Zuker, C.S. *Cell* **1991**, *65*, pp. 219-227.
43. McLaughlin, M. M.; Bossard, M. J.; Koser, P.L.; Cafferkey, R.; Morris, R. A.; Miles, L. M.; Strickler, J.; Bergsma, D. J.; Levy, M. A.; Livi, G. P. *Gene (Amsterdam)* **1992**, *111*, pp. 85-92.
44. Caroni, P.; Rothenfluh, A.; McGlynn, E.; Schneider, C. *J.Biol.Chem.* **1991**, *266*, pp. 10739-10742.
45. Colley, N. J.; Baker, E. K.; Stammes, M. A.; Zuker, C. S. *Cell* **1991**, *67*, pp. 255-263.
46. Schönbrunner, E. R.; Mayer, S.; Tropschug, M.; Fischer, G.; Takahashi, N.; Schmid, F. X. *J.Biol. Chem.* **1991**, *266*, pp. 3630-3635.
47. Tropschug, M.; Barthelmess, I. B.; Neupert, W. *Nature (London)* **1989**, *342*, pp. 953-955.
48. Wiederrecht, G.; Brizuela, L.; Elliston, K.; Sigal, N. H.; Siekierka, J. J. *Proc. Natl. Acad. Sci. USA* **1991**, *88*, pp. 1029-1033.
49. Heitman, J.; Rao Movva, N.; Hiestand, P. C.; Hall, M. N. *Proc. Natl. Acad. Sci. USA* **1991**, *88*, pp. 1948-1952.
50. Ernst, J. F. *DNA* **1988**, *7*, pp. 355-360.
51. Chaudhuri, B.; Hinnen, A. *Protein Synthesis and Targeting in Yeast;* Brown, A.; McCarthy, J.; Sherman, F.; Tuite, M., Eds; NATO-ASI series; Springer-Verlag: Berlin, Germany, in press.
52. Fabre, E.; Nicaud, J,-M.; Lopez, M. C.; Gaillardin, C. *J.Biol.Chem.* **1991**, *266*, pp. 3782-3790.
53. Barr, P.J. *Cell* **1991**, *66*, pp. 1-3.
54. Ellis, R.J.; van der Vies, S.M. *Annu. Rev. Biochem.* **1991**, *60*, pp. 321-347.
55. Winther, J. R.; Sorensen, P. *Proc.Natl. Acad.Sci.USA* **1991**, *88*, pp. 9330-9334.
56. Freedman, R. B. *Cell* **1989**, *57*, pp. 1069-1072.
57. Elliott, S.; Fagin, K. D.; Narhi, L. O.; Miller, J. A.; Jones, M.; Koski, R.; Peters, M; Hsieh, P.; Sachdev, R.; Rosenfeld, R.D.; Rohde, M. F.; Arakawa, T. *J. Protein Chem.* **1990**, *9*, pp. 95-104.
58. Chaudhuri, B.; Helliwell, S. B.; Priestle, J. P. *FEBS Lett.* **1991**, *294*, pp. 213-216.
59. Fuller, R. S.; Brake, A. J.; Thorner, J. *Proc. Natl. Acad. Sci. USA* **1989**, *86*, pp. 1434-1438.
60. Chaudhuri, B; Stephan, C. *FEBS Lett.*, **1992**, *304*, pp. 41-45.

RECEIVED October 26, 1992

Chapter 9

Role of the Protein-Folding Chaperone BiP in Secretion of Foreign Proteins in Eukaryotic Cells

Anne S. Robinson and K. Dane Wittrup

Department of Chemical Engineering, University of Illinois, Urbana, IL 61801

> A mathematical model of protein folding and binding to the chaperone BiP within the eucaryotic secretory pathway has been constructed. Incorporating model parameters reported in the literature, the model results suggest that increasing levels of BiP in the lumen of the ER will lead to more efficient protein secretion, due to competition with nonproductive aggregation of nascent polypeptide chains. We have experimentally measured significant reductions in BiP protein levels in yeast cells secreting foreign proteins. This reduction, brought about by an as yet unknown mechanism, may be responsible for inefficient secretion of foreign proteins by yeast.

The past decade has seen a surge of interest in the process of protein folding and assembly within the cell, with the discovery that folding is assisted *in vivo* by a class of proteins known as chaperones (*1-4*). Since all of the information required to specify the compact folded structure of a protein is present in the amino acid sequence, proteins can be reversibly unfolded and refolded in dilute solutions *in vitro* in the absence of extra factors. However, protein folding *in vivo* after synthesis, or after translocation across membranes, must occur at total protein concentrations up to the millimolar range. At these concentrations, intermolecular aggregation reactions compete with the intramolecular folding process. One role for chaperones within the cell, therefore, is to prevent aggregation of protein folding intermediates.

Chaperones include enzymes such as protein disulfide isomerase and peptidyl prolyl isomerase (*5*). Another type of chaperone binds to folding intermediates, but not folded proteins, and does not appear to cause covalent modification of the protein. Two major families of this type of chaperone have been widely studied, the hsp70 class and the hsp60 class. These two chaperone types differ significantly in structure and function: hsp70 exists as a monomer or dimer (*6*) and binds short peptide stretches in an extended conformation (*7*), while hsp60's form a stable scaffold of two heptamer rings stacked one atop the other (*1*) that appear to recognize elements of secondary structure in unfolded proteins (*7*). Recent work of Hartl et al. (*8*) indicates that hsp70 and hsp60 chaperones may play sequential and complementary roles in protein folding, with hsp70's binding extended polypeptide chains to prevent aggregation, and then passing the polypeptide on to an hsp60 oligomer for completion of folding.

Most compartments of the eucaryotic cell contain hsp70 homologues, while eucaryotic hsp60 homologues have only been identified in mitochondria and chloroplasts (although sequence homology indicates the potential presence of what may be an hsp60-type chaperone in the eucaryotic cytosol (9, 10).) Hsp70's such as DnaK in *E. coli*, yeast SSA, SSB, SSC, SSD, and KAR2, and mammalian hsp73, hsp72, and BiP/GRP78 are well conserved in sequence and function (4). ATP hydrolysis is required for release of unfolded proteins or peptides bound by hsp70 chaperones(11). Sequential binding and release of extended polypeptide chains is a common feature of all hsp70 chaperones, and as such must constitute a significant aspect of their biochemical function. It has been proposed that hsp70's act as reversible molecular detergents which shield exposed hydrophobic patches of unfolded proteins from aggregation (12). Thus, one function of hsp70 chaperones may be directly analogous to that observed for non denaturing detergents in improving the yield of rhodanese refolding *in vitro*(13), or polyethylene glycol in improving refolding of carbonic anhydrase *in vitro* (14) – namely, binding to folding intermediates and hindering formation of protein aggregates.

An hsp70 homolog, IgG heavy chain Binding Protein (BiP), is found in the lumen of the endoplasmic reticulum (ER) of all eucaryotic cells. The KAR2 gene in yeast codes for a BiP homolog, and is an essential gene required for cell growth (15, 16). We will refer to the KAR2 gene product as BiP in this discussion, as this has become the generic term for the ER-resident hsp70 homolog. The first step in the eucaryotic secretory pathway is translocation of the secreted protein across the ER membrane in an extended form, and it has been shown that this translocation process is dependent on the presence of functional KAR2 protein in the ER lumen (17). Translocation intermediates which are artificially jammed in microsomal membranes *in vitro* can be chemically crosslinked with BiP (18). Taken together, these observations indicate that BiP plays a significant role in the early steps of the secretory pathway.

Correct folding and assembly of a protein is a prerequisite for vesicular transport from the ER onward through the secretory pathway (12, 19). Misfolded proteins are retained in the ER, often in association with BiP (20). This has led some to conclude that BiP acts as a proofreading protein, whose chief role is to bind to misfolded proteins in order to prevent their secretion (21-22). BiP, although a soluble protein, is retained in the ER by a receptor-mediated recycling pathway (12), and perhaps by multiple calcium crosslinkages (23). If BiP binds to a protein, the protein must also remain in the ER, if one assumes that transport occurs by passive entrainment in a bulk fluid flow (24).

Dorner and coworkers have shown that the proofreading role for BiP in CHO cells is dominant in certain cases – for example, secretion of von Willebrand factor (vWF), Factor VIII, or tPA mutants lacking glycosylation sites (21-22). Many studies have shown that alteration of glycosylation can lead to misfolding and retention of glycoproteins such as tPA (25-27) , so the requirements for transport of mutant non-glycosylated tPA must differ from those for normal constitutively secreted proteins. Factor VIII is a 2,300 amino acid protein with 25 potential glycosylation sites, and is industrially produced by co-secretion with vWF, which forms a stabilizing complex with Factor VIII (28). The folding and transport of such a large, complex molecule is unlikely to be typical of the majority of secreted proteins. vWF is an adhesive glycoprotein that forms large multimers in the Golgi which are released through a regulated pathway (29). Endothelial cells and megakaryocytes are the only cell types which synthesize vWF *in vivo*. Thus, in each case in which BiP has been shown to inhibit protein secretion (21-22), the protein's native transport pathway is not representative of the majority of constitutively secreted proteins. We are interested in the role of BiP in secretion of less complex and non-mutant proteins, such as antibodies, growth factors, and enzymes.

BiP appears to perform two quite different functions: reversible association with partially unfolded proteins to prevent aggregation, and stable association with misfolded proteins to prevent transport. We believe that the seemingly contradictory roles of ER retention and unfolded protein solubilization can be reconciled by consideration of the observed biochemical functions of BiP in a mathematical framework.

Mathematical Model of BiP Function *in vivo*

We constructed a mathematical model of the processes of folding, aggregation, and binding to BiP which occur when a secretory protein is first translocated across the ER membrane. Our motivation was to predict whether increasing BiP levels in the lumen of the ER would be beneficial or detrimental to protein secretion, based on kinetic rate constants for binding and release of unfolded proteins by hsp70 chaperones reported in the literature. It should be emphasized that the purpose for constructing this model is to predict qualitative trends involved in BiP interactions with secreted proteins. It is quite possible that BiP performs as yet unobserved functions in the lumen of the ER, such as dissociation of translocation complexes after completion of protein translocation. However, given the importance of protein folding in determining secretion kinetics, and the established biochemical evidence that BiP binds to unfolded proteins and releases upon ATP hydrolysis, we believe that these interactions may account for the predominant contribution of BiP to protein secretion.

Mathematical models of the eucaryotic secretory pathway constructed by other investigators have focused on the rates of vesicular transport steps, treating each organelle as a well-mixed compartment (*30-32*). By contrast, the present model includes no information concerning vesicular transport, as the folding steps in the lumen of the ER are likely to be rate-limiting for secretion (*12,19,30*). The process of protein folding *in vivo* has been modeled previously, but the presence of chaperones was not included in those model calculations (*33*).

The proposed steps in the process of protein folding in the ER lumen, and the model equations used to describe the system, are shown in Figure 1. New, unfolded polypeptide chains (U) appear in the ER at the rate R_s, and these chains either aggregate irreversibly with higher order kinetics (in the present model, second order), fold to an aggregation resistant form (F), or form a complex with BiP (C). Transport of the folded protein out of the ER is not included, since the purpose of this model is to focus on the rate-limiting folding step.

The stoichiometry of BiP binding to unfolded nascent chains is taken to be 1:1 in the present model. Palleros et al. observed approximately 1:1 stoichiometry of binding of cytosolic hsp73 to reduced carboxymethylated α-lactalbumin *in vitro* (*34*), and Suzuki et al. immunoprecipitated a truncated T-cell receptor α-chain from CHO cell extracts complexed with BiP at stoichiometric ratios of 1:1 to 2:1 (*20*). Thus, 1:1 binding appears to be a reasonable approximation.

A typical range of values for the second order rate constant for protein-protein binding is 0.5-5.0×10^6 $M^{-1}s^{-1}$, for interactions as diverse as actin polymerization, hemoglobin dimer assembly, and antibody-antigen association (*35, 36*). A second-order rate constant for association of BiP and unfolded proteins (k_b) of 1.0×10^6 $M^{-1}s^{-1}$ was used for base case simulations.

Since ATP hydrolysis stimulates release of BiP-bound proteins, an estimate of the rate of dissociation of the complex can be obtained from the reported ATPase

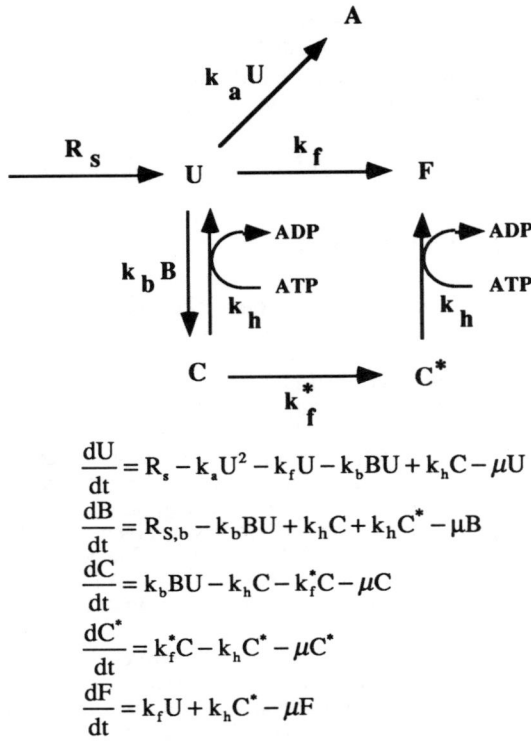

Figure 1. Model for BiP-assisted protein folding in the lumen of the endoplasmic reticulum (ER). Unfolded nascent polypeptide chains (U) form aggregates (A), complexes with BiP (C), or fold to the stable compact form (F). C* represents partially folded protein bound to BiP. Only the correctly folded form (F) is competent for transport from the ER. Feasible ranges for each of the rate constants, with the exception of k_f^*, were found in the literature.

activity(11). The rate of peptide-stimulated ATP hydrolysis by BiP was measured as 0.003 s^{-1} by Flynn, et al. *in vitro*(11). However, accessory factors may accelerate ATP hydrolysis markedly *in vivo*. ATP hydrolysis by DnaK is enhanced in the presence of GrpE and DnaJ to rates as high as 0.35 s^{-1}(37). DnaJ homologues have been found in yeast (38-40), but as yet no eucaryotic homolog of GrpE has been discovered. Thus, a range of 0.003-0.3 s^{-1} for complex dissociation is expected, depending on the extent to which currently unidentified cellular factors accelerate complex dissociation. For the purposes of base case simulations, a value of k_h=0.1 s^{-1} is assumed.

All secreted proteins were lumped together in this model. Rothman has estimated the concentration of nascent chains within the ER at approximately 50 μM (2). Assuming that co-translational translocation occurs at approximately the rate of protein translation in eucaryotes (≈ 2 amino acids/second), and an average secretory protein size is 400 amino acids, then the rate of appearance of new polypeptide chains within the ER (R_s in the model) is 250 nM/s. The rate of synthesis of BiP

($R_{S,b}$) was varied in the simulations to produce differing steady state ratios of chaperone: nascent polypeptide chain.

In our model, aggregation is considered to be irreversible, leading eventually to degradation of the aggregated polypeptides (*41*). However, it has been shown that DnaK can stimulate refolding of aggregated RNase *in vitro* (*42*), and that aggregates of influenza haemagglutinin are disassembled in the ER by an ATP-dependent mechanism which presumably involves BiP (*43*). If BiP is capable of solubilizing some of the aggregated chains for potential folding and secretion, then our assumption of irreversible aggregation will lead to underestimation of the beneficial aspects of BiP. A second order rate constant of 1×10^5 $M^{-1}s^{-1}$ for aggregation of unfolded proteins has been measured *in vitro* (*44*), and is used in this model.

The most speculative component of this model is the assumption that a protein will continue to fold to some extent while bound to BiP. Since BiP binds to peptides as short as seven amino acids(*45*), it is not implausible that the remainder of the protein will assume some elements of secondary and tertiary structure while bound to BiP. In fact, there is experimental evidence that binding of hsp70 chaperones to proteins does not completely block formation of folded structures. Palleros et al. have shown that reduced carboxymethylated α–lactalbumin forms secondary structure (as quantified by CD spectra) while bound to hsp73 (*34*). Hartl et al. have shown that folding intermediates are released from DnaK by GrpE and DnaJ in a form resistant to aggregation – refolding of rhodanese in the presence of DnaK + GrpE + DnaJ is slow, but yields as high as 30% are obtained, compared to undetectable refolding in their absence (*8*). In our model, the parameter k_f^*, which describes folding of the BiP-bound protein, is varied from zero to the rate constant for unassisted folding, k_f. As reviewed by Ptitsyn & Semisotnov, the dominant folding time constant for proteins varies from 1-1000 s (*46*). Since the slowest step in folding is often a proline isomerization, we have chosen a value of k_f=0.01 s^{-1} as typical for unassisted folding of a monomeric secreted protein (*46*).

Steady state solutions of the model equations (for parameter values in the ranges described) were obtained by setting the derivatives to zero, and numerically solving by a variation of Newton's method which uses a finite difference approximation to the Jacobian of the system (*47*). It is of particular interest to determine the fraction of total lumenal protein which is in the correctly folded form (F), and therefore competent for transport, as a function of BiP concentration in the lumen. This secretable fraction is plotted vs. BiP concentration with k_f^* as a parameter in Figure 2, in Figure 3 with k_h as a parameter, and in Figure 4 with the nascent chain concentration as a parameter.

A striking result is that the folded protein fraction increases with increasing BiP concentration for all combinations of parameter values tested, except when $k_f^* = 0$. This implies that the dominant role for BiP is that of a reversible molecular detergent which inhibits aggregation, as long as the BiP-bound protein is capable of folding to some extent to an aggregation-resistant form. Even when k_f^* is 100-fold less than k_f, increasing BiP increases the secretable protein fraction. When k_f^*=0, however, increasing BiP concentration decreases the secretable protein fraction. This may explain the results of Dorner et al., which seemed to indicate that BiP is detrimental to protein secretion. The folding kinetics of Factor VIII, vWF, and mutant non-glycosylated tPA within the lumen of the ER must be very slow, which may correspond approximately to the case k_f^*=0, resulting in a reduction of secretion of these proteins due to their association with BiP.

It is also noteworthy that the model predicts maximum secretion efficiency at BiP: nascent chain ratios above five. The total concentration of ER-resident proteins

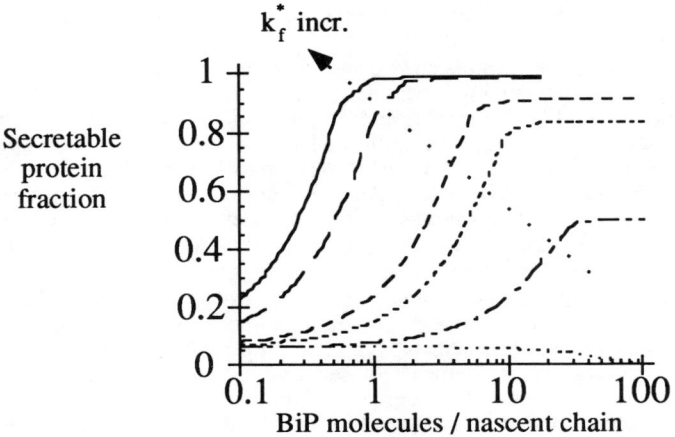

Figure 2. Increasing capability of the BiP-bound protein to fold, represented by increasing k_f^*, results in more efficient secretion. Steady state solutions to model equations were found as described in the text. Other simulation parameters: $k_b = 1 \times 10^6$ M^{-1}s^{-1}, $k_a = 1 \times 10^5$ M^{-1}s^{-1}, $k_f = 0.01$ s^{-1}, $k_h = 0.1$ s^{-1}, nascent chain concentration = 50 μM. At the given nascent chain concentration, BiP concentration is varied on this plot from 5 μM to 5 mM.
k_f^*(s^{-1}) = —— 0.01; — — 0.005; - - - 0.001; ------ 0.0005; —·— 0.0001; ······ 0.0

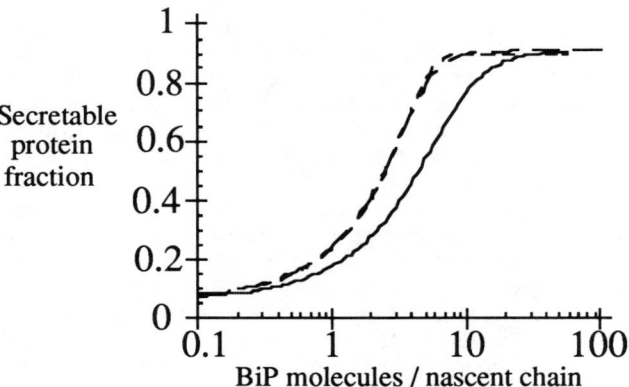

Figure 3. Dependence of secretable fraction on BiP levels is a weak function of the rate of ATP hydrolysis and protein release from BiP. Other simulation parameters: $k_b = 1 \times 10^6$ M^{-1}s^{-1}, $k_f = 0.01$ s^{-1}, $k_a = 1 \times 10^5$ M^{-1}s^{-1}, $k_f^* = 0.001$ s^{-1}, nascent chain concentration = 50 μM. At the given nascent chain concentration, BiP concentration is varied on this plot from 5 μM to 5 mM.
k_h(s^{-1}) = —— 1.0; — — 0.1; - - - 0.01

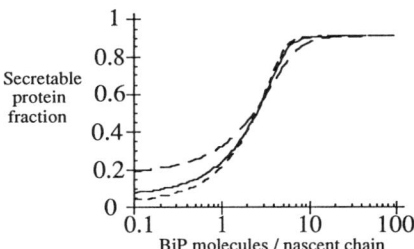

Figure 4. Dependence of secretable fraction on BiP levels is a weak function of the concentration of nascent chains within the ER. Other simulation parameters: $k_b = 1 \times 10^6$ M^{-1}s^{-1}, $k_f = 0.01$ s^{-1}, $k_a = 1 \times 10^5$ M^{-1}s^{-1}, $k_f^* = 0.001$ s^{-1}, and $k_h = 0.1$ s^{-1}. At the given nascent chain concentration, BiP concentration is varied on this plot from 5 μM to 5 mM. Nascent chains (μM) = — — 5; —— 50; - - - 500

has been estimated at approximately 1 mM (48). BiP is one of four major ER-resident soluble proteins (BiP, GRP90, PDI, ERp72), so the concentration of BiP in the ER lumen may actually be in the range of five- to ten-fold over the nascent chain concentration of 50 μM.

Our model results consistently show a beneficial role for BiP in protein secretion, which is robust to changes in all of the model parameters within feasible ranges. This motivated us to experimentally explore the interactions between BiP and foreign proteins secreted in the yeast *Saccharomyces cerevisiae*, an organism for which powerful genetic techniques have been developed (49).

BiP Levels in Yeast Cells Secreting Foreign Proteins

As an initial step in studying the role of BiP in foreign protein secretion in yeast, we determined how native BiP synthesis is influenced by high level foreign protein secretion. BiP is induced by various stimuli which result in the presence of misfolded protein in the ER lumen (15, 16), and if protein misfolding is responsible for inefficient secretion in yeast, one would expect induction of the KAR2 gene which codes for yeast BiP when foreign secreted proteins are overexpressed.

Cell extracts were prepared from strains transformed with multicopy plasmids for constitutive expression and secretion of human erythropoietin, granulocyte colony stimulating factor, platelet derived growth factor, or *Schizosaccharomyces pombe* acid phosphatase (gift of S. Elliott). As shown in Table I, these proteins span a range of structural features in terms of size, glycosylation, and multimeric structure. Cellular levels of BiP were quantified by chemiluminescent slot blot immunoassays, and are shown in Figure 5. Surprisingly, BiP protein was at least five-fold lower in extracts from cells secreting foreign proteins from multicopy plasmids. Similar

Table I. Foreign proteins used in this study

Protein[a]	Oligomer	Glycosylated	Size(aa)
EPO		+	193
PDGF	+		241
GCSF			207
PHO	+	+	435
GCSF-PHO	+	+	548

[a] EPO = human erythropoietin, PDGF = human platelet derived growth factor B chain, GCSF = granulocyte colony stimulating factor, PHO = *Schizosaccharomyces pombe* acid phosphatase, GCSF-PHO = fusion between GCSF and PHO.

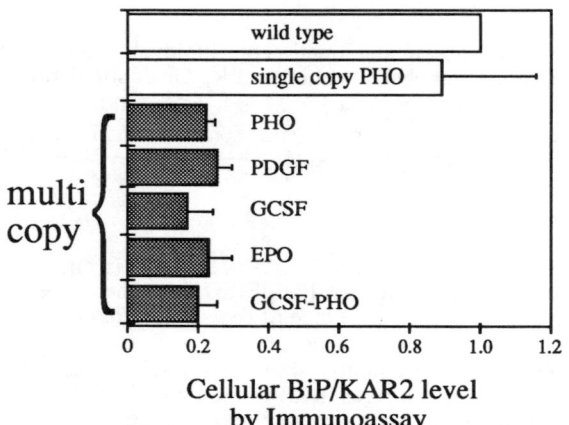

Figure 5. BiP/KAR2 levels in YPH500 (α ura3-52 lys2-801a ade2-101 trp-Δ63 his3-Δ200 leu2-Δ1)cells overexpressing secreted foreign proteins. Protein abbreviations are given in Table I. Extracts from 10 mL of mid-exponential cell culture were prepared by glass bead disruption. Equal amounts of total protein were loaded onto a slot blotting apparatus in serial dilutions. The blots were probed with anti-Kar2 antibody (gift of M. Rose) followed by goat anti-rabbit secondary antibody conjugated to alkaline phosphatase. X-ray film exposed to a chemi-luminescence substrate (Lumi-Phos 530) was used to detect IgG binding. Quantitation of intensities in the linear range of dilutions was performed by densitometric scans.

results were obtained with a yeast strain deficient in vacuolar proteases (BJ5464 α ura3-52 trp1 leu2Δ1 his3Δ200 pep4::HIS3 prb1Δ1.6R can1 GAL), and additional protease inhibitors added to the extracts did not affect the result. Mixing experiments with extracts containing normal levels of BiP protein and extracts with low BiP verified that proteolysis during sample preparation was negligible. Note that expression of *S. pombe* phosphatase from a single copy plasmid did not affect BiP levels, while phosphatase overexpression lowered BiP significantly.

To determine if transcriptional limitations from the weak KAR2 promoter were responsible for the reduction in BiP protein, we attempted to overexpress the KAR2 gene from a galactose inducible promoter. The results are shown in Figure 6.

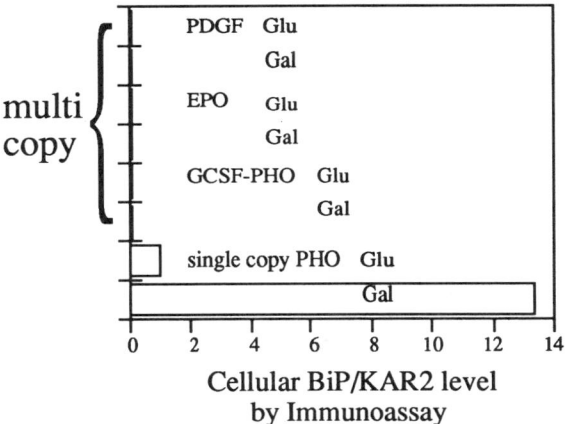

Figure 6. BiP/KAR2 overexpression cannot be induced from a galactose-inducible promoter in cells overexpressing secreted foreign proteins. When YPH500 cells constitutively expressing foreign proteins were transformed with pGALKAR2, galactose induction did not increase BiP levels in the cells. pGALKAR2 is a LEU2-selectable vector for Kar2 overexpression which was subcloned from the URA-based plasmid pMR1341 (gift of M. Rose). Cell extracts and slot blots were performed as described in Figure 5.

Strains which constitutively overexpress secreted proteins cannot be induced to synthesize high levels of KAR2 protein. This result implies that a post-transcriptional limitation is responsible for reduction in BiP.

Since BiP has been shown to play a key role in the secretory process, it is clear that cells with very low BiP levels will not be capable of efficient secretion. In fact, the reduction of cellular BiP which we observe may be directly responsible for the inefficiency of secretion of foreign proteins by yeast. This result is somewhat surprising, and there are several possible explanations. Transcriptional limitations of the KAR2 promoter are unlikely, since transcription of the KAR2 gene from a galactose inducible promoter failed to raise levels of BiP protein. It is unlikely that both the native KAR2 promoter and the GAL promoter are inactive under these growth conditions. Degradation of BiP may be stimulated by foreign protein secretion. If so, this degradation must be non-vacuolar, since BiP reduction is observed in strains deficient in vacuolar proteases.

Our working hypothesis is that the rate-limiting step in synthesis of KAR2 is translocation into the ER. The current model for eucaryotic secretion points to the existence of proteinaceous pores through which secreted proteins are translocated into the ER (50). If this is true, then there are a finite number of sites available for translocation of proteins into the ER. Flooding the cell with mRNA coding for foreign secreted proteins titrates out the available sites of entry to the secretory pathway, reducing the rate of insertion of BiP into the ER. Once BiP levels within the ER drop due to dilution by growth, a futile cycle is initiated, wherein the lack of BiP reduces the overall secretory flux, which in turn reduces the level of BiP, etc. In support of this model, it is noteworthy that lowered levels of BiP were observed for all overexpressed secreted proteins, regardless of structure. Expression of *S. pombe* phosphatase from a single copy plasmid did not lead to an observable reduction in BiP, demonstrating that the mere presence of a foreign protein is not sufficient to lower BiP levels.

Conclusions

We have begun an investigation of the role of the protein folding chaperone BiP in determining the kinetics of secretion of foreign proteins in yeast. A mathematical model of protein folding and chaperone binding kinetics within the secretory pathway predicts that increasing BiP levels should improve secretion efficiency. This prediction contradicts the results of Dorner et al., who found that BiP plays a negative, proofreading role in transport of certain proteins (*21-22*). We believe this inconsistency is due to what is likely to be the unusually slow folding kinetics of the proteins studied. In fact, a 35-fold increase in secretion of active M-CSF occurs upon overexpression of BiP in CHO cells, a result which has been attributed to artifactual alteration of mRNA levels in the BiP-overexpressing cell line (*22*). It is not clear how BiP overproduction would affect M-CSF mRNA synthesis, but an overall stimulation of secretion might lead to enhanced protection of the mRNA from degradation while being translated on polysomes on the surface of the rough ER. In any case, the effect of BiP overproduction on M-CSF secretion is consistent with our model predictions.

We find that levels of BiP in yeast cells overexpressing secreted foreign proteins are significantly reduced. Given the central role for BiP in the secretory process, this reduction may be directly responsible for the inefficient secretion kinetics observed in yeast strains used for overproduction of foreign proteins (*51-53*). We are currently exploring methods for correcting this deficit by overexpression of KAR2.

Acknowledgments

This work was supported in part by NSF PYIA 90-57677 to KDW. ASR is the recipient of a DOD Graduate Fellowship. KAR2 expression plasmids and polyclonal antisera were the kind gift of J. Vogel and M. Rose. Expression plasmids for EPO, GCSF, PDGF, and PHO were the kind gift of S. Elliott (Amgen). D. A. Lauffenburger provided helpful comments on the manuscript.

Literature Cited

1. Ellis, R.J., van der Vies, S.M. *Annu. Rev. Biochem.* **1991**, *60*, 321-347.
2. Rothman, J.E. *Cell* **1989**, *59*, 591-601.
3. Horwich, A.L., Neupert, W., Hartl, F.-U. *TIBTECH* **1990**, *8*, 126-131.
4. *Stress Proteins in Biology and Medicine*; Morimoto, R.I., Tissieres, A., Georgopoulos, C., Eds.; Cold Spring Harbor Laboratory Press: Cold Spring Harbor, NY, 1990; pp. 1-450.
5. Freedman, R.B. *Cell* **1989**, *57*, 1069-1072.
6. Freiden, P.J., Gaut, J.R., Hendershot, L.M. *EMBO J.* **1992**, *11(1)*, 63-70.
7. Landry, S.J., Jordan, R., McMacken, R., Gierasch, L.M. *Nature* **1992**, *6359*, 455-457.
8. Langer, T., Lu, C., Echols, H., Flanagan, J., Hayer, M.K., Hartl, F.U. *Nature* **1992**, *356*, 683-689.

9. Trent, J.D.; Nimmergern, E.; Wall, J.A.; Hartl., F.U.; Horwich, S.L. *Nature* **1991**, *354*, 490-493.
10. North, G. *Nature* **1991**, *354*, 434-435.
11. Flynn, G.C., Chappell, T.G., Rothman, J.E. *Science* **1989**, *245*, 385-390.
12. Pelham, H.R.B., *Annu. Rev. Cell Biol.* **1989**, *5*, 1-23.
13. Tandon, S., Horowitz, P. *J. Biol. Chem.* **1986**, *261*, 15615-15681.
14. Cleland, J.L., Wang, D.I.C., *Bio/Technology* **1990**, *8*, 1274-1278.
15. Rose, M.D., Misra, L.M., Vogel, J.P. *Cell* **1989**, *57*, 1211-1221.
16. Normington, K., Kohno, K., Kozutsumi, Y., Gething, M., Sambrook, J. *Cell* **1989**, *57*, 1223-1236.
17. Vogel, J.P., Misra, L.M., Rose, M.D. *J. Cell Biol.* **1990**, *110*, 1885-1895.
18. Sanders, S.L., Whitfield, K.M., Vogel, J.P., Rose, M.D., Schekman, R.W. *Cell* **1992**, *69*, 353-365.
19. Gething, M., Sambrook, J. *Curr. Op. in Cell Biology* **1990**, *1*, 65-72.
20. Suzuki, C.K.; Bonifacino, J.S.; Lin, A.Y.; Davis, M.M.; Klausner, R.D. *J. Cell Biol.* **1991**, *114(2)*, 189-205.
21. Dorner, A.J., Krane, M.G., Kaufman, R.J. *Mol. & Cell. Biol.* **1988**, *8(10)*, 4063-4070.
22. Dorner, A.J., Wasley, L.C., Kaufman, R.J. *EMBO J.* **1992**, *11(4)*, 1563-1571.
23. Sambrook, J.F. *Cell* **1990**, *61*, 197-199.
24. Wieland, F.T., Gleason, M.L., SErafini, T.A., Rothman, J.E. *Cell* **1987**, *50*, 289-300.
25. Hearing, J., Gething, M.-J., Sambrook, J. *J. Cell Biol.* **1989**, *108*, 355-365.
26. Machamer, C.E., Rose, J. K. *J. Biol. Chem.* **1988**, *263(12)*, 5955-5960.
27. Dorner, A.J., Bole, D.G., Kaufman, R.J. *J. Cell Biol.* **1987**, *105(6)*, 2665-2674.
28. Fox, J.L. *Bio/Technology* **1992**, *10*, 15.
29. Mayadas, T.N., Wagner, D.D. *PNAS USA* **1992**, *89*, 3531-3535.
30. Bibila, T.A., Flickinger, M.C. *Biotech. & Bioeng.* **1991**, *38*, 767-780.
31. Sambanis, A., Lodish, H.F., Stephanopoulos, G. *Biotech. & Bioeng.* **1991**, *38*, 280-295.
32. Park, S., Ramirez, W.F. *Biotech. & Bioeng.* **1989**, *33*, 272-284.
33. Kiefhaber, T., Rudolph, R., Kohler, H.-H., Buchner, J. *Bio/Technology* **1991**, *9*, 825-829.
34. Palleros, D.R.; Welch, W.J., Fink, A.L. *PNAS USA* **1991**, *88*, 5719-5723.
35. Janin, J.; Chothia, C. *J. Biol. Chem.* **1990**, *265(27)*, 16027-16030.
36. Northrup, S.H.; Erickson, H.P. *PNAS USA* **1992**, *89*, 3338-3342.
37. Liberek, K.; Marszalek, J.; Ang, D.; Georgopoulos, C.; Zylicz, M. *PNAS USA* **1991**, *88*, 2874-2878.
38. Caplan, A.J., Douglas, M.G. *J. Cell Biol.* **1991**, *114*, 609-621.
39. Luke, M.M., Sutton, A., Arndt, K.T. *J. Cell Biol.* **1991**, *114*, 623-638.
40. Blumberg, H., Silver, P.A. *Nature* **1991**, *349*, 627-630.
41. Stafford, F.J., Bonifacino, J.S. *J. Cell Biol.* **1991**, *115(5)*, 1225-1236.
42. Skowyra, D.; Georgopoulos, C.; Zylicz, M. *Cell* **1990**, *62*, 939-944.
43. Braakman, I., Helenius, J., Helenius, A. *Nature* **1992**, *356*, 260-262.
44. Zettlmeissl, G., Rudolph, R., Jaenicke, R. *Biochemistry* **1979**, *18*, 5567-5571.
45. Flynn, G.C.; Pohl, J.; Flocco, M.T.; Rothman, J.E. *Nature* **1991**, *353*, 726-730.
46. Ptitsyn, O.B.,Semisotnov, G.V. in *Conformations and Forces in Protein Folding*, Nall, B.T., Dill, K.A., Eds.; American Association for the Advancement of Science: Washington, D.C., 1991.
47. More, J., Garbow, B., Hillstrom, K. *User Guide for MINPACK-1*; Argonne National Laboratory Report ANL-80-74, Argonne, Illinois, August, 1980.

48. Koch, G.L.E. *BioEssays* **1990**, *12*, 527-531.
49. *Guide to Yeast Genetics and Molecular Biology*; Guthrie, C., Fink, G.R., Eds.; Methods in Enzymology v. 194; Academic Press, Inc.: San Diego, CA, 1991.
50. Simon, S.M., Blobel, G. *Cell* **1991**, *65*, 371-380.
51. *Yeast Genetic Engineering*; Barr, P.J., Brake, A.J., Valenzuela, P., Eds.; Butterworths: Boston, MA, 1989.
52. Shuster, J.R. *Curr. Op. in Biotech.* **1991**, *2*, 685-690.
53. Buckholz, R.G.; Gleeson, M.A.G. *Bio/Technology* **1991**, *9*, 1067-1072.

RECEIVED October 26, 1992

Chapter 10

GroEL-Mediated Protein Folding

François Baneyx[1] and Anthony A. Gatenby

Central Research and Development, E. I. du Pont de Nemours and Company, P.O. Box 80402, Wilmington, DE 19880

The molecular chaperone GroEL has been shown to facilitate the refolding of chemically unfolded proteins by a mechanism which depends upon the presence of Mg-ATP and the cochaperonin GroES. In this article, we have used inhibition, proteolysis and fluorescence techniques to show that the adenine nucleotides AMP-PNP, γ-S-ATP and ADP also bind to GroEL and alter the conformation of the tetradecamer. The absolute requirement for the cochaperonin GroES to achieve folding of biologically active Rubisco was further demonstrated using a GroEL single amino acid substitution mutant which interacts with GroES suboptimally. The implications of these results for the mode of action of GroEL are discussed.

In the recent years, the traditional notion that all the information necessary for the proper folding of a polypeptide chain is contained in its amino acid sequence has been challenged by the discovery of molecular chaperones. Molecular chaperones are defined as a class of proteins which assist protein folding without becoming part of the final structure (1) and have been isolated in most of the tissues and cellular compartments examined to date. They include the nucleoplasmins, the heat shock proteins (hsp) 60, 70, and 90, the signal recognition particle as well as the secB and papD proteins in *E. coli* (for recent reviews see references 2-3). Among these polypeptides, the hsp60 and hsp70 families have been particularly well studied, owing to their natural abundance and their multiple roles in the cell. Nevertheless, despite the growing body of information, their function and mechanism of action remain obscure.

The *E. coli* hsp60 homolog is the chaperonin GroEL, a protein which was initially discovered by its requirement for bacteriophage assembly (4). GroEL is organized as a double stack of seven subunits (subunit M_r 57,000) (5-7) and displays a weak ATPase activity fully dependent upon the presence of potassium ions (8-9). A large number of functionally and structurally unrelated proteins are capable of forming a binary complex with GroEL once they have been chemically unfolded, but cannot associate with the chaperonin in their native form. The resulting complexes are usually

---

[1]Current address: Department of Chemical Engineering, BF-10, University of Washington, Seattle, WA 98195

quite stable since they can survive size exclusion chromatography and it has been postulated that they are formed as a result of interactions between hydrophobic patches (*10*). NMR analysis of transferred nuclear Overhauser effects has demonstrated that small peptides and protein fragments adopt a α-helical conformation at the surface of GroEL (*11-12*), suggesting that structural features may also play a role in complex formation. One to two proteins in a non-native form appear to associate with each GroEL tetradecamer (*13-15*). When such binary complexes are supplemented with Mg-ATP and the cochaperonin GroES (organized as a single ring of seven 10 kDa subunits) (*7-8*), bound proteins are released in a folded and active conformation by a process which remains largely unknown. In this paper, we have attempted to shed some light on the mechanism of action the GroEL/GroES system using proteolysis and fluorescence techniques and by comparing the behavior of wild type GroEL to that of a single amino acid substitution mutant, GroEL140.

Interactions between GroEL, Adenine Nucleotides and GroES

Susceptibility to Proteolysis. The binding of ATP and its subsequent hydrolysis by the chaperonin is likely to play a fundamental role in the refolding of proteins associated with GroEL. Non-hydrolyzable ATP analogs such as AMP-PNP and γ-S-ATP (Figure 1) have been reported to support the refolding of GroEL-bound proteins, albeit with a lower efficiency than ATP (*16-18*). In order to show that ATP analogs were able to interact with the ATP binding site of GroEL, a series of inhibition experiments were performed (*19*). Table I shows that all the adenine nucleotides tested inhibited ATP hydrolysis by GroEL. Full inhibition was achieved with γ-S-ATP, while even a 100-fold excess of AMP-PNP only inhibited 60% of the ATP hydrolysis by GroEL. These results suggest that the ATP binding site(s) on GroEL can be very discriminating since ADP binds tighter to the chaperonin than the non-hydrolyzable ATP analog AMP-PNP.

Table I. Inhibition of the ATPase Activity of GroEL by Adenine Nucleotides[a]

Nucleotide	ATPase activity (s^{-1})	Inhibition (%)
ATP	2.10×10^{-2}	0
ATP + AMP-PNP	0.81×10^{-2}	61
ATP + ADP	0.41×10^{-2}	80
ATP + γ-S-ATP	$< 10^{-3}$	100

SOURCE: Adapted from ref. 19.
[a]Assays were performed with 0.1 μM GroEL, 9.375 μM ATP and 1 mM of the adenine nucleotides as described (*19*).

The current body of information suggests that GroEL mediates protein folding in a highly dynamic fashion, harvesting energy from ATP binding and hydrolysis to modify its conformation. Such conformational changes have been demonstrated by protease digestion experiments with the hsp70 homologs Bip (*20*) and DnaK (*21*). Similar experiments were carried out with GroEL (*19*). Figure 2 shows that under the experimental conditions chosen, GroEL is essentially unaffected by the presence of trypsin for up to 30 minutes. However, when ATP is added to the reaction mixture, the half-life of the protein decreases to 22 minutes. This result is indicative of a conformational change induced by ATP. A similar, or in some cases enhanced, susceptibility to proteolytic degradation was observed when ADP, γ-S-ATP, or AMP-

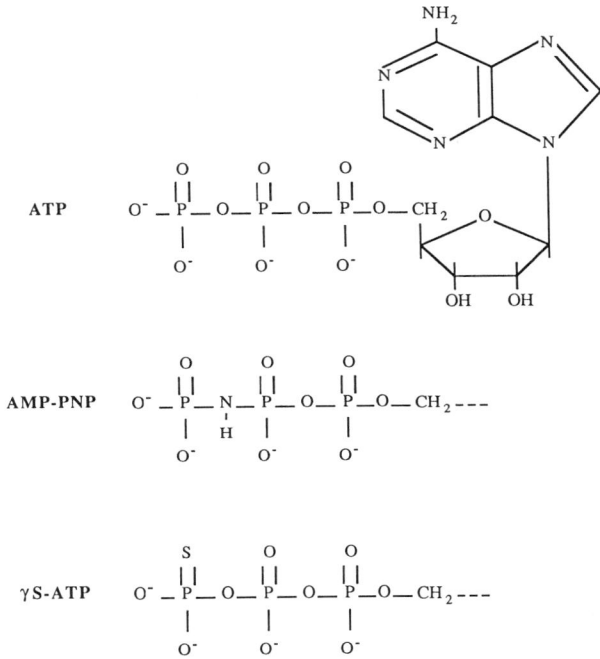

Figure 1. Chemical structure of ATP (adenosine 5' triphosphate) and the non-hydrolyzable analogs AMP-PNP (5' adenylyl imidodiphosphate) and γ-S-ATP (adenosine 5'-O-(3-thiotriphosphate)). Only the relevant domains of the analogs are shown in the figure.

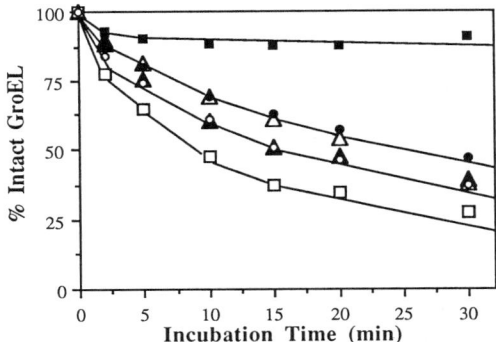

Figure 2. Susceptibility of GroEL to trypsin digestion at 37 °C; effect of adenine nucleotides and GroES. GroEL was incubated with 0.17 mg/ml of trypsin at 37 °C with no additives (■), 2.5 mM ATP (△), 2.5 mM ATP and a 3.8 fold molar excess of GroES (●), 5 mM ADP (○), 5 mM AMP-PNP (□), or 5 mM γ-S-ATP (▲). The reaction was quenched at the times indicated with 10 mM PMSF. Aliquots were resolved by SDS-PAGE and the intensity of the GroEL band quantified by videoscanning. The intensity at time zero was assigned the value 100% (Adapted from ref. 19).

PNP were substituted for ATP. Thus, the binding of adenine nucleotides is sufficient to induce a conformational change in the GroEL tetradecamer, although a subsequent conformational change resulting from ATP hydrolysis cannot be ruled out. In this respect, GroEL behaves like Bip (20), but differently from DnaK, since the latter protein requires ATP hydrolysis to change conformation (21).

In the presence of ATP, GroEL and the cochaperonin GroES form a stable complex consisting of one to two heptamers of GroES per GroEL tetradecamer (8). To determine whether the binding of GroES imparted a further conformational change in GroEL, trypsin digestion was repeated with a 3.8-fold molar excess of $(GroES)_7$ over $(GroEL)_{14}$ in the presence of Mg-ATP (19). Figure 2 shows that addition of GroES did not affect the hydrolysis profile of GroEL.ATP, suggesting that the formation of a complex between GroEL and GroES did not make additional lysine and/or arginine residues available to trypsin. This result was unexpected since electron micrographs of GroEL.GroES complexes formed in the presence of ATP show gross conformational modifications in the GroEL molecule (22). A possible explanation for these results could be that GroES binding generates quaternary structural changes in GroEL.

BisANS Fluorescence. The fluorescent probe bisANS exhibits little intrinsic fluorescence (Figure 3) but becomes highly fluorescent upon binding to hydrophobic domains in proteins. This reporter has been shown to bind to GroEL with a stoichiometry of 2.8 probes per tetradecamer (17). In addition, although bisANS does not interact with GroES, it associates with GroEL.GroES complexes formed in the presence of Mg-ATP. Under these conditions, only 1.5 probes interact with each GroEL tetradecamer (17).

In an attempt to determine whether the conformational changes induced by adenine nucleotide binding affected the hydrophobic patches within the chaperonin, the changes in the fluorescence of GroEL-bound bisANS was examined in the presence of different additives as described in Figure 3. In all cases, addition of adenine nucleotides slightly reduced the maximum intensity of the spectrum (Figure 3a). Since the mere presence of nucleotides did not influence bisANS background fluorescence (data not shown), the decrease in maximum fluorescence intensity may either result from the quenching of probe fluorescence by neighboring amino acids, or by a change in the stoichiometry or strength of binding of bisANS to GroEL as imparted by the binding of adenine nucleotides to the chaperone. ATP and ADP had a comparable quenching effect (5 to 6% decrease in maximum intensity) while AMP-PNP only weakly reduced the fluorescence level. Interestingly, γ-S-ATP, which induces the strongest ATPase inhibition, also yielded the highest quenching (13%). Nevertheless, none of the nucleotides examined induced a shift in the maximum wavelength (493 nm), suggesting that the conformational changes generated by adenine nucleotide binding do not drastically affect the solvent-exposed hydrophobic domains in GroEL.

The fluorescence of bisANS was also examined with samples supplemented with a 3.7-fold molar excess of $(GroES)_7$ over $(GroEL)_{14}$. Figure 3b shows that fluorescence quenchings similar to that reported for the GroEL.GroES.ATP complex (17) were observed with all the adenine nucleotides studied. As was the case in the absence of GroES, the relative order of quenching was AMP-PNP (22% decrease in maximum intensity), ATP (29%), ADP (32%), and γ-S-ATP (36%). Overall, these results confirm the fact that GroES can form a complex with GroEL whether or not the nucleotide present is hydrolyzable. GroES binding did not significantly affect the hydrophobicity of the probe environment since the maximum emission wavelength was not changed.

Using titration experiments, Mendoza et al. (17) were able to show that the dissociation constant of bisANS was essentially unaffected by the binding of GroES to GroEL in the presence of ATP. They suggested that the loss of one bisANS binding site on GroEL could result from a reduced accessibility or physical modification of the exposed hydrophobic domain in GroEL. It is tempting to further speculate that the

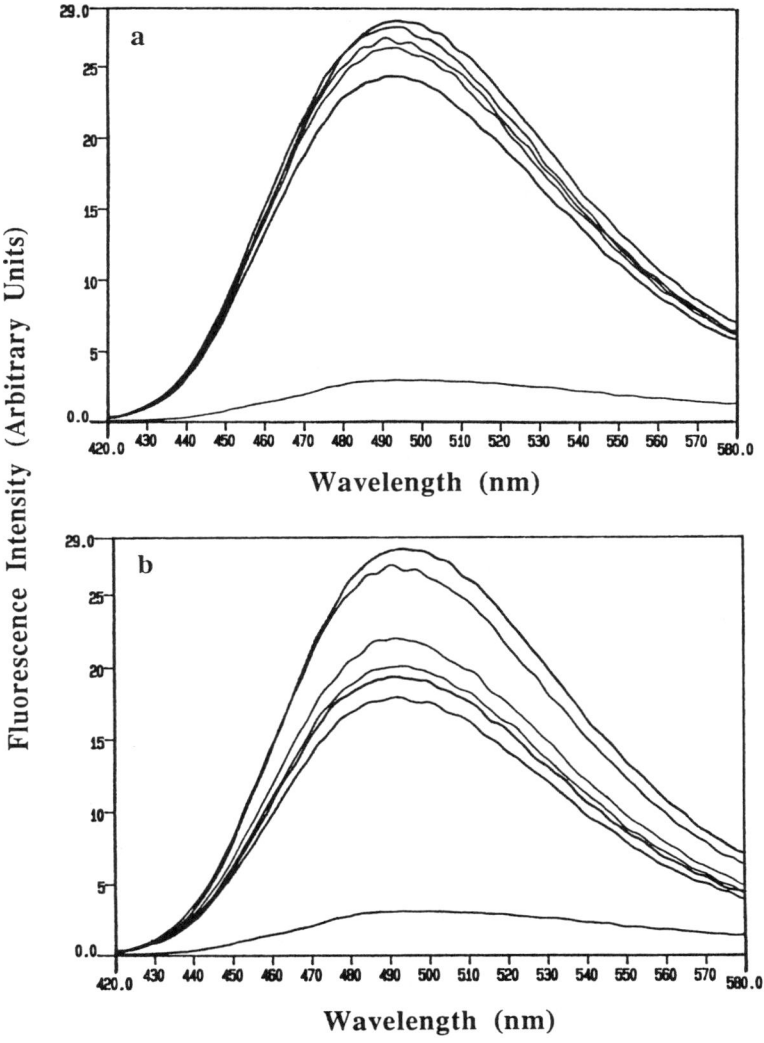

Figure 3. BisANS fluorescence. The fluorescence of bisANS (10 μM final concentration) was measured in the presence of GroEL (1 μM protomer final concentration) and different additives in 100 mM Tris-HCl, pH 7.6, 10 mM KCl, 10 mM $MgCl_2$ at 37 °C using a Perkin-Elmer fluorescence spectrophotometer with excitation at 394 nm and a slit width of 5 nm. Panel a, from top: GroEL, GroEL + 2 mM AMP-PNP, GroEL + 2 mM ATP, GroEL + 2 mM ADP, GroEL + 2 mM γ-S-ATP, and background bisANS fluorescence in buffer. Panel b, from top: GroEL, GroEL + 2 mM ATP, GroEL + GroES + 2 mM AMP-PNP, GroEL + GroES + 2 mM ATP, GroEL + GroES + 2 mM ADP, GroEL + GroES + 2 mM γ-S-ATP, and background bisANS fluorescence in buffer. For all recordings made in the presence of GroES, a 3.7 molar excess of GroES heptamers over GroEL tetradecamers was used, and the background fluorescence of bisANS.GroES substracted.

topological change(s) induced by nucleotide binding is a requirement for the binding of GroES. If hydrophobic interactions are indeed the principal determinant for the binding of non-native proteins to GroEL, the "shrinking" of the hydrophobic patch imparted by GroES binding could weaken the anchoring of the partially unfolded protein and facilitate subdomain folding in a protected environment. The process could continue by successive rounds of ATP hydrolysis coordinated by GroES binding and release, during which the bound protein would progress from a molten globule-like structure to a more compact and native-like conformation (23). At this point, the protein could be released as a result of decreased affinity.

Suboptimal GroES Binding Affects Protein Folding by GroEL

E. coli strains mutant in the groE operon have been isolated on the basis of their inability to support bacteriophage λ morphogenesis (24). Such a mutation (groEL140) was recently mapped to a single amino acid substitution (^{201}Ser -> Phe) in the groEL gene (Georgopoulos, C., CMU Geneva, personal communication). GroEL140 was purified to near homogeneity and compared to wild type GroEL (19). The mutant protein behaved as a tetradecamer and exhibited an ATPase activity approximately 1.5-fold lower relative to the wild type protein. It was capable of binding the adenine nucleotides AMP-PNP, ADP, and γ-S-ATP (in order of increased ATPase activity inhibition). GroEL140 was more sensitive than wild type GroEL to trypsin hydrolysis ($t_{1/2}$ = 30 min) and its susceptibility to hydrolysis was significantly enhanced in the presence of adenine nucleotides (19). Overall, these results indicated that GroEL140 assembled in a looser conformation than GroEL.

The ability of GroEL140 to support the refolding of the dimeric form of ^{35}S-labeled Ribulose bisphosphate carboxylase/oxygenase (Rubisco) from R. rubrum was examined as described (25, 19, for a review see 26). Both wild type and mutant chaperonins could bind comparable amounts of acid unfolded [^{35}S]Rubisco (Figure 4) indicating that the affinity of GroEL140 for non-native proteins was not compromised by the mutation. When ATP was added to the reaction mixture, over 75% of the radioactive material associated with GroEL was released from the chaperonin but only trace amounts folded into a monomer (L1) or the biologically active dimer (L2). Addition of GroES resulted in the recovery of 58% of the radioactivity in the L2 peak and 18% in the L1 peak, while increasing the discharge of the GroEL-bound Rubisco (Figure 4a). Thus, addition of ATP alone to GroEL promotes the discharge of bound Rubisco. This material is however unable to fold properly and, presumably, binds to the column irreversibly (25, 19). The requirement for GroES to obtain active Rubisco dimers is consistent with the hypothesis that GroES couples rounds of ATP hydrolysis with subdomain folding and prevents the "wholesale" dissociation of proteins bound on the surface of GroEL in an inactive form (23).

Figure 4b shows that the mutant GroEL140 only released 18% of the bound Rubisco following treatment with ATP. Discharge was increased to 53% in the presence of GroES, but under these conditions only 14% of the radioactivity properly folded as a L2 specie while 11% migrated to the L1 peak. Since the low level of Rubisco refolding by GroEL140 could only be partially attributed to its lower ATPase activity, we examined the interaction of the mutant chaperonin with GroES in more detail (19). When GroEL140 was mixed with an excess of GroES in the presence of Mg-ATP and resolved on a gel filtration column, no GroES protein could be detected by immunoblotting in the GroEL peak, while a GroES band was clearly visible when the experiment was repeated with wild type GroEL. This result suggested that either GroES was incapable of forming a complex with GroEL140 or that the affinity of GroES for the mutant chaperonin was much lower relative to the wild type, resulting in dissociation of the complex upon size exclusion chromatography. Since GroEL140 could promote low levels of Rubisco refolding, the second possibility seemed more

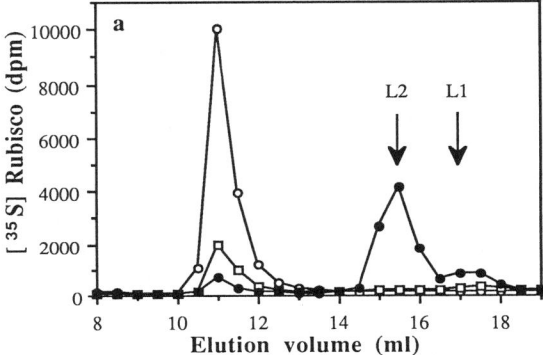
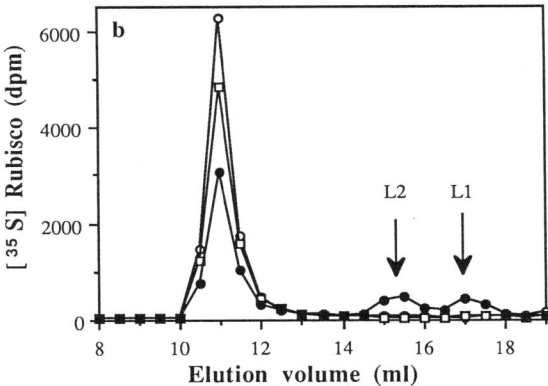

Figure 4. Release and refolding of chaperonin-bound [^{35}S]Rubisco. Acid unfolded [^{35}S]Rubisco was loaded onto GroEL (Panel a) or GroEL140 (Panel b) and incubated with no additives (○), 2.5 mM ATP (□), or 2.5 mM ATP and a 7-fold molar excess of (GroES)$_7$ over (GroEL)$_{14}$ (●). The samples were fractionated on a TSK G3000SW gel filtration column and the fractions counted (19). The positions of the Rubisco monomer (L1) and the biologically active Rubisco dimer (L2) are indicated by arrows (Adapted from ref. 19).

attractive. A suboptimal interaction between GroEL140 and GroES was in fact proven by showing that increasing the molar excess of GroES or the refolding incubation time enhanced the recovery of biologically active Rubisco (*19*).

The above experiments point out that although GroES is not essential to release GroEL-bound Rubisco, it is required for its folding, as anticipated from previous studies (*27-28*). Proteins able to fold rapidly in the absence of a chaperone (e.g. pre β-lactamase [*13*] and DHFR [*16*]) can be released in a folded conformation from GroEL by the action of adenine nucleotides alone. It therefore appears that even though a large percentage (and perhaps all) of the *E. coli* proteins can interact with GroEL (*25*), only a fraction may use the "foldase" activity of the GroEL/GroES system. The interaction may however be necessary for other purposes since GroEL has also been implicated in DNA replication (*29-30*), cell division (*31*), and protein translocation (*32-34*).

Acknowledgements

We are grateful to Saskia van der Vies, Mathhew Todd, Gail Donaldson, and Paul Viitanen for helpful discussions and suggestions. We thank Uwe Bertsch for reading the manuscript.

Literature Cited

1. Ellis, R. J.; Hemmingsen, S. M. *Trends Biochem. Sci.* **1989**, *14*, 339.
2. Ellis, R. J.; van der Vies, S. M. *Annu. Rev. Biochem.* **1991**, *60*, 321.
3. Gething, M. J.; Sambrook, J. *Nature* **1992**, *355*, 33.
4. Georgopoulos, C.; Ang, D. *Seminars Cell Biol.* **1990**, *1*, 19.
5. Hendrix, R. W. *J. Mol. Biol.* **1979**, *129*, 375.
6. Hohn, T.; Hohn, B.; Engel, A.; Wortz, M.; Smith, P. R. *J. Mol. Biol.* **1979**, *129*, 359.
7. Hemmingsen, S. M.; Woolford, C.; van der Vies, S. M.; Tilly, K.; Dennis, D. T.; Georgopoulos, C. P.; Hendrix, R. W.; Ellis, R. J. *Nature* **1988**, *333*, 330.
8. Chandrasekhar, G. N.; Tilly, K.; Woolford, C.; Hendrix, R.; Georgopoulos, C. *J. Biol. Chem.* **1986**, *261*, 12414.
9. Viitanen, P. V.; Lubben, T. H.; Reed, J.; Goloubinoff, P.; O'Keefe, D. P.; Lorimer, G. H. *Biochemistry* **1990**, *29*, 5665.
10. Pelham, H. R. B. *Cell* **1986**, *46*, 959.
11. Landry, S. J.; Gierasch, L. M. *Biochemistry* **1991**, *30*, 7359.
12. Landry, S. J.; Jordan, R.; McMacken, R.; Gierasch, L. M. *Nature* **1992**, *355*, 455.
13. Laminet, A. A.; Ziegelhoffer, T.; Georgopoulos, C.; Plückthun, A. *EMBO J.* **1990**, *9*, 2315.
14. Badcoe, I. G.; Smith, C. J.; Wood, S.; Halsall, D.J.; Holbrook, J.; Lund, P.; Clarke, A. R. *Biochemistry* **1991**, *30*, 9195.
15. Mendoza, J. A.; Lorimer, G. H.; Horowitz, P. M. *J. Biol. Chem.* **1991**, *266*, 16073.
16. Viitanen, P. V.; Donaldson, G. K.; Lorimer, G. H.; Lubben, T. H.; Gatenby, A. A. *Biochemistry* **1991**, *30*, 9716.
17. Mendoza, J. A.; Rogers, E. ; Lorimer, G. H.; Horowitz, P. M. *J. Biol. Chem.* **1991**, *266*, 13044.
18. Fisher, M. T. *Biochemistry* **1992**, *31*, 3955.
19. Baneyx, F.; Gatenby, A. A. *J. Biol. Chem.* **1992**, *267*, 11637.
20. Kassenbrock, C. K.; Kelly, R. B. *EMBO J.* **1989**, *8*, 1461.
21. Liberek, K.; Skowyra, D.; Zylicz, M.; Johnson, C.; Georgopoulos, C. *J. Biol. Chem.* **1991**, *266*, 14491.
22. Saibil, H.; Dong, Z.; Wood, S.; Auf der Mauer, A. *Nature* **1991**, *353*, 25.
23. Martin, J.; Langer, T.; Boteva, R.; Schramel, A.; Horwich, A. L.; -Ulrich Hartl, F. *Nature,* **1991**, *352*, 36.

24. Tilly, K.; Murialdo, H.; Georgopoulos, C. *Proc. Natl. Acad. Sci. USA* **1981**, *78*, 1629.
25. Viitanen, P. V.; Gatenby, A. A.; Lorimer, G. H. *Protein Science* **1992**,*1*, 363.
26. Gatenby, A.A.; Viitanen, P.V.; Lorimer, G.H. *Trends Biotech.* **1990**, *8*, 354.
27. Goloubinoff, P.; Gatenby, A.A.; Lorimer, G.H. *Nature* **1989**, *337*, 44.
28. Goloubinoff, P.; Christeller, J. T.; Gatenby, A. A.; Lorimer, G. H. *Nature* **1989**, *342*, 884.
29. Fayet, O.; Louarn, J. M.; Georgopoulos, C. *Mol. Gen. Genet.* **1986**, *202*, 435.
30. Jenkins, A. J.; March, J. B.; Oliver, I. R.; Masters, M. *Mol. Gen. Genet.* **1986**, *202*, 446.
31. Miki, T.; Orita, T.; Furuno, M.; Horuchi, T. *J. Mol. Biol.* **1988**, *201*, 327.
32. Phillips, G. J.; Silhavy, T. J. *Nature* **1990**, *344*, 882.
33. Kusukawa, N.; Yura, T.; Ueguchi, C.; Akiyama, Y.; Ito, K. *EMBO J.* **1989**, *8*, 3517.
34. Van Dyk, T. K.; Gatenby, A. A.; LaRossa, R. A. *Nature* **1989**, *342*, 451.

RECEIVED October 26, 1992

Chapter 11

Prolyl Isomerizations as Rate-Determining Steps in the Folding of Ribonuclease T1

Lorenz M. Mayr, Thomas Kiefhaber, and Franz X. Schmid

Laboratorium für Biochemie, Universität Bayreuth, D–W–8580 Bayreuth, Germany

Ribonuclease T1 is used as a model protein to elucidate the role of prolyl isomerizations for protein folding. Native ribonuclease T1 contains two *cis* prolyl peptide bonds at P39 and at P55. The *trans* to *cis* isomerizations at these prolyl residues are slow reactions and constitute rate-determining steps in the refolding of this protein. Prior to these final steps, intermediates accumulate during folding that show partially native structure. The replacement of the S54-P55 *cis* prolyl bond by a G54-N55 bond leaves the stability of the protein unchanged. It abolishes, however, a major slow folding reaction and thus simplifies the folding mechanism. On the other hand, replacement of the *cis* proline at position 39 by glycine strongly reduces the stability of the protein and leads to major changes in the folding mechanism. Results for a variant with a tyrosine residue instead of tryptophan at position 59 indicates that the close contact between the residue at position 59 and proline 39, as found in the native protein, is already established in a folding intermediate before the rate-limiting *trans* to *cis* isomerization at P39 takes place.

The folding of a polypeptide chain under native conditions is a spontaneous and reversible process that is directed by the information encoded in the amino acid sequence. The inherent relationship between sequence and folding is presently not understood at the molecular level. In our current work we concentrate on the kinetics of unfolding and refolding of small monomeric proteins. The aim is to detect transient folding intermediates, to characterize their structure and stability, and to define their positions along the folding pathway. In principle, a folding pathway is understood when all intermediates and activated states are arranged in correct order and when their structures are known. In practice, the kinetic approach is confined by several limitations. Only intermediates that are followed by slow steps accumulate and can be studied. Structural investigations are restricted by the transient nature of intermediates, and the high cooperativity of folding limits the number of kinetic steps that can be observed.

Folding is a complex process and it is difficult to establish and to evaluate kinetic mechanisms. The complexity of folding has two major sources. The unfolded state of most proteins is conformationally heterogeneous, thus giving rise to multiple parallel refolding reactions, and partially folded intermediates can be formed transiently on these individual refolding reactions. It is therefore mandatory to use as many different probes as possible to follow folding, to vary the conditions, to employ natural or designed variants for comparison and to draw on thermodynamic and structural information for the molecular interpretation of kinetic results.

An unfolded protein can adopt a tremendous number of different conformations. For all molecules to fold by the same pathway, all conformational transitions in the unfolded state must be very rapid, relative to the overall rate of folding. Not surprisingly, this is generally not the case. In particular, intrinsically slow isomerization processes at Xaa-Pro peptide bonds in unfolded proteins prevent the molecules from rapidly equilibrating conformationally.

Fast and Slow Protein Folding Reactions

The application of fast mixing techniques and sequential mixing procedures revealed kinetic complexity in both protein unfolding and refolding (1-10). The key observation by Garel and Baldwin (8) that both fast and slow phases in the refolding of RNase A produced enzymatically active protein led to the suggestion that part of the kinetic complexity was caused by the co-existence of fast-folding (U_F) and of slow-folding (U_S) species in the unfolded state (eq. 1):

$$N \underset{k_{23}}{\overset{k_{32}}{\rightleftharpoons}} U_F \underset{k_{12}}{\overset{k_{21}}{\rightleftharpoons}} U_S \qquad (1)$$

Slow $U_F \rightleftharpoons U_S$ equilibration reactions in unfolded protein chains after fast unfolding ($N \rightleftharpoons U_F$) of the native protein were a surprise in 1973. A plausible molecular explanation was provided by the suggestion of Brandts et al. (11) that the U_S molecules refold slowly, because they contain incorrect isomers of Xaa-Pro peptide bonds.

Peptide bonds are planar and can be either in the *trans* or in the *cis* conformation. Peptide bonds not involving proline residues are generally in the *trans* state, the *cis* conformation has not been detected in unstructured, linear oligopeptides, and the equilibrium population of the *cis* form is believed to be 0.1 % or less (11-13). Very few non-proline *cis* peptide bonds have been found in native proteins by X-ray crystallography ($14,15$).

The energy difference between the *cis* and the *trans* isomers is very small for peptide bonds between proline and its preceding amino acid, and Xaa-Pro bonds (Fig. 1) typically exist as a mixture of *cis* and *trans* isomers in solution, unless struc-

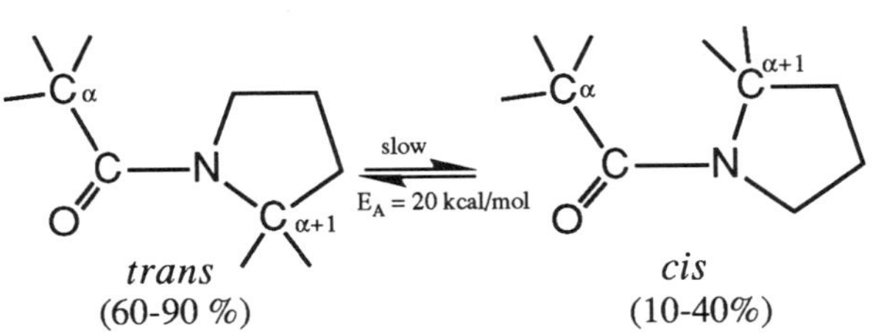

trans
(60-90 %)

slow
E_A = 20 kcal/mol

cis
(10-40%)

Figure 1. Isomerization between the *cis* and the *trans* forms of a Xaa-Pro peptide bond.

tural constraints such as in folded proteins, stabilize one of the two isomers. In the absence of ordered structure the *trans* isomer is usually favored slightly over *cis*. *Trans* contents of 60 - 90 % are frequently found in short linear peptides (*16,17*). The *cis* ⇌ *trans* isomerization involves the rotation about a partial double bond and is an intrinsically slow reaction (time constants around 10 - 100 s are observed at 25°C) with a high activation energy ($E_A \approx$ 85 kcal/mol).

Approximately 7 % of all prolyl peptide bonds in native proteins with known three-dimensional structure are *cis* (*14*). The conformational state of each peptide bond is usually well defined, being either *cis* or *trans* in every molecule, depending on the structural constraints imposed by the chain folding. Rare exceptions exist: *cis/trans* equilibria at particular Xaa-Pro bonds have been detected in native staphylococcal nuclease (*18*), in insulin (*19*) and in calbindin (*20*). After unfolding (N ⟶ U_F), prolyl bonds become generally free to isomerize slowly as in small oligopeptides, thus leading to an equilibrium mixture of unfolded protein molecules with different prolyl isomers. The chains with the correct set of isomers, U_F molecules, usually refold rapidly. Chains with at least one incorrect isomer (the U_S molecules) refold more slowly. The re-isomerizations of the wrong prolyl bonds are slow steps in folding. Non-native isomers usually do not block refolding, therefore chains with incorrect isomers can rapidly form partially native structure prior to prolyl peptide bond re-isomerization (*21-26*).

The formation of ordered structure and the re-isomerization of incorrect prolyl peptide bonds are probably not independent events in folding, but intimately coupled processes. Two aspects of the problem are worth noting. As already mentioned, unfolded protein chains with incorrect prolyl bonds can start to fold. At some point, however, correct isomers are required for folding to continue and thus the final steps of folding are often limited in rate by prolyl re-isomerization. The formation of ordered structure around the incorrect prolyl bonds early in folding can modify the kinetic properties of the subsequent isomerizations. This interdependence of folding and isomerization depends probably on two major factors: on the location of the Xaa-Pro bonds in the protein structure and on the conditions used for folding.

Our interest in the interrelationship between folding and prolyl isomerization is twofold. (i) Most proteins contain prolyl residues. The deceleration of folding by incorrect isomers and slow *cis/trans* isomerizations are therefore ubiquitous phenomena in protein folding. They could pose a problem for folding in the cell, since the newly formed protein chains are arrested for an extended time in partially folded states and are thus sensitive to aggregation. (ii) The importance of the various prolyl residues for the overall folding process might strongly differ. A correlation of certain kinetic events in folding with defined prolyl residues would improve our understanding of the molecular mechanisms of folding. The characterization of folding intermediates with such prolyl residues in the incorrect isomeric state would yield valuable information on the role of individual, proline-containing chain segments for the structure and for the stability of folding intermediates.

Structure and Stability of Ribonuclease T1.

Structure. In our studies of the interrelationship between prolyl isomerization and protein folding we use ribonuclease T1 (RNase T1) from *Aspergillus oryzae* as a model protein. RNase T1 is a small single-domain protein with 104 residues. The amino acid sequences of RNase T1 and of several related microbial RNases (*27*) are known and the three-dimensional structures of the apo-enzyme and of several complexes with nucleotides have been solved at high resolution (*28-31*). The structure of RNase T1 (Fig. 2) consists of an α-helix of 4.5 turns and two antiparallel ß-sheet structures, which are composed of two (β_1-β_2) and of five (β_3-β_7) strands, respectively. These elements of secondary structure are connected by extended loop regions. Two disulfide bonds form a small (2-10) and a large (6-103) covalently linked loop.

The protein contains four prolyl peptide bonds. Two of them (W59-P60 and S72-P73) are *trans* and the other two (Y38-P39 and S54-P55) are *cis* in the native protein. Both *cis* Pro residues connect the end of loop structures with ß-strands. P39 is located near the active site and is buried in the hydrophobic core of RNase T1; P55 is located near the surface. We will show later that isomerizations at these two *cis* prolines are rate-determining steps in the folding of RNase T1.

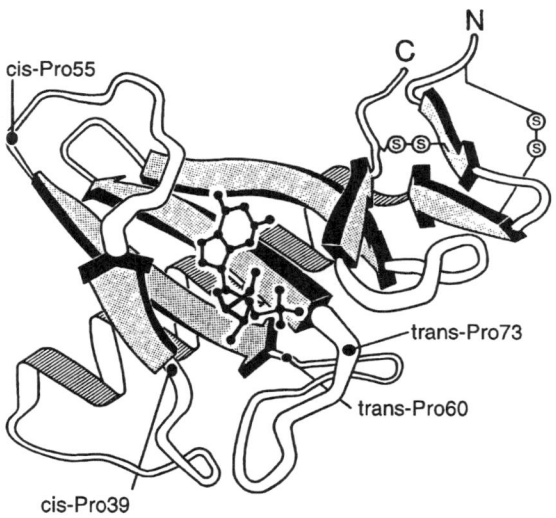

Figure 2. Backbone structure of ribonuclease T1. The location of the *cis* and *trans* prolyl peptide bonds are indicated. Figure courtesy of Dr. U. Heinemann.

Stability. Unfolding of RNase T1 by heat or by chemical denaturants is reversible under a wide variety of conditions (*32-37*). This is a necessary prerequisite for a meaningful interpretation of kinetic results. Unfolding transitions of the wild-type protein are well approximated by a simple N $\rightleftharpoons$ U two-state model. Identical transition curves are obtained when unfolding is followed by different probes, and the calorimetric enthalpy of unfolding coincides with the unfolding enthalpy derived from the van't Hoff relation (*36*). The stability of RNase T1 is strongly dependent on pH with a maximum near pH 5 (*37*), and it can be increased by the addition of salts, such as NaCl or $CaCl_2$ (*36,38,39*). Reduction of the two disulfide bonds results in a strong destabilization of the protein. It can, however, attain a folded, enzymatically active conformation at low temperature and in the presence of high concentrations of NaCl (*38,40*). This allows us to study the folding of RNase T1 in the presence as well as in the absence of disulfide crosslinks.

Kinetic Models for Unfolding and Slow Refolding of RNase T1.

In contrast to the apparent simplicity of the two-state equilibrium unfolding transitions, the unfolding and refolding kinetics of RNase T1 are complex processes. The unfolding reaction is followed by slow isomerizations of the denatured protein chains, and refolding is composed of at least three phases with rates and amplitudes that depend on the folding conditions as well as on the probes that were used to monitor folding (*25,41*). In this review we will first present our current models for the unfolding and refolding of RNase T1 and then describe the experimental evidence that has led to these models. Also, results for several designed variants with substitution at or near proline residues are included (42-45). These variants were constructed to examine the predictions of our model, to simplify the folding mechanism, and to elucidate molecular aspects of the folding mechanism of RNase T1. A major result of these studies was that an alternative replacement of either P55 or P39 had strikingly different effects on the stability and on the folding kinetics of RNase T1.

Model for Unfolding. The kinetic mechanisms to describe the unfolding and refolding of RNase T1 are shown in schemes I and II, respectively. In the model for unfolding (scheme I) we assume that the actual unfolding reaction (N ⟶ U_F) is followed by two slow isomerizations of prolyl peptide bonds, and that these two isomerizations occur at the proline residues P39 and P55. Both are *cis* in N and in U_F, but isomerize largely to the more favorable *trans* state in the unfolded protein. Contributions from the *trans* prolines 60 and 73 are not considered. At equilibrium 3.5 % of all unfolded molecules are in the U_F state with P39 and P55 in the correct *cis* state (25,41). Furthermore, we assume that a *cis/trans* equilibrium of about 20/80 is established at both P39 and P55. This suggests that the species with two incorrect isomers ($U_{39t}{}^{55t}$) predominates in the unfolded state (60-70%). The species with single incorrect isomers ($U_{39t}{}^{55c}$ and $U_{39c}{}^{55t}$) are populated to approximately 10-20% each.

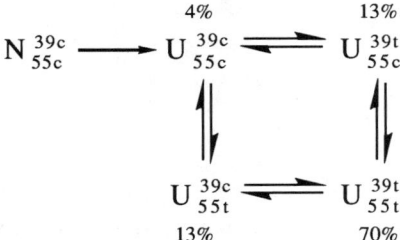

Scheme I. Kinetic model for the unfolding and isomerization of RNase T1. This model is valid for unfolding only. The superscript and the subscript indicate the isomeric states of prolines 39 and 55, respectively, in the correct, native-like *cis* (c) and in the incorrect, non-native *trans* (t) isomeric states. As an example, $U_{55c}{}^{39t}$ is an unfolded species with Pro55 in the correct *cis* and Pro39 in the incorrect *trans* state. The two isomerizations are independent of each other, therefore the scheme is symmetric with identical rate constants in the horizontal and vertical directions, respectively. The given percentages for the individual unfolded species are estimates only.

Model for Refolding. In the model for refolding (scheme II) we assume that the 3.5% molecules with the correct isomers refold rapidly (U_F ⟶ N). The two slow-folding species with one incorrect prolyl isomer each ($U_{39t}{}^{55c}$ and $U_{39c}{}^{55t}$) and the species with both prolines in the incorrect isomeric state ($U_{39t}{}^{55t}$) can regain rapidly most of their secondary structure and presumably part of their tertiary structure in the milliseconds range (the U_i ⟶ I_i steps, scheme II). Then slow folding steps follow that involve the isomerizations of the incorrect prolyl isomers. A peculiar feature of the folding model in scheme II is that the major unfolded species with two incorrect isomers can enter two alternative folding pathways (the upper or the lower pathway in scheme II), depending upon which isomerization occurs first. The distribution of refolding molecules on these two pathways is determined by the relative rates of isomerization at the stage of the rapidly formed intermediate $I_{39t}{}^{55t}$.

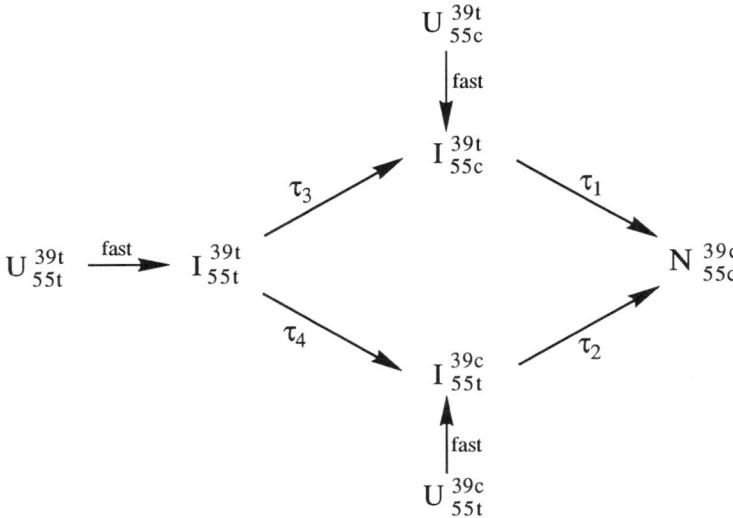

Scheme II. Kinetic model for the slow refolding of RNase T1 under strongly native conditions. U stands for unfolded species, I for intermediates of refolding, and N is the native protein. The superscript and the subscript indicate the isomeric states of prolines 39 and 55, respectively, in the correct *cis* (c) and the incorrect *trans* (t) isomeric states.

Experimental Basis of the Folding Models. In the following discussion we describe the experimental approach that was used to establish and to test the mechanisms for unfolding (scheme I) and for refolding (scheme II).

Complex Refolding of RNase T1. In the initial approach to characterize the refolding of RNase T1, we measured the kinetics of folding by different structural and functional probes and varied the conditions of both unfolding and refolding (25,41). The structural probes included the changes of absorbance, fluorescence, and of amide circular dichroism (CD) during refolding. The return of the RNase function was measured by activity assays. Refolding experiments were carried out at different pH values and in the presence of varying concentrations of stabilizing salts (such as NaCl) and destabilizing additives (such as GdmCl or urea). In all refolding experiments rapid changes with variable amplitudes were observed within the dead time of mixing (1-20 s). The kinetics of slow refolding were complex and in most cases they could be described as a sum of two first-order reactions. These two processes were labelled the "intermediate" and the "very slow" refolding reactions. Their derived rate constants and amplitudes were dependent on both the solvent conditions and the probe used to follow folding.

Slow Formation of U_S Molecules after Unfolding. Part of the kinetic complexity of refolding clearly originates from slow isomerizations in the unfolded protein, as outlined in scheme I. Slow refolding experiments carried out after varying times of unfolding (41) indicated that the species that are responsible for the slow phases of refolding are produced by slow chain isomerizations after the N $\longrightarrow$ U_F reaction in the unfolded state. These equilibrations occur with time constants of approximately 1000s at 0 °C and show activation energies of 88 kJ/mol, a value typical of prolyl isomerizations.

Assays for the Formation of Native Molecules during Refolding. Spectral differences between the unfolded and the native state of a protein are convenient and sensitive probes to follow folding reactions. Unfortunately, the mechanistic interpretation of such folding kinetics is difficult. As an example, biphasic kinetics could originate from the parallel refolding of two unfolded species, from the sequential folding reaction of a single species or a combination of both. To discriminate between these alternatives, a probe is desirable that monitors selectively the appearance of the fully folded native protein. The regain of the function of a protein could be such a probe. It has been found in several instances, however, that folding intermediates can already show properties of the native protein, such as catalytic activity.

The formation of native molecules during refolding can be measured without interference from partially folded intermediates by a method that is based on the energetics of folding (46). The completely folded species, N, is separated from all partially folded species by a high barrier of activation energy. Therefore, only N molecules unfold slowly, whereas all intermediates (including native-like, enzymatically active intermediates [46]) that have not yet crossed this final barrier unfold much more rapidly. Accordingly, the formation of native protein can be measured by unfolding assays. Folding is initiated in these experiments by dilution to the desired refolding conditions at time zero in a test tube. Then, after various time intervals of refolding, samples are withdrawn and transferred to unfolding conditions. The amplitude of the resulting slow unfolding reaction is a measure for the amount of completely refolded molecules that was present at the time when folding was interrupted.

The analysis of RNase T1 refolding by this technique (25,41) gave three salient results. (i) 3.5 % of the native molecules are formed within the dead time of the experiment. They are produced by the $U_F \longrightarrow N$ reaction. (ii) Slow refolding occurs on two major pathways. About 30 % of the molecules show sigmoidal folding kinetics, which indicates that two slow sequential steps are involved in their refolding. (iii) About 65 % of the RNase T1 refold on a pathway that is limited in rate by a very slow process with time constants of 6500 s (at pH 5, 10 °C) and 3000 s (at pH 8, 10 °C). Apparently, the slow folding events are dominated by the $U_{39t}{}^{55t}$ species. The sequential reaction reflects the lower branch in scheme II and the very slow reaction reflects the upper branch, which is limited in rate by the very slow re-isomerization at P39. Folding reactions that originate from the minor species $U_{39c}{}^{55t}$ and $U_{39t}{}^{55c}$ contribute presumably as well, but could not identified unambiguously.

Rapid Formation of Intermediates. The rapid formation of partially folded intermediates on all folding pathways is a distinctive feature of the folding model in scheme II. Although only 3-4 % of fast-folding U_F molecules are present, 40-50 % of the changes in absorbance and fluorescence (depending on the folding conditions) occur within the dead time of manual mixing (2-20 s). This constituted early evidence for folding intermediates. More decisive results were obtained by CD spectroscopy. The native-like amide CD spectrum of RNase T1 was regained in less than 20 s (25). This time was required to record the first spectrum of the refolding protein. This initial spectrum was more strongly negative than the CD of native RNase T1. We attribute this difference not to the formation of excess secondary structure early in folding, but to contributions of aromatic residues that have not yet developed at this stage of folding. Stopped-flow mixing experiments corroborated and extended this result. At 225 nm the regain of the native-like CD value is complete within 15 ms (45). The transient accumulation of folding intermediates could be measured directly by an assay that is similar to the assay for native molecules (25). In summary, these findings indicate that well ordered intermediates are formed on all folding pathways of RNase T1. They display essentially native-like secondary structure and partial tertiary structure, as judged by absorption and fluorescence spectroscopy.

Nature of the Slow Folding Steps. In the folding mechanism (scheme II) we assumed that all slow steps that constitute the inner rhombus of the model are prolyl isomerizations. This assignment was initially based on the finding that all slow folding species were produced after unfolding by slow prolyl isomerizations and should thus contain incorrect isomers (25,41). An excellent tool to identify the nature of slow folding reactions is provided by peptidyl-prolyl *cis/trans* isomerase. Indeed, all slow folding reactions of RNase T1 were accelerated by this enzyme, albeit with different efficiency (47,48). Catalysis of the very slow process was fairly poor, indicating that P39 is not well accessible for prolyl isomerase at the stage of the intermediate I_{39t55c}. Catalysis of the intermediate phase was much better, and, moreover, when the concentration of prolyl isomerase was increased this phase itself became clearly biphasic. As we had suspected earlier, this kinetic phase is composed of several reactions with similar rates, and an average time constant is measured in the absence of prolyl isomerase.

Even more revealing results were obtained when the formation of native molecules was followed in the presence of increasing concentrations of prolyl isomerase (25). Both steps of the sequential folding pathway were efficiently catalyzed, demonstrating that both involve prolyl isomerizations, as assumed in our model. In addition, a shift from the upper to the lower pathway occurred in these experiments. The simplest explanation for this shift involves the I_{39t55t} molecules with two incorrect isomers. They can enter either pathway, depending on which prolyl bond isomerizes first. Prolyl isomerase appears to catalyze *trans* ──→ *cis* at P39 more efficiently than at P55 in these molecules and thus leads to a preference for the lower pathway in scheme II.

In summary, these data support the kinetic mechanism in scheme II in its major aspects. (i) Two prolyl isomerizations create four unfolded species, and the dominant species contains two incorrect prolyl isomers. (ii) Extensive structure formation occurs early in refolding on all folding pathways. (iii) The slow steps of refolding are limited in rate by the re-isomerization of the incorrect prolyl bonds and occur on two major pathways. The choice of pathway seems to be determined by the rank order of isomerization. The mechanism in scheme II is complex, but clearly a simplification. It is valid only under strongly native conditions and contributions from minor species (e. g., with incorrect isomers at the *trans* prolines 60 and 73) are not included.

Role of Individual Prolines Probed by Directed Mutagenesis.

The kinetic results obtained for wild-type RNase T1 suggest that prolyl isomerizations, presumably at the *cis* prolines P39 and P55 are responsible for the complexity of slow refolding. We reasoned therefore that amino acid replacements at or near these proline residues should help in the molecular assignment of the observed folding reactions. Ideally, the replacement of a *cis* proline by another amino acid should (i) increase the amount of fast-folding species (U_F, scheme I), (ii) abolish the sequential reactions that involve two consecutive isomerizations, and (iii) lead to a general simplification of slow refolding. Replacements of *cis* prolines may, however, change the stability and the mechanism of folding and thus complicate the easy comparison with the folding of the wild-type protein.

In our experiments we used three variants of RNase T1 with substitutions at P55, at P39 and at W59, which is in close spatial contact to P39 in the native protein. Although strikingly different results were obtained with these three variants they enabled us to understand several molecular aspects of the mechanism of RNase T1 folding.

Folding of the (S54G,P55N) Variant of RNase T1. P55 is part of a surface loop with few contacts to other regions of the protein. In the closely related RNase C2 (from *Aspergillus clavatus*) the residues S54 and P55 that form a *cis* peptide bond in RNase T1 are replaced by G54 and N55, respectively. The flanking sequences on both sides are identical in the two proteins (*27*). To study the influence of P55 on the folding mechanism, we used a variant of RNase T1 with such a G54-N55 sequence. By these "evolutionary guided" replacements we hoped to obtain a mutant protein, in which the *cis* S54-P55 bond is replaced by a normal peptide bond, but which is still very similar to the wild-type protein in its structure and stability.

These expectations were borne out by the results (*42*). The structure and the stability of RNase T1 did not change to a significant extent after the substitution of the *cis* S54-P55 peptide bond by a normal G54-N55 bond. The kinetic results were clearcut. The fraction of U_F molecules increased more than fourfold from 3.5 to 16 % after the replacement of the *cis* S54-P55 bond, indicating that isomerization of this bond is indeed a source for slow folding molecules, and that P55 is predominantly in the incorrect *trans* state in unfolded wild-type RNase T1. The fourfold difference in U_F is well explained by a *trans/cis* ratio of about 80/20 at P55, as found typically in small proline-containing peptides. A similar *trans/cis* ratio exists also at P39, as suggested by the presence of 84% U_S molecules in (S54G,P55N)RNase T1. The combination of these isomer distributions leads to the presence of more than 96% slow-folding molecules in the wild-type protein.

In the mutant, the dominant species with two incorrect isomers ($U_{55t}{}^{39t}$) and the species with an incorrect P55 only ($U_{55t}{}^{39c}$) no longer exist. Consequently all reactions originating from these species are absent, thus leading to the disappearance of the steps that constitute the "intermediate" phase in fluorescence and lead to the sigmoidal formation of native protein. The consequence is an enormous simplification of the mechanism for slow refolding. Four slow steps are required to explain the folding of wild-type U_S molecules (scheme II). In the mutant, this complexity is reduced to a single slow proline-controlled step (presumably at P39) that limits the refolding of all U_S molecules.

A second remarkable feature is the rate of re-isomerization of the remaining prolyl bond (presumably P39, see below) during refolding. It is strongly dependent on the isomeric state of P55. In the presence of an incorrect *trans* P55, P39 re-isomerizes with a time constant of about 190 s (at pH 8, 10°C; cf. lower branch in scheme II). However, when P55 isomerizes first to the native *cis* state (upper branch in scheme II) or is absent (as in the mutant) then *trans* ⟶ *cis* isomerization of P39 is strongly decelerated and proceeds with a time constant of about 3000 s. This reaction can be accelerated by adding "structure-breaking" salts, such as GdmCl, for both wild-type and mutant RNase T1 (*25,43*). This is very unusual. It suggests that the incorrect *trans* isomer of P39 is "locked" in a folded intermediate structure, which retards its isomerization. The intermediate can be destabilized either by an additional incorrect isomer (at P55) or else by unfavorable folding conditions.

The characterization of the intermediates in the folding of wild-type RNase T1 was complicated by the multitude of intermediates with different combinations of incorrect prolyl isomers (cf. scheme II). The replacement of the S54-P55 bond in (S54G,P55N)RNase T1 removed a major source of heterogeneity and only a single folding intermediate (presumably with an incorrect *trans* P39) is present. It is separated from the native state by a very slow step and could be studied at leisure. This intermediate resembles the native protein in the aromatic CD spectrum. It shows about 40 % enzymatic activity and the accessibility of W59 to quenching by acrylamide is as low as in the native protein. The amide CD spectrum is restored in the milliseconds time range (*45*). Activity and quenching measurements have longer dead times and therefore we do not know whether activity and solvent accessibility change in the same time range also.

Folding of the W59Y Variant of RNase T1. The very slow refolding reaction of RNase T1 is unexpectedly slow in comparison to the proline limited refolding kinetics of other proteins. A tentative explanation was that the incorrect *trans* isomer at P39 becomes trapped in a native-like folding intermediate and is thus sterically prevented from reaching the native *cis* state. P39 is closely packed against the tryptophan residue 59 in native RNase T1. Such a close packing might possibly restrict the isomerization at P39 during folding and we therefore decided to replace W59 by a smaller aromatic residue, tyrosine.

This substitution does not influence the stability of the protein to a significant extent (*43*). Refolding of the W59Y variant, however, is much faster than that of the wild-type protein. It is essentially complete after 35 min (at pH 8 and 10 °C) compared to 250 min for the wild-type protein under the same experimental conditions. This acceleration originates from an apparent loss of the major very slow phase and it is found with all probes that were used to monitor folding. Of course, the mutation of a nonproline residue, such as W59, should not lead to the disappearance of a slow reaction or to a redistribution of fast and slow folding species. Indeed, the folding mechanism (scheme II) is apparently not changed. The amount of fast folding molecules is not affected in the variant and the appearance of native molecules in the time course of refolding still exhibits a lag phase, indicating that two prolyl isomerizations occur as in refolding of the wildtype protein. A major clue for understanding the molecular effect of the W59Y mutation on folding came again from experiments with prolyl isomerase (*43*). In the presence of this enzyme the slow refolding of the W59Y mutant can clearly be separated into two distinct kinetic phases. One phase is well accelerated by the enzyme, and thus resembles the intermediate phase in the wild-type protein. The other phase is only poorly catalyzed and corresponds to the very slow reaction of the wild-type protein. Apparently, this reaction is still present in the folding of the W59Y variant; but it is about tenfold accelerated and coincides with the intermediate phase under most conditions. From these results we conclude that during refolding of the wild-type protein P39 and W59 are already in close contact at the stage of a folding intermediate before the rate-limiting reisomerization of the Y38-P39 bond. This close packing retards the isomerization of P39. The retardation is relieved when the structure of this folding intermediate is destabilized, and, as a consequence, this folding reaction is accelerated, when the GdmCl concentration is increased. In the W59Y variant no retardation is found, since, after replacement of W59 by a less bulky tyrosine residue, more space is available for P39 to reisomerize. The "obstructive" effect of W59 on refolding is found only under strongly native conditions, where ordered structure is formed rapidly. Under conditions that do not favor the formation of intermediates (above 1.7 M GdmCl) the slow refolding kinetics of the W59 and the Y59 variants of RNase T1 become identical. The W59Y replacement is thus a "conditional folding mutant" of RNase T1.

Folding of the P39G Variant of RNase T1. The structural environments of P39 and P55 differ strongly in the native protein. P39 is in a rigid part of the polypeptide chain, it is almost inaccessible to solvent, and its neighbouring residues, Y38 and H40, are part of the active site (*28,29*). This region is conserved in all homologous eukaryotic RNases (*27*) and thus no evolutionary guidance for the construction of a mutated variant was available. We decided to replace P39 by a glycine residue, since this small and flexible amino acid might be most easily accommodated at a geometrically restrained position. The resulting P39G variant is only marginally stable and its folding kinetics differ strongly from the kinetics of the wild-type protein and also of the S54G,P55N variant. As we will see, the altered kinetic properties are not easily explained by the proline model for protein folding.

The backbone structure of RNase T1 seems not to be affected by the P39G mutation. The amide CD is unchanged and the variant shows the same enzymatic activity as the wild-type protein. The fluorescence of W59 is decreased in the P39G variant. The close proximity of the positions 39 and 59 in the folded protein is probably responsible for this effect.

The P39G mutation strongly destabilizes the protein. The midpoint of the thermal unfolding transition is lowered by 17 degrees and the free energy of stabilization is decreased from 42 kJ/mol (wild-type protein) to 20 kJ/mol (P39G variant) at pH 5, 25°C (Mayr and Schmid, unpublished results). Despite this quantitative difference in stability the dependence on pH (with a maximum near pH 5) and on NaCl concentration is the same for wild-type and (P39G)RNase T1. The enormous loss of stability explains the serious problems that we encountered in the production and purification of (P39G)RNase T1. This variant could not be produced by using the isolation procedure of the wild-type protein, but only after reducing the growth temperature from 37°C to 28 °C and performing all purification steps at 4°C as rapidly as possible.

Unfolding and refolding of (P39G)RNase T1 are reversible reactions. The rate of unfolding of the mutant protein is 50-fold increased. Slow refolding, as monitored by fluorescence can be approximated by a biphasic process and resembles the refolding of the wild-type protein. The use of unfolding assays revealed that the refolding of (P39G)RNase T1 still involves two sequential slow steps, similar to the wild-type protein. The surprising conclusion at this point was that unlike the case of the S54G,P55N variant, where the substitution of a *cis* proline led to the disappearance of a major slow phase, the substitution at P39 did not simplify the slow folding kinetics of RNase T1.

The refolding of the P39G variant has, however, several properties that are distinct from the folding of the wild-type protein, and are not expected from the simple proline model for protein folding. (i) Contrary to expectation and contrary to the results for the P55 variant, the amount of fast-folding U_F molecules is not increased, but is decreased virtually to zero. (ii) The species that give rise to the slowest phase of refolding are not formed slowly after unfolding (cf. scheme I), but are present immediately after the fast N $\longrightarrow$ U_F reaction. The other slow refolding reaction, however, which leads to the sigmoidicity of refolding, is correlated with a slow isomerization that occurs in unfolded (P39G)RNase T1. (iii) The unusual acceleration of the very slow refolding reaction of the wild-type protein upon increasing the GdmCl concentration from 0.2 to 2.0 M is not observed for the P39G variant. (iv) Only the faster of the two consecutive slow folding reactions is catalyzed by prolyl isomerases. The very slow reaction is not accelerated (Mayr and Schmid, unpublished results).

These results suggest that only the faster of the two slow folding reactions of (P39G)RNase T1 involves prolyl isomerization as the rate-limiting step. It is identical in rate with the intermediate phase in the refolding of the wild-type protein and thus we asssume that it reflects the trans $\longrightarrow$ *cis* isomerization at P55. The product of this reaction is, however, not the fully folded protein, but an intermediate that has to undergo a second, very slow reaction before folding is complete. This very slow reaction of the P39G variant is a novel reaction. It is not related with the very slow phase of folding of the wild-type protein and it is apparently not a prolyl isomerization. It is not created by a slow isomerization of the unfolded chain, it cannot be catalyzed, and all molecules have to undergo this reaction during refolding.

In summary we note two major effects of the P39G substitution. (i) The stability of the protein is reduced by more than 20 kJ/mol, and (ii) there is a novel very slow refolding reaction that is not observed in the folding of the wild-type protein. What is the molecular nature of this slow refolding reaction, and is it related with the strong destabilization by the P39G mutation? The replacement of a *cis* Tyr-Pro bond by a Tyr-Gly bond could have two different consequences. (i) The *cis* bond at P39 is replaced by a *trans* bond at G39 in the variant. This difference in main chain geometry would distort the folded state, thus leading to the observed strong decrease in stability and also to a high barrier for the final step of refolding. (ii) The *cis* conformation of the 38-39 bond is preserved in the folded P39G RNase T1. For peptide bonds not involving proline, the *trans* form is 10-20 kJ more stable than the *cis* form (*12,13*), and such an increment of free energy would be required to stabilize the *cis* form in the folded protein. In this case, the novel very slow folding reaction of the P39G variant could reflect the inherently unfavorable *trans* ⟶ *cis* isomerization of the Y38-G39 bond. This bond is probably more than 99% *trans* in the unfolded protein, and therefore all molecules would have to undergo this isomerization. The activation energies of *cis* ⇌ *trans* isomerizations are near 85 kJ/mol for both normal peptide bonds and peptide bonds preceding proline. Regarding this high activation barrier and the unfavorable *cis/trans* equilibrium, a *trans* ⟶ *cis* isomerization of a non-proline peptide bond could indeed be a very slow reaction in protein refolding, but not in unfolding, since the *trans* isomer is strongly favored in the unfolded state.

The experimental data are not yet sufficient to exclude one of the above alternatives. In addition, both effects could be involved. A native-like chain conformation around P39 might be necessary as a template for the correct folding and docking of other chain regions. A change in the sequence and/or the main chain conformation could thus reduce both the stability and the rate of folding. Clearly, experimental handles are needed to identify non-proline peptide bond isomerizations in unfolding and refolding reactions.

Concluding Section

Our results indicate that the *trans* to *cis* isomerizations at P39 and at P55 largely determine the mechanism of slow folding of RNase T1. Contributions from the *cis* to *trans* isomerizations at P60 and P73 could not be detected, possibly since they are correlated with small amplitudes only. It is evident that structure formation and prolyl isomerization steps are mutually interdependent during refolding. Rapid folding decelerates the re-isomerization at P39 by the formation of premature native-like structure that involves an interaction of P39 with W59. Strikingly different results were obtained for RNase T1 variants where the *cis* prolines were replaced. The substitutions at positions 54 and 55 led to a clear result: a slow folding reaction was abolished and the mechanism of folding simplified as expected from the kinetic model. The replacement of P39 drastically destabilized the protein and led to a new, very slow event in folding that was not observed in the folding of wild-type RNase T1. This shows that the *cis* bond at P39 is of utmost importance for stability and folding. The change in the folding mechanism, however, impairs a straightforward comparison of the folding kinetics and the identification of the P39-limited event in the folding of the wild-type protein. The molecular origin of the new slow reaction in the folding of the P39G variant is unclear at present. It could originate from a *trans* to *cis* isomerization at G39. The isomeric state of the Y38-G39 bond in native (P39G)RNase T1 is not yet known.

We observed an interesting interrelationsship between folding and prolyl isomerization steps. The original proposal (*11*) that prolyl isomerization precedes folding appears to be a limiting case that holds only for strongly unfavorable conditions, such as near or within the unfolding transition of a protein. Under more strongly native conditions protein chains can rapidly acquire extensive ordered structure still in the presence of incorrect prolyl isomers. Later in folding, the correct chain conformation is required and further folding is limited in rate by the re-isomerization of the incorrect prolyl isomers. The extent of rapid structure formation depends on the location of the incorrect prolyl bonds and on the folding conditions. Prefolding can lead to acceleration of prolyl isomerization as in RNase A (*21,22*) but also to deceleration, as in the case of P39 in RNase T1.

Evidently, the rapid formation of partially folded intermediates is not always of advantage for folding, since protein may become trapped in an incorrect structure (in this case a structure with an *trans* isomer at P39). A fairly low stability of intermediate structures may be an important factor to warrant correct and rapid folding. Such a marginal stability is probably also important to maintain the folding protein chains in a state that remains accessible for folding catalysts, such as protein disulfide isomerase and prolyl isomerase.

Acknowledgments. We thank Ulrich Hahn and his coworkers for samples of purified protein and for supplying *E. coli* strains that produce variants of RNase T1. This work was supported by grants from the Deutsche Forschungsgemeinschaft, the Fonds der Chemie, and the Hoechst AG.

Literature Cited

1. Epstein, H. F.; Schechter, A. N.; Chen, R. F.; Anfinsen; C. B. *J. Mol. Biol.* **1971**, *60*, 499-508.
2. Ikai, A.; Tanford, C. *Nature* **1971**, *230*, 100-102.
3. Ikai, A.; Tanford, C. *J. Mol. Biol.* **1973**, *73*, 145-163.
4. Ikai, A.; Fish, W. W.; Tanford, C. *J. Mol. Biol.* **1973**, *73*, 165-184.
5. Tanford, C.; Aune, K. C.; Ikai, A. *J. Mol. Biol.* **1973**, *73*, 185-197.
6. Tsong, T. Y.; Baldwin, R. L.; Elson, E. L. *Proc. Nat. Acad. Sci. U. S. A.* **1971**, *68*, 2712-2715.
7. Tsong, T. Y.; Elson, E. L.; Baldwin, R. L. *Proc. Nat. Acad. Sci. U. S. A.* **1972**, *69*, 1809-1812.
8. Garel, J. R.; Baldwin, R. L. *Proc. Nat. Acad. Sci. U. S. A.* **1973**, *70*, 3347-3351.
9. Garel, J. R.; Nall, B. T.; Baldwin, R. L. *Proc. Nat. Acad. Sci. U. S. A.* **1976**, *73*, 1853-1857.
10. Hagerman, P. J.; Baldwin, R. L. *Biochemistry* **1976**, *15*, 1462-1473.
11. Brandts, J. F.; Halvorson, H. R.; Brennan, M. *Biochemistry* **1975**, *14*, 4953-4963.
12. Ramachandran, G. N.; Mitra, A. K. *J. Mol. Biol.* **1976**, *107*, 85-92.
13. Jorgensen, W. J.; Gao, J. *J. Am. Chem. Soc.* **1988**, *110*, 4212-4216.
14. Stewart, D. E.; Sarkar, A.; Wampler, J. E. *J. Mol. Biol.* **1990**, *214*, 253-260.
15. Herzberg, O.; Moult, J. *Proteins: Struct. Funct. Genet.* **1991**, *11*, 223-229.
16. Cheng, H. N.; Bovey, F. A. *Biopolymers* **1977**, *16*, 1465-1472.
17. Grathwohl, C.; Wüthrich, K. *Biopolymers* **1981**, *20*, 2623-2633.
18. Evans, P. A.; Dobson, C. M.; Kautz, R. A.; Hatfull, G.; Fox, R. O. *Nature* **1987**, *329*, 266-268.
19. Higgins, K. A.; Craik, D. J.; Hall, J. G.; Andrews, P. R. *Drug Design and Delivery* **1988**, *3*, 159-170.
20. Chazin, W. J.; Kördel, J.; Drakenberg, T.; Thulin, E.; Brodin, P.; Grundström, T.; Forsén, S. *Proc. Nat. Acad. Sci. U. S. A.* **1989**, *86*, 2195-2198.
21. Cook, K. H.; Schmid, F. X.; Baldwin, R. L. *Proc. Nat. Acad. Sci. U. S. A.* **1979**, *76*, 6157-6161.
22. Schmid, F. X.; Blaschek, H. *Eur. J. Biochem.* **1981**, *114*, 111-117.
23. Goto, Y.; Hamaguchi, K. *J. Mol. Biol.* **1982**, *156*, 891-910.
24. Kelley, R. F.; Richards, F. M. *Biochemistry* **1987**, *26*, 6765-6774.
25. Kiefhaber, T.; Quaas, R.; Hahn, U.; Schmid, F.X. *Biochemistry* **1990**, *29*, 3061-3070.
26. Nall, B. T., In *Protein Folding* Gierasch, L. M.; King, J., Eds.; AAAS Press: Washington, DC, 1990, pp. 198-207.
27. Heinemann, U.; Hahn, U. In *Protein-Nucleic Acid Interaction* Saenger, W.; Heinemann, U., Eds.; Macmillan: London, UK, 1989; pp. 111-141.
28. Heinemann, U.; Saenger, W. *Nature* **1982**, *299*, 27-32.

29. Arni, R.; Heinemann, U.; Tokuoka, R.; Saenger, W. *J. Biol. Chem.* **1988**, *263*, 15358-15368.
30. Kostrewa, D.; Choe, H.-W.; Heinemann, U.; Saenger, W. *Biochemistry* **1989**, *28*, 7592-7600.
31. Martinez-Oyanedel, J.; Choe, H.-W.; Heinemann, U.; Saenger, W. *J. Mol. Biol.* **1991**, *222*, 335-352.
32. Thomson, J. A.; Shirley, B. A.; Grimsley, G. R.; Pace, C. N. *J. Biol. Chem.* **1989**, *264*, 11614-11620.
33. Shirley, B. A.; Stanssens, P., Stayaert, J.; Pace, C. N. *J. Biol. Chem.* **1989**, *264*, 11621-11625.
34. Pace, C. N. *Trends Biochem. Sci.* **1990**, *15*, 14-17.
35. Pace, C. N.; Heinemann, U.; Hahn, U.; Saenger, W. *Angew. Chemie, Int. Edition* **1991**, *30*, 343-360.
36. Kiefhaber, T.; Schmid, F.X.; Renner, M.; Hinz, H.-J.; Hahn, U; Quaas, R. *Biochemistry* **1990**, *29*, 8250-8257.
37. Pace, C. N.; Laurents, D. V.; Thomsom, J. A. *Biochemistry* **1990**, *29*, 2564-2572.
38. Oobatake, M.; Takahashi, K.; Ooi, T. *J. Biochem.* 1979, 86, 55-62.
39. Pace, C. N.; Grimsley, G. R. *Biochemistry* **1988**, *27*, 3242-3246.
40. Pace, C. N.; Grimsley, G. R.; Thomson, J. A.; Barnett, B. J. *J. Biol. Chem.* **1988**, *263*, 11820-11825.
41. Kiefhaber, T.; Quaas, R.; Hahn, U.; Schmid, F.X. *Biochemistry* **1990**, *29*, 3053-3060.
42. Kiefhaber, T.; Grunert, H.-P.; Hahn, U.; Schmid, F.X. *Biochemistry* **1990**, *29*, 6475-6479.
43. Kiefhaber, T.; Grunert, H.-P.; Hahn, U.; Schmid, F.X. *Proteins: Structure, Function and Genetics* **1992**, *12*, 171-179.
44. Kiefhaber, T.; Kohler, H.-H.; Schmid, F.X. *J. Mol. Biol.* **1992**, *224*, 217-229.
45. Kiefhaber, T.; Schmid, F. X.; Willaert, K.; Engelborghs, Y.; Chaffotte, A.-F. *Proteins* **1992**, *1*, in press.
46. Schmid, F. X. *Biochemistry* **1983**, *22*, 4690-4696.
47. Schönbrunner, E. R.; Mayer, S.; Tropschug, M.; Fischer, G.; Takahashi, N.; Schmid, F. X. *J. Biol. Chem* **1991**, *266*, 3630-3635.
48. Fischer, G.; Wittmann-Liebold, B.; Lang, K.; Kiefhaber, T. Schmid, F. X. *Nature* **1989**, *337*, 476-478.

RECEIVED October 26, 1992

Chapter 12

Kinetic Control of Protein Folding by Detergent Micelles, Liposomes, and Chaperonins

P. M. Horowitz

Department of Biochemistry, University of Texas Health Science Center, San Antonio, TX 78284-7760

Recovery of active proteins after unfolding, or after recombinant expression, is often kinetically limited by aggregation of transient folding intermediates with exposed hydrophobic surfaces. Folding intermediates of the enzyme rhodanese (EC 2.8.1.1) can be trapped and efficiently reactivated by interactions with detergent micelles, liposomes or the chaperonin proteins, groEL and groES from *E. coli*. These systems now permit study of the conformation of bound intermediates, and we present an overview of recent work in this area. In each case, the trapped intermediates have characteristics of molten globules. Limited proteolysis indicates that specific regions on the intermediates are protected by interaction. Control over detergent- or liposome- assisted refolding can be exerted by varying the lipid compositions, and in all cases the folding efficiency can be modified by general solution conditions. These approaches have implications for understanding the roles of accessory components in the practical control of protein folding and the recovery of recombinant proteins.

Problems with the Native Folding Paradigm: The Role of Intermediates

The amino acid sequence of a protein determines its three dimensional structure (*1*). Facile formation of native structure is most likely to occur in dilute, homogeneous solutions. However, even under these conditions, in many cases, the native structure is not easily regained after it has been disrupted by denaturation or after it has been perturbed by altered solution conditions (*2*). One reason for this difficulty is that kinetic competition often leads to aggregates that refold so slowly that the process appears irreversible (*3*). Thus, although the native structure may be thermodynamically most stable, it is formed so slowly that there is no useful formation of functional protein. Important practical issues depend on the distinction between the most stable state and the states that, although transient, are significantly populated on a biologically important time scale.

These problems are particularly relevant inside the cell, where a protein finds itself in an environment that is complex, concentrated, interactive, and dynamic. Interestingly, what we would normally call the native structure of an isolated protein is often not compatible with many biological requirements. Thus, the state that a protein adopts in dilute solution may not be capable of interacting with membranes, or it may not be capable of assembling properly into macromolecular assemblies. The enzyme rhodanese is being used increasingly to study many of these issues (4). Rhodanese is synthesized on cytoplasmic ribosomes and transported into the matrix of the mitochondrion (5). Rhodanese is a good model for several reasons: first, it is difficult to refold after denaturation (6); second, it is unstable in solution (7); and third, it forms inclusion bodies when expressed in E. coli (8).

One way to understand and solve the problems caused by conformational *cul-de-sacs* is to invoke the idea that protein folding occurs through intermediates that have transient and exposed interactive-surfaces. The native structure is expected to be formed in such a way that it minimizes the exposure of these interactive surfaces. Intermediates could form in several ways: from the denatured protein; during vectorial synthesis on ribosomes; or by structural perturbations of the folded protein. Then, these intermediates could kinetically partition into various paths ranging from what are normally considered desirable ends such as folding, assembly or membrane interactions, on the one hand, to less positive outcomes such as degradation and turnover, aggregation or enhanced chemical reactivity such as oxidation. This leads to the hypothesis that control of intermediates can direct the fates of proteins.

Overview of the Enzyme Rhodanese

The enzyme rhodanese has been increasingly used to study properties of folding intermediates (4). As noted above, rhodanese is a mitochondrial matrix protein that is synthesized in the cytoplasm. The catalytic function of rhodanese is to transfer sulfur atoms (9), and this ability has been suggested to be involved in the synthesis and the control of iron-sulfur centers that are important electron carriers in the mitochondrion (10). The protein as normally isolated is a single polypeptide chain and it is monomeric (11). The sequence of rhodanese from the cDNA is very similar to the sequence derived for the protein itself (12,13), so there has been very little, if any, processing of the polypeptide chain after mitochondrial import. This conclusion is supported by comparisons of the mature protein with protein that is produced directly under the direction of the rhodanese mRNA (14). The active site of rhodanese is in the region of contact between the two domains into which the protein is folded (11).

The sequence of the first 25 amino acids from the N-terminus has consensus characteristics of a mitochondrial import signal (12). Specifically, this region contains basic residues embedded in a hydrophobic sequence with a capacity to form an amphiphilic alpha helix. Unlike most imported proteins, this leader sequence is not cleaved after import. Thus, the mature protein contains all the information for folding, targeting and mitochondrial import. In addition, the N-terminal sequence is helical on the surface of rhodanese (15). A synthetic peptide corresponding to this sequence is random in solution, but it binds to the chaperonin or heat shock protein, cpn60, in an alpha helical conformation (16). The role of cpn60 in facilitating folding will be discussed below.

Hydrophobic surfaces are important in the structure and function of rhodanese (11). Extensive hydrophobic interactions in the interdomain region stabilize the two domain structure, and hydrophobic interactions occur between the N-terminal sequence and the body of the N-terminal domain (11). Rhodanese has been proposed to be able to partition on and off the membrane in course of *in vivo* function (17), and these interactions may be related to modulated exposure of hydrophobic surfaces.

Difficulties in Refolding Denatured Rhodanese

Until recently, it was very difficult to get refolding of rhodanese after *in vitro* denaturation (*18*). The most common result when refolding was attempted was that the protein precipitated. This effect was particularly clear when the solubility of rhodanese was measured as a function of the concentration of the denaturant, guanidinium hydrochloride (*19*). In this case, there was a critical concentration of guanidinium hydrochloride at which almost all the rhodanese precipitated from solution. This precipitation was proposed to result from aggregation of folding intermediates with exposed hydrophobic surfaces. This exposure could be demonstrated by using fluorescent probes such as 1,1'-bi(4-anilino)naphthalene-5,5'-disulfonic acid, bisANS. These probes are virtually non-fluorescent in aqueous solution, but become strongly fluorescent when bound to hydrophobic surfaces (*20*). Using the clue that there were structured species that might aggregate by interactions between perturbation-induced hydrophobic surfaces, it was found that precipitation could be prevented completely by including non-denaturing detergents such as lauryl maltoside in the incubation mixture (*21*).

Detergent Assisted Refolding is Possible and Reveals Intermediates

Urea induced unfolding is reversible if detergents are used, and, although the process is fully reversible, it is not described by a two-state model (*22*). Analytically, the transition curves for the denaturation can be fitted by two two-state transitions in which the one at lower urea concentrations corresponds to the formation of a folding intermediate.

If the samples, at each of the urea concentrations above, are tested within 10 minutes, rather than 24 hours required to ensure equilibrium, it is found that the enzyme quickly adopts a non-native conformation that only slowly relaxes to the equilibrium conformer (*22*). For example, when diluted to low urea concentrations, denatured rhodanese rapidly forms a state in which the fluorescence wavelength maximum is at 345 nm which is between that for the native enzyme (335 nm) and the fully unfolded protein (355 nm). This result suggests that the initial state formed on dilution has partially buried tryptophan residues. Samples at the low urea concentrations slowly reached their equilibrium values after continued incubation for several hours. These type of data indicate that intermediates form that are kinetically stable.

Detergent Micelles Capture Incompletely Folded Rhodanese and Permit Studies of Intermediate-Like Species

The detergents that are used are only effective above their critical micelle concentrations (*23*). Gel filtration chromatography demonstrates that reactivatable folding intermediates form transient, inactive complexes with the detergent micelles (*24*). Gel filtration studies using micelles containing cardiolipin are particularly informative (*24*). Micelles of lauryl maltoside containing cardiolipin permit refolding, but the folding process is slowed relative to micelles containing only lauryl maltoside. For example, 30 min after the start of refolding, most of the rhodanese coelutes in gel filtration with the much larger cardiolipin-containing micelles. The enzyme activity is very low, and the small amount of activity that is observed, elutes at a position characteristic of the native rhodanese. After a further 29 hours of incubation, a much larger amount of activity is observed, and a much larger amount of protein has dissociated from the micelle. The results indicate that the micelles can bind an inactive form of rhodanese, and full reactivation follows release of the protein from the micelles. Experiments like these demonstrate that rhodanese forms complexes with increased efficiency when the micelles contain the mitochondrial, anionic phospholipid, cardiolipin. Although rhodanese has a net negative charge, its N-terminal sequence is positively charged, and this sequence may interact with charged phospholipids in membrane transport. Overall, the presence of cardiolipin increases the residency time of the protein in the micelle. The net effects are to increase the yield and decrease the rate of folding. The same general effects are seen with micelles of lauryl maltoside alone, but the process is considerably faster. Therefore, it is possible to exert some control over the folding reaction by the choice of the type of lipid and the overall composition of the micelles.

The observation that rhodanese can bind to micelles as an inactive intermediate that is only active after release is similar to effect of chaperonin proteins (see below). This micelle system, then, provides an opportunity to observe the properties of structures that are related to folding intermediates. The folding intermediate(s) that is captured in the detergent lauryl maltoside is a case in point (22). The fluorescence maximum of the captured state is at 345 nm compared with the native enzyme at 335 nm or the denatured enzyme at 355 nm. Studies of the ability of solutes to quench the intrinsic fluorescence of captured rhodanese indicate that the solvent exposure of tryptophan residues is between that of the native and the denatured enzyme. The UV-CD spectra in the region of 222 nm indicate that the intermediate retains a considerable amount of regular secondary structure. On the other hand, the near UV-CD signal in the region of 280 nm is dramatically decreased in the intermediate, indicating relaxed tertiary interactions around the tryptophan residues. The tryptophan residues in the intermediate are not fully mobile as monitored by fluorescence polarization. These properties of the rhodanese folding intermediate are similar to those normally used to define what have been called "molten globule states" or "compact folding intermediates" in the folding of proteins (25).

Chaperonin Proteins can Protect Interactive Folding Intermediates and Participate in Biologically Related Folding Processes

Protein folding in a biological system occurs in a complex environment, and one idea that has been proposed to understand how protein folding can be controlled in the cell invokes the properties of proteins called chaperones that are suggested to interact with folding intermediates to control interactions (26). These issues have been discussed well for the process of protein import from the cytoplasm to the mitochondrion (27). A fully folded protein can't enter the mitochondrion, and import is dependent on maintaining the target protein in a partially folded, import competent state. Protein assistants, at various stages of the process, are proposed to interact with a partially folded protein to help keep it in the import competent conformation. One of the proteins involved in the process of mitochondrial protein trafficking is hsp60 (heat shock protein with 60 kDa subunits) which is found in the mitochondrial matrix. This protein is a close homolog of the chaperonin, cpn60 (also called groEL), that is found in *E. coli*. Cpn60 is the chaperonin that has been used to influence the folding and interactions of rhodanese.

A model for the effect of cpn60 can be described that contains several features (27). Cpn60 is an oligomer composed of 14 60kD subunits that are assembled into a structure resembling a double doughnut composed of 2 stacked, 7-member rings. This structure binds non-native conformers. Even though interactive surfaces on the passenger protein may be exposed, they can be protected sterically by being part of a macromolecular complex. Trapped structures of this type that can be formed with other polypeptide chains may also allow assembly at these interactive surfaces prior to release of the bound polypeptide thus providing the appropriate sequencing of interactions. This may be applicable to the orderly assembly of oligomeric proteins. For the protein rhodanese, the release step requires additional factors including ATP and a second oligomeric protein, cpn10 (20). This process allows ordered release to permit folding.

Detergent-assisted Folding and Chaperonin-assisted Folding Share Common Features

The correspondence between the detergent assisted folding and models, such as the one outlined above, for chaperonin-protein interactions were so similar, that we were led to test the influence of chaperonins on rhodanese folding (20). The E. coli chaperonins, cpn60 and cpn10, were found to be very effective at facilitating folding of rhodanese which requires a combination of cpn60 + cpn10 + ATP (20). Comparisons of the kinetics of folding in the presence of the detergent lauryl maltoside with the cpn system indicates that the activity will reach the same level in both cases but the rate is much slower with the detergents (20). In addition, under appropriate conditions, there is some folding even when there are no additives. Interestingly, folding with the chaperonins is slower than that observed in this so-called spontaneous case, but there is a significant increase in efficiency. This observation is the basis of the idea that chaperonins may generally function by directing folding in the sense of increasing efficiency, rather than speed. Cpn60, in the absence of the other components, can trap partially folded rhodanese and arrest spontaneous folding (28). If cpn10 and the other components are added back later, folding continues and the activity reaches the levels expected if all components had been present from the beginning. Additionally, light scattering experiments show that cpn60 prevents the aggregation that would occur on attempted refolding (27). This is similar to the effects of detergents (19).

The complex between cpn60 and rhodanese is very tight, and the complex can be isolated in a form that has a high efficiency for reactivation of the bound rhodanese (27,28). This is important to ensure that properties of the bound protein reflect species that are competent to subsequently fold. This system, then, gives us experimental access to a cpn60-bound intermediate. In terms of properties that have been investigated, such as intrinsic fluorescence, controlled tryptic digestion and the binding of hydrophobic probes, the cpn60 bound intermediate is like the detergent trapped intermediate (27,29). The fluorescence properties show that: a) the maximum of the fluorescence spectrum is at an intermediate position; b) quenching of tryptophan is intermediate between fully exposed and buried residues; and c) studies with fluorescence probes suggest that the bound intermediate has increased exposure of hydrophobic surfaces. These data show that there is folding on cpn60 to a molten-globule like structure having some native-like characteristics. The N-terminus of rhodanese is not prevented by interactions with cpn60 from being proteolyzed (29).

In view of the importance of hydrophobic surfaces in determining the fates of rhodanese folding intermediates, we were interested in whether there were complementary surfaces on cpn60. The fluorescence spectra of the probe, bisANS, whose fluorescence is enhanced when it binds to hydrophobic surfaces indicate that there are hydrophobic sites on cpn60 (20). Fluorescence titration indicates that only a few bisANS molecules are required to saturate the cpn60 oligomer. This indicates a small hydrophobic region is formed on the oligomer when the subunits assemble. It is possible that this region is the hole in doughnut structure. No hydrophobic sites are present on cpn10. It is possible that, by modulating hydrophobic surfaces, cpn60 can alternately bind and release intermediates that would otherwise aggregate by interactions of their own hydrophobic surfaces.

Rhodanese Can Bind to Liposomes as a Reactivatable, Partially Folded Intermediate

All these results indicate that the rhodanese intermediate(s) bound to cpn60 have characteristics expected for a partially folded protein that is in a state that can interact with membranes; so we tested the ability of rhodanese intermediates to interact with phosphatidylserine liposomes (*30*). The interaction was monitored by measuring the release of carboxyfluorescein that was trapped in liposomes. Rhodanese was diluted from urea and allowed to refold for varying times before introduction of liposomes. At early times after initiating refolding, rhodanese efficiently disrupts liposomes. The longer the delay to allow refolding, the less effective is the release, and native enzyme or fully folded enzyme has no effect. In addition to any effect on liposome disruption, rhodanese becomes tightly, but transiently associated with the liposomes (*31*). The binding can be directly detected using an ultracentrifugation based liposome flotation assay. The binding was found to depend on the state of rhodanese and on the nature of the lipid.

Fluorescence reveals that the spontaneous release of rhodanese from the liposomes is associated with changes in the intrinsic protein fluorescence, and the complex with liposomes is stable for at least 48 hrs with CL (*31*). When large unilamellar vesicles are made with phosphatidylcholine, the protein is initially bound to the liposomes only if presented as a partially folded species. The interaction is characterized by a fluorescence spectrum indicating an enhanced quantum yield. The native enzyme is not bound to the liposomes. The bound form is slowly released, and, after 48 hours, its fluorescence spectrum has significantly changed from the originally bound form to give a final spectrum characteristic of unbound rhodanese. When the vesicles contain cardiolipin, rhodanese is bound to the liposome in a kinetically stable form. It is inactive, and there is no release or reactivation over 48 hours.

Rhodanese that is bound as an inactive intermediate can be refolded and reactivated by liposome disruption (*31*). If rhodanese is first captured as a partially folded intermediate using cardiolipin containing liposomes, there will be no spontaneous reactivation during the experiment. However, if the liposomes are disrupted by the addition of the detergents lauryl maltoside or Triton X100, there is considerable reactivation. Importantly, Triton X100 will not, by itself, support refolding of unfolded rhodanese. These results support the hypothesis that the role of the detergent in this type of experiment is not to refold the protein, but to release the enzyme from a liposome-bound state in a form that is able to quickly adopt an active conformation. Thus, in this case, reactivation is due to liposome disruption and not detergent assistance.

Summary and Conclusions

The essential features that form the basis for our experimental approach are contained in the hypothesis that there is sequential folding of the polypeptide chain. If we follow folding from the unfolded polypeptide, domain coalescence precedes domain association giving extensive, exposed hydrophobic surfaces. In competition with folding to a soluble, active conformation are kinetically competing interactions leading either to aggregation, or complex formation with chaperonins, or binding to detergent micelles or liposomes. It is the ordered release of these partially folded states, or the release of these intermediates at low concentrations, that give increased yields of active rhodanese. In this picture, the N-terminus of rhodanese folds onto the surface of the N-terminal domain and assumes a helical conformation. The binding of this N-terminal sequence involves significant hydrophobic interactions with the surface of the N-terminal domain. This last interaction contributes stability. Thus, interactions involving the N-terminus can have profound consequences on the global structure of rhodanese. Chaperonins, such as cpn60, can associate with the interactive intermediates and have several consequences. They can: a) protect against heat shock, b) prevent aggregation that competes with proper folding; c) maintain non-native conformers that may be required for modulated membrane association or membrane transport; or c) they may even be necessary to present the protein for ordered proteolysis or oxidation. Overall, the chaperonins or the other assistants do not make folding faster in the rhodanese system. Their roles, in general, may be to raise the energy of activation for competing processes to prevent inappropriate kinetic trapping, or they may arrest all folding to stabilize intermediates and permit additional interactions that would not be easy from the native state.

(Supported by Welch Grant AQ723 and NIH grants GM 25177 and ES05729).

Literature Cited

1. Kane, J.F.; Hartly D.L. *Trends Biotechnol.* **1988**, *6*, 95.
2. Jaenicke, R., *Prog. Biophys. Mol. Biol.* **1987**, *49*, 117-237.
3. Mitraki, A., and King, J., *Bio/Technology* **1989**, 7, 690-696.
4. Horowitz, P.M. In *Conformation and Forces in Protein Folding*; Nall, B.T. and Dill, K.A., Eds.; AAAS: Washington, D.C., **1991**; pp 185-189.
5. Boggaram, M., Horowitz, P.M. and Waterman, M.R., *Biochem. Biophys. Res. Commun.* **1985**, *130*, 407-411
6. Horowitz, P.M. and Simon, D., *J. Biol. Chem.* **1986**, *261*, 13887-13891
7. Aird, B.A. and Horowitz, P.M. *Biochim. Biophys. Acta* **1988**, *956*, 30-38.
8. Miller, D.M., Kurzban, G.P., Mendoza, J.A., Chirgwin, J.M., Hardies, S.C. and Horowitz, P.M. *Biochim. Biophys. Acta* **1992**, *1121*, 286-292.
9. Westley, J., *Adv. Enzymol. Rel. Areas Mol. Biol.* **1973**, *39*, 327-368.
10. Ogata, K. , Dai, X. and Volini, M., *J. Biol. Chem.* **1989**, *264*, 2718-2725.
11. Ploegman, J.H., Drent, G., Kalk, K.H., Hol, W.G.J., Heinrikson, R.L., Keim, P., Weng, L. and Russel, J., *Nature* **1978**, *273*, 124-129.
12. Miller, D.M., Delgado, R., Chirgwin, J.M., Hardies, S.C. and Horowitz, P.M., *J. Biol. Chem.* **1991**, *266*, 4686-4691.
13. Russel, J., Weng, L., Keim, P.S., and Heinrikson, R.L. *J. Biol. Chem.* **1978**, *253*, 8102-8108
14. Boggaram, M., Horowitz, P.M. and Waterman, M.R. *Biochem. Biophys. Res. Commun.* **1985**, *130*, 407-411.
15. Hol, W.G.J. *Adv. Biophys* **1985**, *19*, 133.
16. Landry, S.J., and Gierasch, L.M. *Biochemistry* **1991**, *30*, 7359-7362.
17. Ogata, K. and Volini, M. *J. Biol. Chem.* **1990**, *265*, 8087-8093.
18. Horowitz, P.M. and Simon, D. *J. Biol. Chem.* **1986**, *261*, 13887-13891
19. Horowitz, P.M. and Criscimagna, N.L. *J. Biol. Chem.* **1986**, *261*, 15652-15658.

20. Mendoza, J.A., Rogers, E., Lorimer, G.H. and Horowitz, P.M. *J. Biol. Chem.* **1991**, *266*, 13044-13049.
21. Tandon, S. and Horowitz, P.M. *J. Biol. Chem.* **1986**, *261*, 15615-15618.
22. Horowitz, P.M. and Criscimagna, N.L. *J. Biol. Chem.* **1990**, *265*, 2576-2583.
23. Tandon, S. and Horowitz, P.M. *J. Biol. Chem.* **1987**, *262*, 4486-4491.
24. Zardeneta, G. and Horowitz, P.M. *J. Biol. Chem.* **1992**, *267*, 5811-5816.
25. Ptitsyn, O.B.J. *Protein Chem.* **1987**, 6, 272- 280.
26. Ellis, RT.J. *Semin. Cell Biol.* **1990**, 1, 1-9.
27. Osterman, J., Horwich, A.L., Neupert, W., and Hartl, F.-U. *Nature* **1989**, *341*, 125-130.
28. Mendoza, J.A., Rogers, E., Lorimer, G.H. and Horowitz, P.M. *J. Biol. Chem.* **1991**, *266*, 13587-13591.
29. Mendoza, J.A. and Horowitz, P.M. **1992**, *J. Biol. Chem.*, in press.
30. Mendoza, J.A., Grant, E. Jr., and Horowitz, P.M. **1992**, *J. Prot. Chem.*, in press.
31. Zardeneta, G. and Horowitz, P.M. **1992**, *Eur. J. Biochem.*, in press.

RECEIVED October 26, 1992

IN VITRO PROTEIN FOLDING

Chapter 13

Comparison of Amino Acid Helix Propensities (s-values) in Different Experimental Systems

A. Chakrabartty and R. L. Baldwin

Department of Biochemistry, School of Medicine, Stanford University, Stanford, CA 94305−5307

Three years ago, work started on measuring the α-helix propensities of the amino acids in different short peptide systems (1 - 8). As in earlier experiments with synthetic polypeptides (9,10), the helix propensity of an amino acid is taken as its s-value, where s is the propagation parameter of the Zimm-Bragg theory of α-helix formation (11). Although the results found with short peptides are still preliminary, the helix propensities are evidently different from those measured earlier with random-sequence polypeptides by the "host-guest" technique (9,10). The rank order of helix propensities in the different peptide systems is similar; however, one consistent set of s-values has not been produced. Here we compare the results of these substitution experiments and comment on possible reasons for the disagreements.

Experimental Systems for Measurement of Helix Propensities

The major differences between the short peptide and host-guest systems are in the sizes of the molecules studied and in the use of defined sequences and chain lengths in short peptide systems. With the exception of the dimeric coiled-coil system of O'Neil and DeGrado (3), the results are evaluated by applying helix-coil transition theory. The s-values obtained for the uncharged non-aromatic amino acids from these systems are summarized in Table I.

Host-Guest System of Scheraga and Coworkers (9,10). This system consists of a series of random copolymers which vary in their degree of polymerization and contain chiefly hydroxybutyl-L-glutamine (host residue); they are doped with varying small amounts of another amino acid (guest residue). Homopolymers of the host residues form water-soluble helices which undergo thermal helix-coil transitions. The transition curves are altered when guest residues are present. The host-guest determination of s(Ala) was found to be similar to that determined previously in a different block copolymer system *(12)*. Analysis of an entire series of host-guest copolymers, containing each of the twenty amino acids, has yielded the s-values of all the amino acids. Except for proline and glycine, the s-values are all close to 1 and the differences between them are small. The helix nucleation constants were also found to be very small. A major consequence of this finding is that short helices are predicted not to form, unless they are otherwise stabilized by long-range interactions, such as side-chain interactions or tertiary interactions between helices as are found in proteins *(13)*. Based on this conclusion, Scheraga and coworkers modified Zimm-Bragg theory by including parameters which consider long-range interactions *(14, 15)*. The values of the interaction parameters were determined by fitting the modified theory to various published data, and estimates of helix contents, using the best fit values, did reproduce several of the experimental results *(14, 15)*.

An opposing viewpoint has arisen, however, from the work of Marqusee *et al (16)*. Marqusee *et al. (16)* demonstrated stable helix formation by 16-residue peptides comprised chiefly of Ala with a few interspersed Lys residues and no apparent stabilizing side-chain interactions. The helix content in this series of peptides decreases as the number of Lys residues increases (see their data for 1 M NaCl, where this effect can probably not be attributed to electrostatic repulsion between Lys[+] residues). Their data suggest that helix formation in these short peptides is stabilized by Ala residues alone, and not by Lys[+] residues or any long-range interaction. They concluded from this finding that the host-guest technique underestimates the s-value of Ala because there are helix-stabilizing interactions, probably hydrophobic interactions, among the sidechains of the host residues *(17)*, which are not accounted for in the host-guest analysis *(16)*.

Scheraga and coworkers recently proposed an alternative explanation*(18)* for the data of Marqusee *et al. (16)*, which was based on Monte Carlo simulation of helix formation in peptides whose sequences are the same as those of Marqusee *et al. (16)*. They attribute the stability of these short helices to the presence of Lys residues, which interfere with hydration of the backbone atoms in nonhelical conformations and thereby increase the helix content *(18)*. Furthermore, Scheraga and coworkers contend that 1 M NaCl is not sufficient to screen electrostatic repulsion in peptides of high charge density, and the helix-inducing effect of Lys is offset by the electrostatic repulsion in the peptides with high Lys content*(18)*.

Collaborative experiments between the laboratories of Drs. John Stewart (University of Colorado) and R.L. Baldwin (Stanford University) investigating possible sidechain interactions between the hydroxybutyl-L-glutamine residues are under way (Padmanabhan, S; York, E.J.; Stewart, J.M.; Baldwin, R.L. unpublished results), and the results will hopefully resolve this issue.

Table I. Helix Propensities Measured in Different Experimental Systems

Amino acid	Published s values and calculated s values from published data (at 0°C)			
	A	B	C	D
Ala	1.07	2.19		1.99
Leu	1.14	1.55	0.85	1.70
Met	1.20	1.41	0.64	0.87
Gln	0.98	1.20	0.56	0.61
Ile	1.14	1.02	0.37	0.44
Val	0.95	0.93	0.18	0.20
Ser	0.76	0.86	0.25	
Thr	0.82	0.79	0.18	
Asn	0.78	0.73	0.18	
Gly	0.59	0.57	0.06	0.02

A   Published values from Scheraga and coworkers (9).
B   Published values from Kallenbach and coworkers(4,6), pH 7, 0.01M KF.
C   Calculated values from data of Stellwagen and coworkers (8) using Lifson-Roig theory (19), pH 7, 0.01 M KCl.
D   Published values for Gly from Baldwin and coworkers (5), calculated values for Leu, Met, Ile, and Val from data of Padmanabhan et al. (1), and calculated values for Gln and Ala from unpublished data of Chakrabartty and Baldwin, pH 7, 1.0 M NaCl.
Note: While the published data of O'Neil and DeGrado (3) yield values of $\Delta\Delta G°$ or ratios of s-values in their system, the absolute values of s cannot be obtained.

Peptide System of Baldwin and Coworkers (1,5). Our system is a helical reference peptide based on the design of Marqusee et al. (16) (sequence: Ac-YKAAAAKAAAAKAAAAK-amide). We substitute one or more of the Ala residues with another residue and measure the change in helix content by circular dichroism (CD), and apply Lifson-Roig helix-coil theory (19) to calculate the s-value of the substituted residue.

We chose Ala as a reference because it is a strong helix-former and its sidechain is small, nonpolar, and unable to participate in significant sidechain interactions (16). To minimize long-range charge interactions amongst Lys residues, they are placed 5 residues apart, causing them to spiral around the helix and be well separated in space. In addition, the CD measurements are made in 1.0 M NaCl so that charge interactions are screened. The single Tyr at the N-terminus of the peptide enables us to determine peptide concentration with high accuracy by tyrosine absorbance.

Our reference peptide contains more than one type of residue. To facilitate calculation of the s-value of a substituted residues, however, we treat the reference peptide as a homopolymer and calculate its average s value which is determined chiefly by Ala (12 residues), with a small contribution from Lys (3 internal residues; only the internal residues contribute to the average s when Lifson-Roig theory is used). The s-value and nucleation parameter of our reference peptide have been determined by both CD *(5,20)* and hydrogen exchange methods *(21)*, and similar values have been obtained by each technique.

We decided first to determine the s-value of Gly because it has been shown to be a strong helix breaker *(22)*: thus, the effect of a Gly-substitution should be large and easily measured. In addition, we substituted every Ala in the reference peptide with Gly in a series of individual peptides, thus obtaining the entire substitution curve, not just one point. The determined value of s(Gly) is very small, 100-fold smaller than s(Ala), and it is significantly different from values obtained in most other systems (see Table I). We have also calculated the s-values of Leu, Met, Val, and Ile from the data of Padmanabhan *et al.(1)*. These values, which are reported in Table I, are tentative and will be redetermined by synthesizing an entire series of substituted peptides.

Recently we made a discovery which significantly affects the interpretation of substitution experiments based on CD measurements. All of the substitution experiments reported use peptides that contain either Tyr or Trp to determine peptide concentration. We found that aromatic sidechains of these residues, even when they are present at the frayed ends of helical peptides, contribute to far-UV CD. For example, in Table II we report large differences in the ellipticities at 222 nm for three helical peptides which contain Tyr, Trp, or Ala at the N-terminus, but are otherwise identical in sequence.

Table II. Effect of Aromatic Residues on CD Measurements

Sequence	Mean residue ellipticity at 222 nm (deg cm^2/dmol)*
Ac-YKAAAAKAAAAKAAAAK-amide	-25,000
Ac-AKAAAAKAAAAKAAAAK-amide	-30,000
Ac-WKAAAAKAAAAKAAAAK-amide	-32,000

* Conditions: 0°C, pH7.0, 1.0 M NaCl and 1 mM each of sodium phosphate, sodium borate, and sodium citrate.

The helix contents of these peptides are nearly identical when measured by amide-proton exchange using NMR *(21)* (Chakrabartty, A.; Padmanabhan, S; Baldwin, R.L. unpublished results). This apparent contradiction is reconciled if the differences in CD are caused by aromatic sidechain contributions to the CD spectrum and not by differences in helix content. Similar effects are not observed with non-aromatic uncharged amino acids (Chakrabartty, A; Baldwin, R.L. unpublished results). The aromatic contribution to ellipticity at 222 nm is comparatively minute in unstructured peptides (Chakrabartty, A.; Baldwin, R.L. unpublished results). Thus, the effect we observe requires the presence of a helix and its magnitude may vary with helix content.

It is well known that aromatic groups found in cores of globular proteins *(see 23)* or in conformationally constrained diketopiperazines *(24)* can produce large far UV-CD bands, unlike the minor bands of unconstrained aromatic groups. Interactions of aromatic groups with one another and with peptide groups produces the unusual far UV-CD spectrum of helical poly-L-tyrosine *(see 23)*. The aromatic groups in the short peptides mentioned above, however, are present at the frayed end of a peptide helix, a position with very few structural restraints. The aromatic CD in these peptides, therefore, is somewhat of a mystery.

The aromatic contribution to peptide CD complicates the proper interpretation of the substitution experiment and it will affect the reported s-values. Consequently, it is necessary that it either be corrected or eliminated. A method of achieving this goal, without giving up the use of Tyr residues for determining peptide concentrations is being developed (Chakrabartty, A.; Baldwin, R.L., unpublished results).

Peptide System of Stellwagen and Coworkers *(2,8)*. Stellwagen and coworkers have analyzed the helix-forming tendencies of all 20 amino acids using the helical *(i,i+4)* E,K peptide designed by Marqusee and Baldwin *(25)* (sequence: Ac-YEAAAKEAAAKEAAAKA-amide). Ala9 was substituted with each of the other 19 amino acids and helix contents were determined by CD.

The major difference between this system and ours is the presence of three *(i,i+4)* ion pairs between Glu and Lys which contribute to helix stability. Nevertheless, their observed change in helix content produced by Ala→Gly substitution, in terms of actual ellipticities at 222 nm, is essentially the same as ours. Furthermore, upon using their data and Lifson-Roig theory (which does not take account of helix stabilizing sidechain interactions) to calculate s-values for uncharged non-aromatic amino acids, we find very good agreement, with the exception of Leu, with our values (Table I). We are not yet able to evaluate the contribution of ion pairs, when they are present in configurations such as those in the above reference peptide. The problem lies in the difficulty of assigning s-values to charged residues and also in the evaluation of the charge-helix dipole interaction produced by a charged residue *(see 26)*. The solution is being pursued by collaborative efforts between the laboratories of Drs. J.A. Schellman (University of Oregon) and R.L. Baldwin (Stanford University).

As regards the large disparity between the s-values of Leu, note that in these substitution experiments the relationship between the s-value of the substituent and helix content is nonlinear; at helix contents close to 1, a small error in helix content translates into a large error in s-value. Whenever a strong helix former, such as Leu, is examined, a small error in estimating helix content will result in a large error in s. This is the most likely reason for the disparity.

Peptide System of Kallenbach and Coworkers (4,6). This system employs a helical reference peptide (sequence: succinyl-YSEEEEKKKKEEEEKKKK-amide) in which the arrangement of Glu and Lys residues enables the formation of $(i, i+4)$ ion pair interactions among 16 charged residues. Ion pair interactions are critical for helix formation in this peptide because an isomeric peptide with an altered sequence (succinyl-YSEEKKEEKKEEKKEEKK-amide), which precludes any $(i, i+4)$ Glu-Lys ion pair interactions, shows no significant helix formation at neutral pH.

The helix propensities of uncharged non-aromatic amino acids were determined by inserting three copies of each amino acid into the reference peptide at positions X (succinyl-YSEEEEKKKKXXXEEEEKKKK-amide). The helix contents of the peptides were anaylzed by CD and s-values were calculated (see Table I) using their own helix-coil theory, which incorporates $(i, i+4)$ sidechain interactions but is otherwise similar to classical helix-coil theories.

Their rank order of helix-forming tendencies, in terms of actual ellipticities at 222 nm, *(4)* is similar to ours *(1,5)* (with the exception of Leu) as well as to Stellwagen's *(2,8)*, but is significantly different from host-guest results *(9)*. In terms of s-values, Kallenbach's values correlate both with ours and with the values we calculate from Stellwagen's data but not with Scheraga's host-guest values (Table III).

Table III. Correlations Between s-values from Different Laboratories

y-axis	x-axis		
	Stellwagen	Kallenbach	Scheraga
Baldwin	R = 0.90	R = 0.95	R = 0.52
	m = 1.20	m = 1.36	m = 1.67
	b = 0.06	b = -0.89	b = -0.96
Stellwagen		R = 0.98	R = 0.80
		m = 0.80	m = 1.03
		b = -0.44	b = -0.60
Kallenbach			R = 0.71
			m = 1.73
			b = -0.50

R, correlation coefficient; m, slope; b, y –intercept.

The correlation between Kallenbach's values and ours or Stellwagen's is good (R = 0.95 and 0.98, respectively) and the slopes of the linear regression lines are close to 1. If we disregard our value of s(Leu) the correlation is even better (R = 0.98; m = 1.2; b = -0.85). There appears, however, to be a constant offset of approximately 0.7 between s-values of Kallenbach's and ours or Stellwagen's (Figure 1).

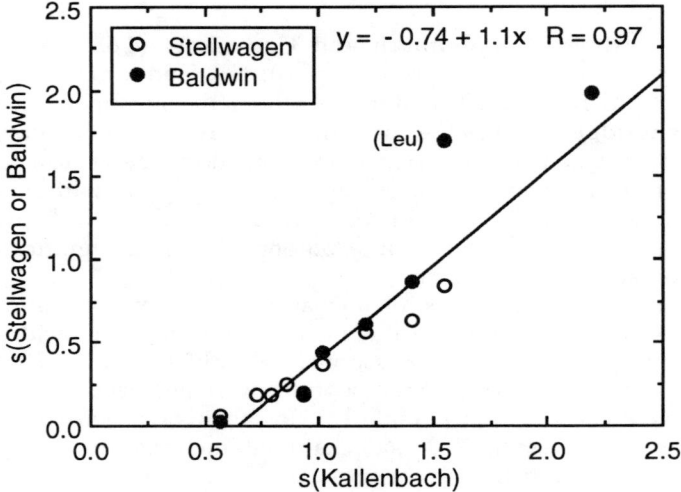

Figure 1. Correlation of s-values of Kallenbach and coworkers with s-values of Baldwin and coworkers and also the calculated s-values from the data of Stellwagen and coworkers using Lifson-Roig theory. The linear regression line is a composite fit of Baldwin and Stellwagen values against Kallenbach values. The Baldwin value for s(Leu) was omitted from the analysis.

The cause of this offset lies at the heart of the disagreement between the s-values determined in different peptide systems. The obvious difference between these peptide systems is the amount of helix stabilization imparted by sidechain interactions. While our Ala-Lys peptide system possesses no evident sidechain interactions, helix formation in Kallenbach's system is entirely dependent on such interactions, and sidechain interactions make some contribution to the helical stability of Stellwagen's system. The predominance of sidechain interactions in Kallenbach's system motivated them to modify classical helix-coil theory to include an *(i,i+4)* sidechain interaction parameter.

To investigate the consequence of this modification, we analyzed their ellipticity data using unmodified Lifson-Roig theory (i.e. with no interaction parameters) and compared the results with their published s-values (Figure 2). It should be noted that when the helix nucleation parameter is small (as in this case), the Lifson-Roig theory produces virtually identical results to Kallenbach's theory if the interaction parameter is removed, and if only small peptides (n ≈ 20) are considered.

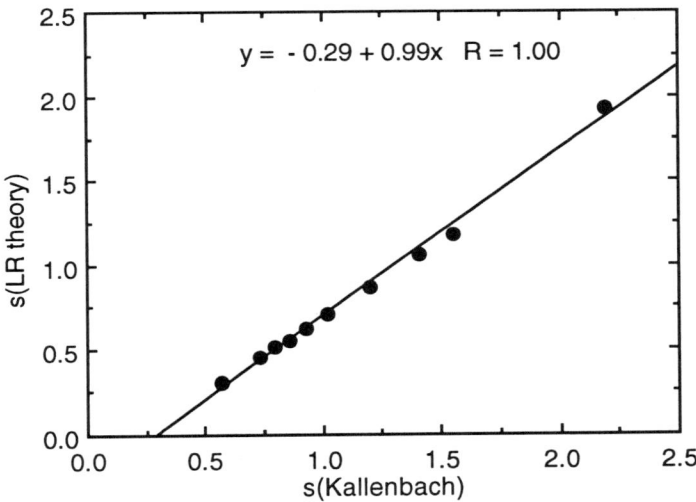

Figure 2. Correlation of the s-values of Kallenbach and coworkers with the s-values calculated from their data using Lifson-Roig theory.

The striking feature of Figure 2 is that omission of the interaction parameter produces a similar, but smaller, offset to that in Figure 1. The similarity between Figures 1 and 2 suggest that the value of the interaction parameter in Kallenbach's theory is a likely cause of the offset in Figure 1. Thus, one set of consistent s-values may be produced if the Kallenbach interaction parameter is reduced in magnitude. While it is essential that stabilizing $(i,i+4)$ ion pair interactions be explicitly accounted for in the analysis, as Kallenbach and coworkers have done, there may also be both stabilizing and destabilizing interactions in the $(i,i+3)$, $(i,i+2)$ and $(i,i+1)$ configurations in Kallenbach's peptide helix. The allowed geometries and strengths of these interactions are not well established. Kallenbach's measurements were made in 0.01 M KF where electrostatic repulsion between like charges may also seriously affect helical stability. Therefore, it is not yet possible to conclude whether $(i, i+4)$ interactions will dominate over the others, making it permissible to ignore the other contributions. If the contributions of these other interactions are significant, then adjustment of the interaction parameter and/or addition of new interaction parameters will be necessary.

Dimeric Coiled-Coil System of O'Neil and DeGrado (3). O'Neil and DeGrado used a system in which one peptide is in equilibrium between helical dimeric and random coiled monomeric states. They chose one site in the peptide to substitute all 20 amino acids. This site is surrounded by small neutral amino acids and is solvent-exposed in the dimeric helical state. The monomer/dimer two-state equilibrium constant of the system is affected by the substituted amino acid, and this constant is related to the helix-forming tendency of the substituted amino acid. Their results, in terms of $\Delta G°(Xaa) - \Delta G°(Gly)$, correlate well with those of Kallenbach and coworkers (4,6).

It is not possible to obtain directly s-values from the results of their two-state analysis. There exists, however, a more complex ensemble-of-states model of Skolnick and Holtzer (27,28) which can be used to obtain s-values. While the two state analysis considers only nonhelical monomer and completely helical dimer states, the ensemble-of-states model considers the possible existence of partially helical monomers and dimers, and completely helical monomers as well. This analysis is currently being performed by DeGrado's laboratory (E.I. du Pont de Nemours and Company) in collaboration with Dr. A. Holtzer (Washington University), and independently by Dr. H. Qian (University of Oregon). Therefore, it is too soon to comment on the results obtained in this system.

Preformed Helix Nucleus System of Kemp and Coworkers (7). By synthesizing a conformationally constrained analog of Ac-Pro-Pro, Kemp and coworkers have partially immobilized three carbonyl groups in an orientation which mimics that of the three peptide carbonyls of a single turn in a helix. This helical turn analog, when attached to the N-terminus of a peptide, behaves as a helix nucleus or template which facilitates helix formation in the attached peptide. Such helices exhibit a very broad helix-coil transition expected from the non-cooperative process caused by the large nucleation parameter of the template(see 26). By attaching the template to short polyalanine peptides (n = 4 to 6) and monitoring the conformations of the conjugates by NMR they have estimated the s-value of Ala to be around 1, close to the host-guest result but considerably different from results of substitution experiments in short peptides (see Table I). It will be interesting to find out how the behaviour of their system changes as the major amino acid is varied; experiments of this kind have not yet been reported.

Investigation of Helix Propensities in Proteins

The relative helix-forming tendencies of Ala and Gly were investigated in a protein system by Fersht and coworkers *(29)* using mutational experiments in solvent-exposed sites of barnase (RNase Ba from *Bacillus amyloliquefaciens*). They observe a strong position dependence different from that observed in the peptide substitution experiments *(5)*. Ala→Gly mutations stabilize the protein when the mutation site is close to the ends of the protein helix, but they are destabilizing when the mutation site is at in an interior position.

Fersht and coworkers' explanation of the stabilizing effect of the mutation at the helix ends is that the Ala sidechain prevents H-bonding of solvent to the backbone atoms of the terminal helical turns and mutation removes the sidechain, thus allowing solvent access to the backbone atoms. The destabilizing effect of the mutation in the middle of the protein helix, on the other hand, was attributed to the reduction in the amount of buried hydrophobic surface in the protein caused by mutation.

In peptide systems, fraying at the ends of isolated helices obscures the observation of end effects, such as the one observed by Fersht and coworkers. Effects in the helix interior, similar to the one mentioned above, are observed in peptide systems, however, and burial of hydrophobic surface is one of the factors which contributes to the s-value of an amino acid. The greater flexibility of the peptide backbone in the random coil when Gly is present also contributes. A basic difference between the protein and peptide systems is that, in the absence of sidechain interactions, the substitution effect in an isolated helix is entirely regular and uniform in the helix interior, and gradually diminishes at positions close to either end of the helix, whereas it depends on the surrounding context, provided by the remainder of the protein, in the protein system.

Another possible aspect of protein helix formation which cannot be detected in peptide systems, and which may contribute to the differences in stability of the barnase mutants, involves desolvation of backbone atoms of the protein helix*(30,31)*. It has been postulated that during folding, formation of tertiary structure from secondary structure requires desolvation of H-bonded backbone atoms; a significant positive enthalpy is predicted to be associated with this desolvation *(30,31)*. Consequently, mutation-induced changes in stability may include contributions from this effect.

Predicting Helix Propensity by Molecular Dynamics and Monte Carlo Simulation.

The substitution experiment in short peptides is well suited for molecular dynamics simulation, because of the small size of the peptide helix, and the availability of experimental data makes it easy to test the calculations. Hermans and coworkers (32) recently performed such a simulation on a 14-residue polyalanine helix in which one of the residues in the middle is replaced with α-amino isobutyric acid, α-amino n-butyric acid, Val, Gly, D-Ala, t-Leu, or Pro. Their calculated values of $\Delta\Delta G°$ for Ala → α-amino isobutyric acid, Val, and Gly correlated very well with the values of O'Neil and DeGrado (3) and reasonably well with those of Lyu et al. (4). Besides reproducing experimental data, the simulations provide insight into the structural determinants of the s-value such as backbone flexibility of the random coil, in the case of Gly and Pro, and absence of a Cβ-helix backbone contact in the case of Gly (32).

Monte Carlo simulations have also been used to probe the thermodynamic and structural origins of peptide helices. The Monte Carlo simulations of Vila et al. (18) have been discussed under the section *Host-Guest System of Scheraga and Coworkers*. Creamer and Rose (33) have used Monte Carlo simulations to calculate the difference in configurational entropy of apolar residues between the unfolded state (modelled by a flexible Ala-X-Ala tripeptide) and the helical state (modelled by a rigid Ala_4-X-Ala_4 helix). Their computed values of TΔS for Ala, Leu, Met, Phe, Ile, Val correlated well with $-[\theta]_{222}$ values of Padmanabhan et al. (1) and with $\Delta\Delta G°$ values of O'Neil and DeGrado (3) and Lyu et al. (4).

Concluding Remarks

The s-values which have been obtained so far, in different short peptide systems, disagree numerically, but produce the same rank order, and are quite different from host-guest results. The s-values obtained in different systems bear, however, a simple relation to each other, and the numerical disagreement may be eliminated when the helix-stabilizing effect of ion pair interactions is better understood.

The field is active, and new developments can be expected rapidly. The use of so many different and ingenious experimental systems allows a critical test of the elegant theories of the helix-coil transition which were formulated three decades ago.

Acknowledgement

We thank Drs. Andrew Doig, Nancy Ng, and Marty Scholtz for helpful discussions and suggestions, and Drs. J. Hermans and H.A. Scheraga for preprints of their manuscripts. A.C. is the recipient of post-doctoral fellowship from the Medical Research Council of Canada.

Literature Cited

1. Padmanabhan, S.; Marqusee, S.; Ridgeway, T.; Laue, T.M.; Baldwin, R.L. *Nature* **1990**, 344, 268.

2. Merutka, G.; Lipton, W.; Shalango, W.; Park, S.-H.; Stellwagen, E *Biochemistry* **1990**, 29, 7511.
3. O'Neil, K.T.; DeGrado,W.F. *Science* **1990**, 250, 646.
4. Lyu, P.C.; Liff, M.I.; Marky, L.A.; Kallenbach, N.R. *Science* **1990**, 250, 669.
5. Chakrabartty, A.; Schellman, J.A.; Baldwin, R.L. *Nature* **1991**, 351, 586.
6. Gans, P.J.; Lyu, P.C.; Manning, M.C.; Woody, R.W.; Kallenbach, N.R. *Biopolymers* **1991**, 31, 1605.
7. Kemp, D.S.; Boyd, J.G.; Muendel, C.C. *Nature* **1991** 352, 451.
8. Stellwagen, E.; Park, S.-H.; Shalango, W.; Jain, A. *Biopolymers* **1992**, 32, 1193.
9. Sueki, M.; Lee, S.; Powers, S.P.; Denton, J.B.; Konishi, Y.; Scheraga, H.A. *Macromolecules* **1984**, 17, 148.
10. Wojcik, J.; Altmann, K.-H.; Scheraga, H.A. *Bioploymers* **1990** 30, 121.
11. Zimm, B.H.; Bragg, J.K. *J. Chem. Phys.* **1959** 31, 526.
12. Ingwall, R.T.; Scheraga, H.A.; Lotan, N.; Berger, A.; Katchalski, E. *Biopolymers* **1968** 6, 331.
13. Scheraga, H.A. *Proc. Natl. Acad. Sci. U.S.A.* **1985** 82, 5585.
14. Vasquez, M.; Pincus, M.R.; Scheraga, H.A. *Biopolymers* **1987** 26, 351.
15. Vasquez,M.; Scheraga, H.A. *Biopolymers* **1988** 27,41.
16. Marqusee, S.; Robbins, V.H.; Baldwin, R.L. *Proc. Natl. Acad. Sci. U.S.A.* **1989** 86, 5286.
17. Lotan, N.; Yaron, A.; Berger, A. *Biopolymers* **1966** 4, 365.
18. Vila, J.; Williams, R.L.; Grant, J.A.; Wojcik, J.; Scheraga, H.A. *Proc. Natl. Acad. Sci. U.S.A.* **1992**, 89, 7821.
19. Lifson, R.; Roig, A.J. *J. Chem. Phys.* **1961**, 34, 1963.
20. Scholtz, J.M.; Qian, H.; York, E.J.; Stewart, J.M.; Baldwin, R.L. *Biopolymers* **1991**, 31, 1463.
21. Rohl, C.A.; Scholtz, J.M.; York, E.J.; Stewart, J.M.; Baldwin, R.L. *Biochemistry* **1992**, 31, 1263.
22. Strehlow, K.G.; Baldwin, R.L. *Biochemistry* **1989** 28, 2130.
23. Adler, A.J.; Greenfield, N.J.; Fasman, G. *Methods Enzymol.* **1973** 27, 675.
24. Snow, J.W.; Hooker, T.M.; Schellman, J.A. *Biopolymers* **1977** 16, 121.
25. Marqusee, S.; Baldwin, R.L. *Proc. Natl. Acad. Sci. U.S.A.* **1987**, 84, 8898.
26. Scholtz, J.M.; Baldwin, R.L. *Annu. Rev. Biophys. Biophys. Chem.* **1992** 21, 95.
27. Skolnick, J.; Holtzer, A. *Macromolecules* **1982** 15, 303.
28. Skolnick, J.; Holtzer, A. *Macromolecules* **1982** 15, 812.
29. Serrano, L.; Neira, J.-L.; Sancho, J.; Fersht, A.R. *Nature* **1992** 356, 453.
30. Ben-Naim, A. *J. Phys. Chem.* **1991** 95, 1437.
31. Yang, A.-S.; Sharp, K.A.; Honig, B. *J. Mol. Biol.* **1992** in press.
32. Hermans, J.; Anderson, A.G.; Yun, R.H. *Biochemistry* **1992** 31, 5646.
33. Creamer, T.P.; Rose, G.D.*Proc. Natl. Acad. Sci. U.S.A.* **1992**, 89, 5937.

RECEIVED October 26, 1992

Chapter 14

Single-Step Solubilization and Folding of IGF-1 Aggregates from *Escherichia coli*

Judy Y. Chang and James R. Swartz

Department of Cell Culture and Fermentation, Research and Development, Genentech, Inc., 460 Point San Bruno Boulevard, South San Francisco, CA 94080

>A significant amount of IGF-1 can be expressed using the LamB signal sequence in E. coli (1). However, more than 90% of the IGF-1 protein forms insoluble protein aggregates (refractile particles), presumably in the periplasm. These protein aggregates are not detected by IGF-1 radioimmunoassay (RIA) and exhibit no appreciable bio-activity. A very simple, one step protein folding protocol was developed for the aggregated IGF-1. The inactive protein is solubilized using low concentrations of urea and DTT in an alkaline buffer. The protein is solubilized and folded in the same solution. Therefore, a denaturant removal or dilution step is not required. Recovery of correctly folded, active protein is dramatically affected by protein concentration and solvent conditions.

Many recombinant proteins expressed in bacteria are unable to fold properly and often accumulate to form inactive aggregates called refractile particles. Refractile particles can be formed in the periplasm (1-3) or cytoplasm (2,4), following expression of the protein with or without a signal peptide, respectively. Insulin-like Growth Factor 1 (IGF-1) is a single chain protein of 70 amino acids with three disulfide bridges in the molecule. These disulfide bonds are: Cys18-Cys61, Cys6-Cys48 and Cys47-Cys52. We use a periplasmic protease deficient W3110 host carrying a plasmid, which contains the alkaline phosphatase promoter and the LamB signal peptide from E. coli and the human IGF-1 structural gene, to produce Insulin-like Growth Factor 1 in E. coli. The gene product has the N-terminal sequence corresponding to authentic IGF-1, is presumably accumulated in the periplasmic space, and can accumulate to as much as 10 - 15 % of the cell's protein. A small portion of the secreted IGF-1 (< 10% of the total) appears in the growth medium. This material is soluble, RIA detectable, and exhibits bio-activity. However, the majority of the IGF-1 accumulates within the E. coli periplasm and forms insoluble protein aggregates (1), refractile particles. The cell associated IGF-1 is insoluble, not detected by an IGF-1 radioimmunoassay, and does not exhibit bioactivity. In the process of characterizing these refractile particles, we developed a very simple, one step protein folding protocol to produce active IGF-1.

Isolation of IGF-1 Aggregates

IGF-1 refractile particles can be isolated easily by centrifugation after cell breakage.

Four grams of cells (wet weight) from a 10 liter fermentation of E. coli carrying the

IGF-1 secretion plasmid was resuspended in 100 ml of cell lysis buffer containing 25 mM Tris, pH 7.5 plus 5 mM EDTA, with the presence of 0.2 mg/ml lysozyme. Cells were sonicated at 4°C for 5 min. Cell lysates were centrifuged at 12,000 x g for 10 min. The distribution of IGF-1 protein in the whole cell, supernatant, and pellet fractions was examined using a Coomassie blue stained 4-20% Tricine SDS-PAGE gel under reducing conditions (Fig. 1). For the whole cell lysate, approximately 10% of the total E. coli protein is IGF-1. The supernatant fraction reveals a protein pattern similar to that of whole cell lysate, except that very little IGF-1 is detected. The pellet fraction (refractile particles) contains IGF-1 as the dominant protein suggesting that the centrifuged pellet is predominantly IGF-1 refractile particles.

IGF-1 in the various fractions was also examined with a Coomassie blue stained, non-reduced SDS-PAGE gel (data not shown). Under the non-reduced conditions, no significant IGF-1 migrating near the IGF-1 standard was detected . However, numerous high molecular weight, faint bands appeared, suggesting that the majority of cell associated IGF-1 is in disulfide linked aggregates. Visual observation as well as the SDS-PAGE results indicated that the aggregated form is completely dissociated upon heating in the presence of SDS and a reducing reagent.

Solubilization of IGF-1 Aggregates

To solubilize the IGF-1 aggregates, the effects of two chaotropic reagents and a reducing reagent were tested (Table I). Refractile particles at approximately 1.5 mg/ml were resuspended in 25 mM Tris, pH 7.5 and 5 mM EDTA. Two chaotropic reagents, urea and GuCl, and a reducing reagent, DTT, were supplemented, at various concentrations, alone or in combination. The solubilization of the refractile particles was examined by observing the clearing of the refractile particle suspension. The turbidity/solubility observation was confirmed by centrifuging the solution to remove the insoluble protein and then examining the resulting supernatant and pellet fractions on Coomassie blue stained, reduced SDS gels. IGF-1 folding was followed by the increase in radioimmunoassay recognizable titers. The turbidity observations and the soluble protein concentration results agreed well with the gel analysis. These results showed that the refractile particles are not soluble or are only slightly soluble with a chaotropic reagent alone (Table I), even in the presence of 8 M urea or 6 M GuCl. With the addition of 10 mM DTT, the refractile particles become completely soluble (Table I). The control experiment with 10 mM DTT alone revealed that the aggregates are not solubilized by a reducing reagent alone. After solubilization, samples were examined using a radioimmunoassay to detect immuno-recognizable IGF-1 and the resulting value was used to calculate the folding yield. Very little folding was detected, even for the completely solubilized samples. However, we were delighted that some folding was detected in the sample with 8 M urea plus 10 mM DTT. Since urea plus DTT worked well to solubilize the refractile particles and this solubilization protocol produced RIA recognizable IGF-1, we decided to use urea in further studies.

Initial IGF-1 Folding

Traditionally, to recover active protein from refractile particles, strong denaturing conditions, such as 8 M urea (5) or 6 M GuCl (6,7), are required for refractile particle solubilization. The completely unfolded protein is then folded after decreasing the concentration of the denaturant by dialysis or by dilution (5,7). Encouraged by the RIA recognizable activity detected in the solubilization experiments, we set out to test the idea of using a single solubilization/folding step to avoid the dialysis or the dilution step. In addition, if some degree of correct or beneficial structure is already present in the aggregated IGF-1 formed in the periplasm, then subsequent in vitro

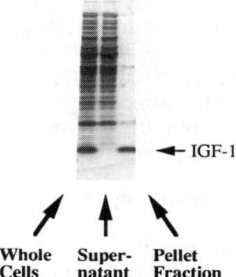

Fig. 1 Reduced, SDS Page gels of the refractile particle preparations.

Table I. Solubilization of IGF-1 aggregates

Chaotrope	DTT (mM)	solubility	Folding Yield (%)
-	0	-	0.07
-	10	-	0.04
2 M urea	0	-	0.08
4 M "	0	-	0.11
6 M "	0	-	0.13
8 M "	0	-	0.15
8 M "	10	++++	1.15
2 M GuCl	0	-	0.19
4 M "	0	-	0.16
6 M "	0	-	0.13
6 M "	10	++++	0.11

folding may be improved by a more gentle solubilization of the aggregates by using a minimal concentration of denaturant and reducing reagent. To test the idea, we examined the minimal concentration of urea and DTT required for IGF-1 solubilization and folding. IGF-1 refractile particles at approximately 1.5 mg/ml were resuspended in 100 mM sodium acetate, pH 8.2, with various concentrations of urea and DTT and incubated at 23°C for 3 hours. IGF-1 folding yield was monitored by radioimmunoassay (RIA). IGF-1 solubilization and folding are dependent on the DTT and the urea concentrations (Fig. 2). 2 mM appeared to be the optimal DTT concentration for all urea concentrations except with 1M urea. A deleterious effect of high DTT concentration, 4 mM, was observed only when the urea concentration was low, 1 M. At 1 M urea, solubilization of the IGF-1 aggregates was incomplete. Increasing DTT concentration for samples with 1 M urea resulted in a yield decrease, presumably because of excess reducing agent without complete chaotropic solubilization. At 4 M urea, very little RIA recognizable IGF-1 was detected, regardless of the DTT concentrations. High folding yields were obtained for 2 M urea and the folding yield reached a maximum with 2 M urea and 2 mM DTT. The results indicated that it is important to have simultaneous disulfide bond reduction and decrease of noncovalent forces to allow one step IGF-1 solubilization and folding.

Characterization of the Folded IGF-1

This RIA detectable IGF-1 was examined using reverse phase HPLC. IGF-1 after folding was applied to a Vydac C-18 column which had been equilibrated with 0.1% trifluoroacetic acid in 25% acetonitrile (pH = 2). Elution was conducted at 40°C with a flow rate of 2 ml/min and with a 28.5-29.5% linear gradient of acetonitrile. The IGF-1 elution profile reveal two major IGF-1 peaks (Fig. 3). One co-migrated with the authentic IGF-1 standard, the other migrated with a misfolded IGF-1 form. The ratio of the correct form to the misfolded form is approximately 2 to 1. The peak area of the correct form accounts for all of the RIA titer. This result suggests that when IGF-1 becomes RIA recognizable, it has acquired a correctly folded conformation. Additional IGF-1 variants were also detected in the more hydrophobic region of the HPLC. These are IGF-1 dimers and oligomers along with IGF-1 monomers of unknown nature.

To characterize the disulfide linkage and to check the general amino acid sequence of the folded protein, folded IGF-1 was isolated using a preparative HPLC at pH 7.0 and subjected to HPLC peptide mapping analysis after V8 proteinase digestion (8). The fraction co-migrating with the authentic IGF-1 standard was analyzed. The HPLC peptide mapping profile is identical to that of authentic IGF-1 (data not shown). The result suggested that this form contained correct disulfide linkages. In addition, this fraction was detected by RIA and exhibited full bioactivity (Fig. 4). The fraction co-migrating with the misfolded IGF-1 form was also examined. The HPLC peptide profile indicated that two disulfide linkages were formed incorrectly at Cys6-Cys47 and Cys48-Cys52. In addition, this misfolded form was not significantly detected by either the RIA assay or the bioassay.

Optimization of IGF-1 Folding

To increase the folding yield, several protein folding parameters were investigated.

Effect of pH. To study the effect of pH and buffers on IGF-1 solubility and folding, refractile particles at 1.5 mg/ml were resuspended into a solution containing 100 mM NaCl, 2 M urea, and 2 mM DTT and buffered with 100 mM of various reagents. The pH of each buffer was varied systematically within its effective pH range. Samples were allowed to fold for 5 hours at 23°C. IGF-1 folding was

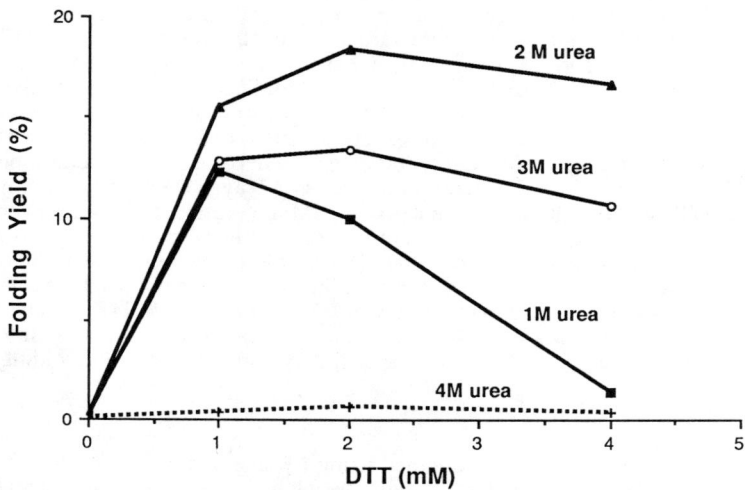

Fig. 2 Initial optimization of urea and DTT for IGF-1 aggregate solubilization and folding. The standard deviation of the RIA is approximately 8%.

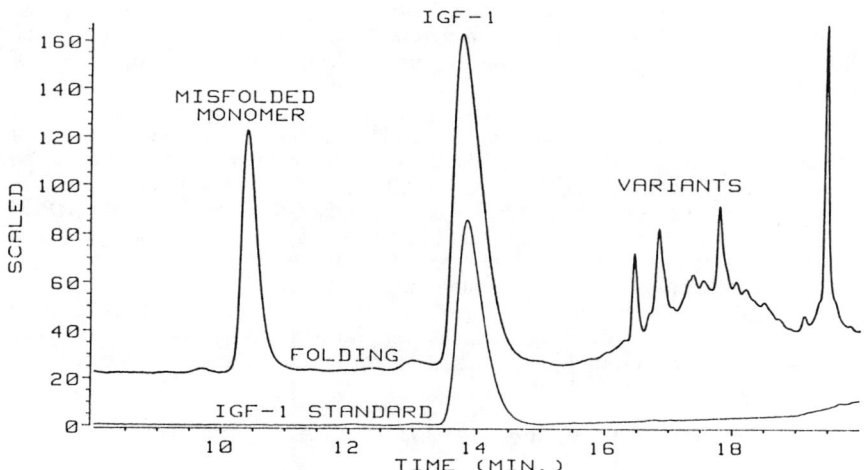

Fig. 3 HPLC analysis of folded IGF-1.

followed by the RIA assay. A dramatic pH effect was observed. The solubility of the IGF-1 protein increased with increasing pH, regardless of the buffer used (data not shown). At pH 9 and above, almost all IGF-1 protein was solubilized. This is expected since the pKa's of our reductant, DTT, are at alkaline pH. High pH will increase the formation of thiolated anions which are involved in disulfide exchange. RIA detectable titers increased very drastically as the pH was increased between pH 7.5 and 10.5 (Fig. 5). Various buffers at the same pH exhibited different effects on IGF-1 folding. We selected Capso at pH 10.5 as the buffer of choice.

Folding Kinetics. The kinetics of IGF-1 folding were examined. IGF-1 refractile particles at approximately 1.5 mg/ml were resuspended in 100 mM Capso, pH 10.5, with 2 M urea, 2 mM DTT, 100 mM NaCl and incubated at 23°C. Correctly folded IGF-1 was monitored at time intervals by HPLC using a Vydac C18 column as described. Total IGF-1 protein, after solubilization in 50 mM Tris buffer, pH 8.0, containing 6 M urea, 5 mM EDTA, and 10 mM DTT, was also assayed by HPLC. Two 4.6x50 mm PLRP-S columns linked in series were used and the denatured IGF-1 was eluted at a flow rate of 1 ml/min with a linear 32-45% acetonitrile gradient containing 0.1% trifluoroacetic acid. In subsequent experiments, HPLC based methods were used for both total and folded IGF-1. Results (Fig. 6) revealed that relatively little correctly folded IGF-1 was present in the 0.5 hr and 1 hr samples. After this point, the concentration of folded IGF-1 continued to increase with folding duration and appeared to plateau after approximately 5 hours. Subsequent analysis was done with 5 hour samples except where noted. Results in Fig. 2 which were done with 3 hour samples, may not represent complete IGF-1 folding. However, the validity of the relative effects indicated by the 3 hour time points was supported by subsequent studies.

Solvent Effects. To further improve IGF-1 folding, various effectors cited in the literature, such as the addition of polyethylene glycol (*9,10*) of various molecular weights (300-10,000) and at various concentrations (1% to 9%), glycerol (10%-40%), triton (0.05%-0.5%), NaCl (*11*), and other solvents (*11,13*) were tested. Most of these agents produced no apparent effect except for the addition of MeOH and EtOH. The effect of MeOH is shown in Fig. 7. With 20% MeOH, the yield of the correct IGF-1 form increased from 33% to 46% and the ratio of the correct to misfolded form improved from approximately 2.4 to 4.1. This is consistent with Tamura's (*13*) result that adding MeOH to folding buffer improves the formation of the authentic IGF-1 form. The kinetics of folding also accelerated in the presence of methanol. The folding yield plateaued at approximately 2 hours instead of 5 hours (data not shown).

The effect of increased salt concentration was also examined. HPLC results revealed that by increasing NaCl from 100 mM to 1 M, no apparent effect on the yield of folding or on the ratio of the correct to misfolded form was obtained (Fig. 7). However, by using 1M NaCl combined with 20% MeOH, the ratio of correct to misfolded forms further improved to 4.5 - 4.7 and the yield was improved to 48% (Fig. 7). Although folding with MeOH plus 1M NaCl is only slightly superior to folding with MeOH plus 0.1 M NaCl, it is known that salts effectively neutralize the electrostatic forces in proteins and minimize unfavorable repulsive interactions during folding of the polypeptide chain (*14*). We, therefore, decided to include 20% MeOH and 1 M NaCl in the folding cocktail.

Temperature Effects. The effect of temperature on IGF-1 folding was examined. IGF-1 was allowed to fold using Capso buffer, pH 10.5, containing 1 M NaCl, 20% MeOH, 2 M urea, and 2 mM DTT at several temperatures. The formation of correctly folded IGF-1 was followed by HPLC analysis. Fig. 8 shows the effect of

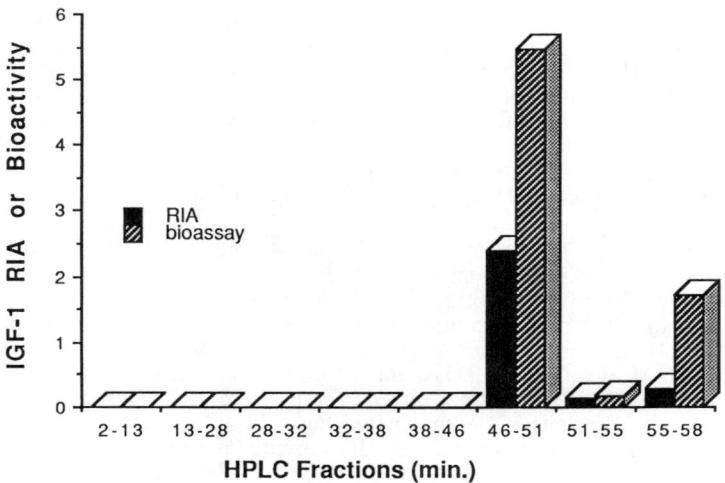

Fig. 4 Bioactivity and immunoactivity of folded IGF-1. Various IGF-1 forms from the folded IGF-1 refractile particles were isolated using a Waters-C4 preparative HPLC at pH 7.0. The correctly folded (fraction 46-51), the misfolded (fraction 51-55), and the regenerate fraction (fraction 55-58) were collected and analyzed by bioassay and radioimmunoassay for IGF-1. The bioassay for IGF-1 activity measures the ability of IGF-1 to enhance the incorporation of tritiated thymidine, in a dose-dependent manner, into the DNA of BALB/c 3T3 fibroblasts.

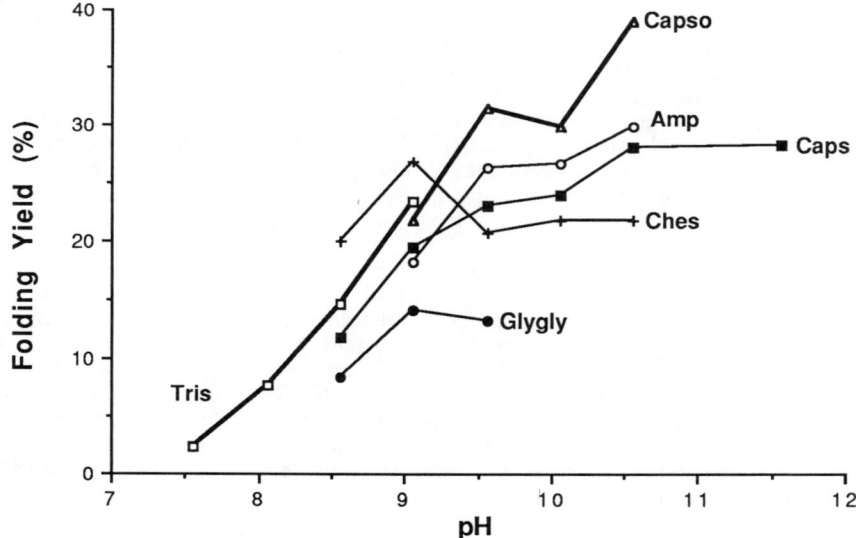

Fig. 5 Effect of pH and buffers on IGF-1 folding. The pKa's of buffers are: Tris, 8.3; Glygly, 8.4; Ches, 9.3; Capso, 9.6; Amp, 9.7; and Caps, 10.4.

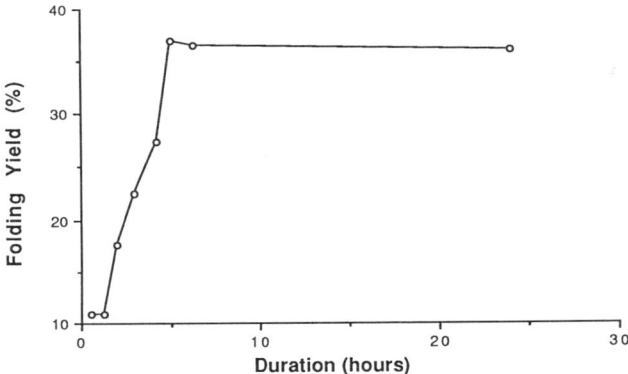

Fig. 6 Kinetics of IGF-1 folding.

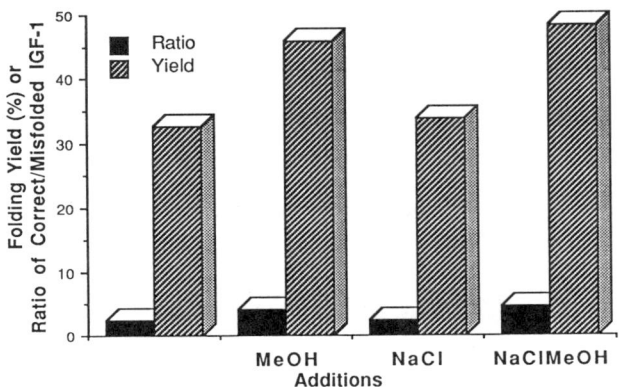

Fig. 7 **Effect of NaCl and Methanol on IGF-1 folding.** IGF-1 refractile particles at approximately 1.5 mg/ml were resuspended in 100 mM Capso, pH 10.5, containing 100 mM NaCl, 2 M urea, 2 mM DTT, and with or without 20% Methanol and with or without additional 900 mM NaCl. IGF-1 was allowed to fold at 23°C for 5 hours. Correctly folded and misfolded IGF-1 were measured by HPLC using a Vydac C-18 column.

temperature on the maximal folding yield obtained at each temperature. Similar IGF-1 levels were obtained for folding at 15, 23, and 37 °C. Only at 4°C was a significantly lower IGF-1 level obtained. For convenience, we use 23°C, room temperature, as the temperature of choice.

Effect of IGF-1 Aggregates Concentration. It is known that protein folding yield is dependant on protein concentration. To minimize the volume required for the folding reaction, we investigated the effect of high IGF-1 concentrations. In one set of experiments, the concentrations of DTT and IGF-1 refractile particles were varied simultaneously while keeping the molar ratio of DTT to IGF-1 protein constant at 10. Solubility results revealed that the majority of the IGF-1 protein was solubilized, regardless of added DTT or IGF-1 aggregate concentrations. Therefore, as the concentration of added DTT and IGF-1 aggregates increased, the soluble IGF-1 concentration also increased (data not shown). IGF-1 folding was measured by HPLC analysis and results are shown in Fig. 9. The yield of the correctly folded IGF-1 form increased with increasing IGF-1 concentration and then plateaued at an initial refractile particle concentration of 1.5 mg/ml. The folding yield of IGF-1 was nearly constant at IGF-1 aggregate concentrations up to 4.5 mg/ml. The results suggest that by using higher concentrations of DTT and refractile particles simultaneously to increase the soluble IGF-1 concentration, one can minimize the folding volume and maintain a high specific folding yield.

The other two sets of experiments were performed by varying the concentration of IGF-1 refractile particles and keeping a constant DTT concentration at 1 mM or 2 mM (Fig. 9). The optimal concentration for IGF-1 folding was obtained with a refractile particle concentration of 1.5 mg/ml. The reduction in the folding yield at higher aggregate concentrations is consistent with the observation of elevated protein aggregation under these conditions.

We also observed a lowered folding yield in the control samples while studying the effect of IGF-1 concentration. As yet, we have no satisfactory explanation for the variation. However, our subsequent data supported the superior performance of folding with a fixed DTT to IGF-1 ratio.

Conclusions

We have developed a very simple, one step *in vitro* protein folding protocol to produce bio-active IGF-1. Our approach was to optimize the concentration of the denaturants so that protein solubilization and folding occurred in one step, and the denaturant removal or dilution step before protein folding could be eliminated. The optimized folding condition uses 2 M urea. By maintaining a DTT to IGF-1 molar ratio of 10, a relatively high protein concentration (4.5 mg/ml) can be used. The pH and choice of buffer exert dramatic effects on folding yield. We obtained the best results with Capso at pH 10.5. One molar NaCl and 20% MeOH were added to improve the formation of native IGF-1. For ease of operation, we chose 23°C, room temperature. Two to five hours of incubation is required to obtain maximal folding yield with these conditions. Using this protocol, we can take advantage of E. coli secretion to produce large quantities of bioactive IGF-1.

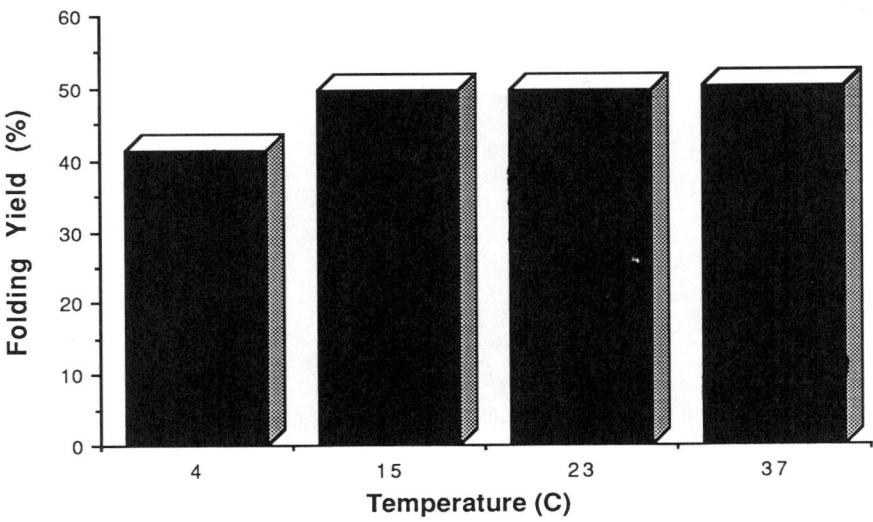

Fig. 8 Effect of temperature on IGF-1 folding. The data represent the maximal yield obtained at a given temperature.

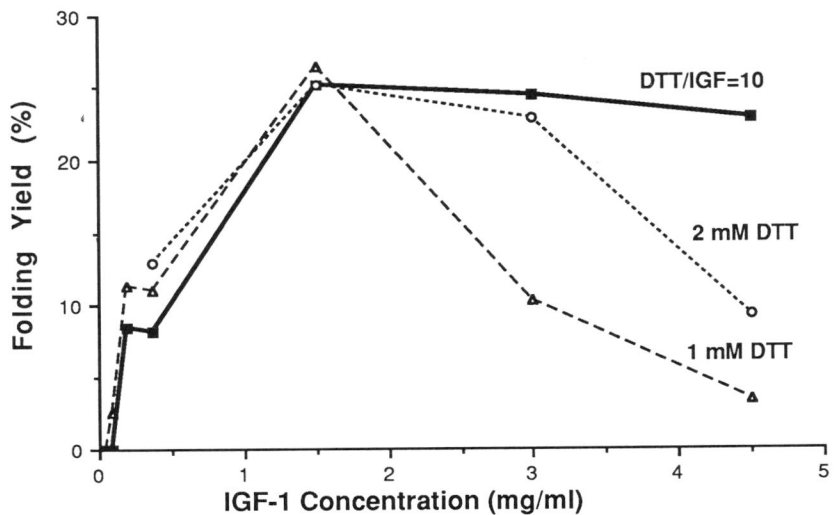

Fig. 9 Effect of DTT and IGF-1 concentration on folding. IGF-1 refractile particles were resuspended in 100 mM Capso, pH 10.5, containing 1 M NaCl, 2 M urea, and 20% MeOH. IGF-1 was allowed to fold at 23°C for 5 hours and analyzed.

Acknowledgements

This project was supported by numerous colleagues at Genentech, Inc. We would especially like to thank Nancy MacFarland for helping in strain construction, Dan Yansura and Laura Simmons for plasmid construction, Chuck Olson for HPLC assay advice, Marian Eng and Victor Ling for disulfide mapping analysis, Karl Clauser and Kathy O'Connell for Mass Spec analysis, Assay Services for RIA and Bioassay services, and Fermentation Operations for performing fermentations.

References

1. Wong, E., Seetharam, R., Kotts, C. E., Heeren, R. A., Klein, B. K., Braford, S. R., Mathis, K. J., Bishop, B. F., Siegel, N. R., Smith, C. E., and Tacon, W. C., *Gene* **1988**, *68*, 193-203.
2. Bowden, G. A., Paredes A. M., and Georgiou, G., *Bio/Tech.* **1991**, *9*, 725-730.
3. Silhavy, T. J., Benson, S. A., and Emr, S. D., *Microbiol. Rev.* **1983**, *47*, 313-344.
4. Taylor, G., Hoare, M., Gray, D. R., and Marsion, F. A. O., *Bio/Tech.* **1986**, *4*, 555-557.
5. Kumagai, I., Takeda, S., Hibino, T., and Miura, K., *Protein Engineering* **1990**, *3*, 449-452.
6. Winkler, M. E. and Blaber, M., *Biochem.* **1986**, *25*, 4041-4045.
7. Sarmientos, P., Duchesne, M., Denefle, P., Boiziau, J., Fromage, N., Delporte, N., Parker, F., Lelievre, Y., Mayaux, J., and Cartwright, T., *Bio/Tech.* **1989**, *7*, 495-501.
8. Canova-Davis, E., Eng, M., Mukku, V., Reifsnyder, D. H., Olson, C. V., and Ling, V. T., *Biochem. J.* **1992**, *285*, 207-213.
9. Cleland, J. L., Hedgepeth, C., and Wang, D. I. C., *J. Biol. Chem.* **1992**, *267*, 13327-13334.
10. Cleland, J. L., Builder, S. E., Swartz, J. R., Winkler, M., Chang, J. Y., and Wang, D. I. C, *Bio/Tech.* **1992**, *10*, 1013-1019.
11. Damodaran, S., *Biochem. Biophy. Acta* **1987**, *914*, 114-121.
12. Snyder, G. H., *J. Biol. Chem.* **1984**, *259*, 7468-7472.
13. Tamura, K., Hamada, H., Ishii, Y., Koyama, S., and Niwa, M., *11th American Peptide Symposium;* Abstract, **1989**.
14. Eagland, D., *in Water Relations of Foods;* Duckworth, R. B.; Academic Press: New York, NY, **1975**; p. 73.

RECEIVED October 26, 1992

Chapter 15

Role of Disulfide Bonds in Folding of Recombinant Human Granulocyte Colony Stimulating Factor Produced in *Escherichia coli*

Hsieng S. Lu, Christi L. Clogston, Lee Ann Merewether, Linda O. Narhi, and Thomas C. Boone

Amgen Inc., Amgen Center, Thousand Oaks, CA 91320

> Oxidative folding of recombinant human granulocyte stimulating factor produced in *E. coli* follows a sequential pathway: $I_1 \xrightarrow{fast} I_2 \xrightarrow{slow} N$, where I_1 and I_2 are the two folding intermediates and N, the folded, biologically active form. Based on their difference in hydrophobicity, these forms at various folding stages can be separated and quantified by reverse-phase HPLC. The folding can be accelerated by increasing the folding temperature or by adding copper sulfate as a catalyst. Isolation and characterization of the intermediates revealed that I_1 represents the partially folded but fully reduced intermediate form; I_2 has partially folded structure containing a single Cys^{36}-Cys^{42} disulfide bond; and N, the final folded form, has two disulfide bonds. I_1 and I_2 are less stable and conformationally different from form N. Our studies showed that in the folding process disulfide bond formation enables the reduced granulocyte colony stimulating factor to fold into a structurally correct and biologically active form.

It has been known that elucidating a protein folding pathway lies in measuring the structural properties of intermediate protein species, since they are usually short-lived. As described by Creighton and others (1-4), a model that includes an oxidative refolding of a disulfide-reduced protein allows one to investigate the pathway of protein folding in detail. The folding intermediates can be kinetically trapped by chemical modification at different disulfide-reduced states during folding. The resulting stabilized intermediates can then be structurally characterized to establish the role of disulfide bond formation in the folding of biologically active proteins. We have employed this similar approach to study the folding of recombinant human granulocyte colony stimulating factor produced in *E. coli*.

Granulocyte colony-stimulating factor (G-CSF) is one of the hemopoietic growth factors which play an important role in the stimulation, proliferation, and differentiation of hematopoietic progenitors, and are also required for activation of the mature cell functions (5,6). G-CSF is capable of supporting neutrophil proliferation *in vivo* and *in vitro* (7-10). The human G-CSF gene has been cloned and characterized (11,12). Large quantities of recombinant human G-CSF (hG-CSF) produced in genetically engineered *E. coli* have been successfully used in human clinical studies to treat neutropenic patients in a variety of clinical situations (13-17).

E. coli-produced recombinant hG-CSF is a 175 amino acid polypeptide chain containing an extra Met (at position -1) at its N-terminus (Fig. 1; trade name: Neupogen^{TM}). The molecule also contains a free cysteine at position 17 and two intramolecular disulfide bonds, Cys^{36}-Cys^{42} and Cys^{64}-Cys^{74} (18). The two disulfide bonds form two small loops which are separated by 21 amino acids. We have produced recombinant hG-CSF by *E. coli* direct expression at high expression levels. Like other bacteria-derived recombinant proteins, hG-CSF produced intracellularly in *E. coli* tends to precipitate, forming inclusion bodies. Since the insoluble form in the inclusion bodies is disulfide-reduced and not properly folded, the protein at this stage is not biologically active. Recovery and isolation of the biologically active hG-CSF thus requires implementation of procedures including solubilization of the inclusion bodies followed by a folding and oxidation step prior to any chromatographic separation. Practically, the identification of an optimized folding and oxidation procedure as well as detailed understanding of the folding pathway is beneficial to the overall manufacturing process. To accomplish this goal, we conducted the kinetic studies on the *in vitro* folding of reduced hG-CSF prepared from solubilized inclusion bodies. Biological and physicochemical characterization of the isolated folding intermediates and Cys->Ser analogs made by site-directed mutagenesis revealed that formation of disulfide bonds plays an important role during folding of hG-CSF. Some of the detailed experimental procedures not described here can be referred to in an earlier report (19).

RESULTS

Detection of hG-CSF Folding Intermediates and Folding Kinetics by reverse-phase HPLC. Due to the difference in hydrophobicity, the native, recombinant hG-CSF and its reduced, denatured form are well separated by reverse-phase (RP-) HPLC (19). The difference in the chromatographic elution times among different hG-CSF forms has allowed us to detect folding intermediates and to study folding kinetics by RP-HPLC.

Recombinant hG-CSF was produced by *E. coli* direct expression. After cell breakage, the inclusion bodies containing the recombinant protein were collected and extensively washed in 0.2 M Tris-HCl, pH 8.0 containing 1 mM EDTA. Approximately 35% of the proteins in inclusion bodies is rhG-CSF. The level of the impurity at this stage appears to have little effect on the folding studies.

In our earlier study, solubilization of hG-CSF from the inclusion bodies using Sarkosyl (laurylsarkosine) was found to be suitable (20). Similar results were also obtained by solubilization using 6 M guanidine hydrochloride (GdnHCl) or other denaturants as demonstrated in other recombinant proteins. However, GdnHCl and urea were not selected for the studies because both denaturants seem to cause hG-CSF to precipitate during folding studies. As presented here, the solubilization

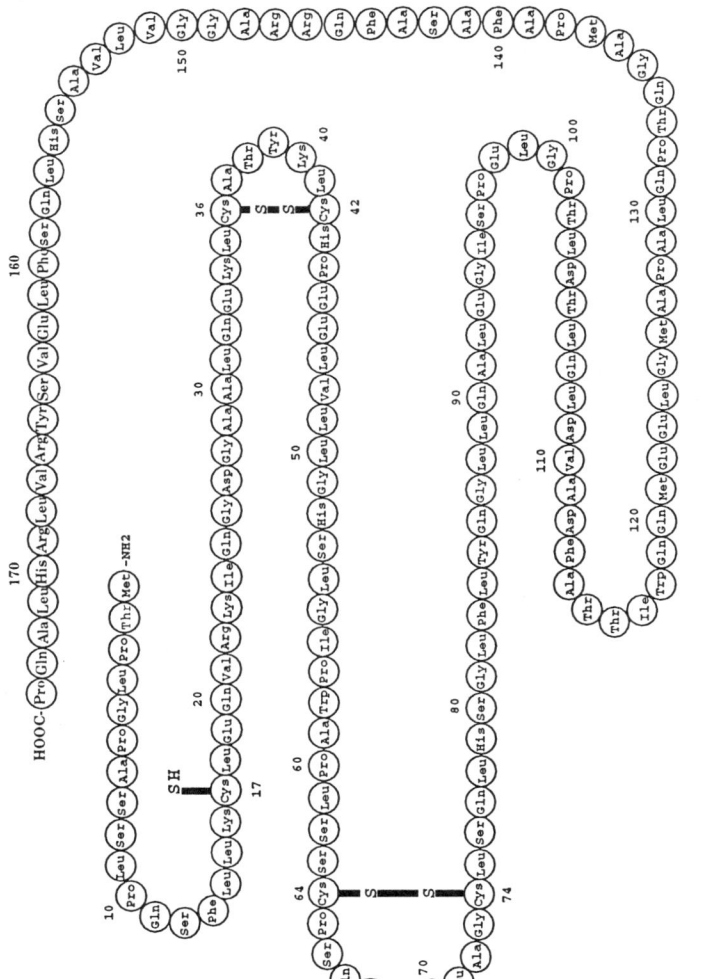

Figure 1. Covalent structure of recombinant methionyl human G-CSF. Two disulfide bonds are assigned at positions 36 and 42, 64 and 74, respectively. Cys[17] is a free thiol.

of the collected inclusion bodies was performed in 2% Sarkosyl and 50 mM Tris-HCl, pH 8.0 for 20 min. at 25° C. The protein concentration was kept at 1-2 mg/ml. The subsequent folding studies were performed at 25° C, 37° C, and at 25° C in the presence of 40 μM copper sulfate. Aliquots of different samples were taken at different time points for RP-HPLC. The folding rate was expressed by percent disappearance of the intermediates and formation of final oxidized form per second. Fig. 2 shows typical chromatographic profiles of the solubilized hG-CSF samples prepared at different folding times during incubation at 25° C in the presence of copper sulfate. The populations of the three major hG-CSF-related species, I_1, I_2, and N, change dramatically as a function of time. I_1 is defined as the intermediate form representing solubilized rhG-CSF. It is chromatographically different from the reduced, unfolded rhG-CSF in HPLC and will be discussed later. I_2 is the intermediate converted from I_1, and N is the folded, biologically active form. At 20 min, I_1, the starting reduced hG-CSF, contains 67% of the total population with the remainder being the I_2 form. At 1, 2, and 4 h, the generation of I_2 has proceeded further with the concomitant disappearance of I_1 and appearance of the final oxidized form N. Under such conditions, folding of N reaches its maximum in 12 h and is greater than 95% complete. Table 1 lists the initial first order folding rate for the conversion. The initial folding rate for conversion of I_1 into I_2 was estimated to be 1.9 x10^{-2} s^{-1} and for conversion of I_2 into N to be approximately 3.1 x10^{-3} s^{-1}.

Shown in Fig. 3 are the folding kinetics of hG-CSF at 25° C and 37° C without copper sulfate. The initial rate for I_2 formation at both temperatures are similar (1.0 x10^{-2} s^{-1}) but slower than the folding in the presence of copper sulfate (Table 1). However, at 25° C the generation of completely oxidized form N is relatively slow (6.6 x10^{-4} s^{-1}). In this case, I_2 persists much longer and approximately 20 h is required to reach half-maximal folding from I_2 to N versus 4.5 h in copper sulfate. At 37° C the folding from I_2 to N is faster than at 25°C without copper sulfate, but biphasic kinetics, which takes place at approximately 8 h, becomes apparent. This has decreased the folding rate of the completely oxidized form (reaching only 80% after 23 h folding). The folding studies thus suggest that copper sulfate catalysis is crucial for completion of hG-CSF oxidative folding. Folding at elevated temperature (37°C) is not recommended.

G-CSF Folding Intermediates Are Disulfide-reduced. Trapping experiments of the intermediates with an alkylating agent, [^{3}H]-C_2-iodoacetic acid, in 0.3 M Tris-HCl, pH 8.3 (20 min at 25° C) were performed to confirm if the intermediate forms detected by RP-HPLC contain disulfide reduced species. After alkylation, the resulting carboxymethylated derivatives contain more negatively charged groups than the oxidized form and are separable by cationic exchange HPLC. Fig. 4 illustrates the separation of the alkylated I_1 and I_2 derivatives from N by sulfoethyl polyaspartamide silica-based columns (4.6 x 200 mm, 5 μ) in 20 mM NaOAc, pH 5.4 by a 30-min linear gradient from 0 to 0.2 M NaCl at a flow rate of 1 ml/min. Table 2 lists the labeling results for N and the purified intermediates. In the absence and presence of the denaturant, GdnHCl, N gives 0 and 1 mole of label per mole of protein, respectively, suggesting that there is one free cysteine probably residing at a more hydrophobic environment. The fully reduced and denatured G-CSF gives 5 moles of label. The trapped I_1 and I_2 give 3.74 and 1.9 moles of label per moles of protein, respectively (Table 2). These results indicate that 4 and 2 free cysteinyl residues are not forming disulfide bonds since they are available for the labeling.

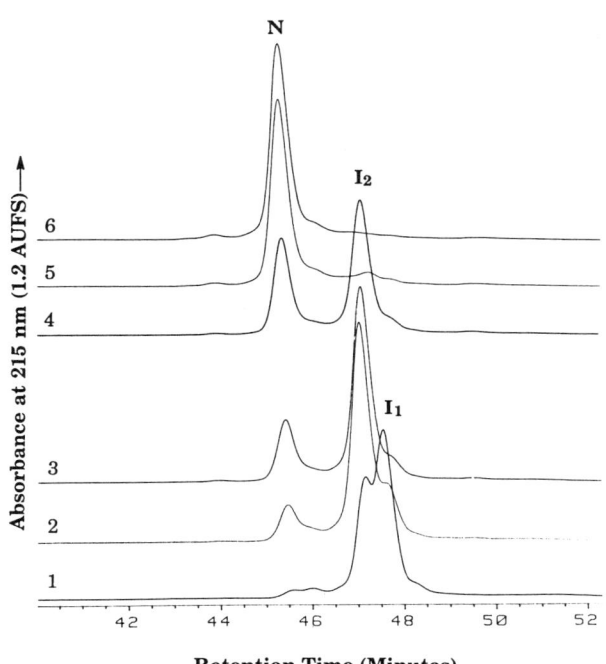

Figure 2. Folding of hG-CSF at 25°C in $CuSO_4$ monitored by RP-HPLC at different times. Chromatograms 1-6, 20 min, 1, 2, 4, 8, and 12 h, respectively. Reproduced with permission from reference 19. Copyright 1992 The American Society for Biochemistry & Molecular Biology.)

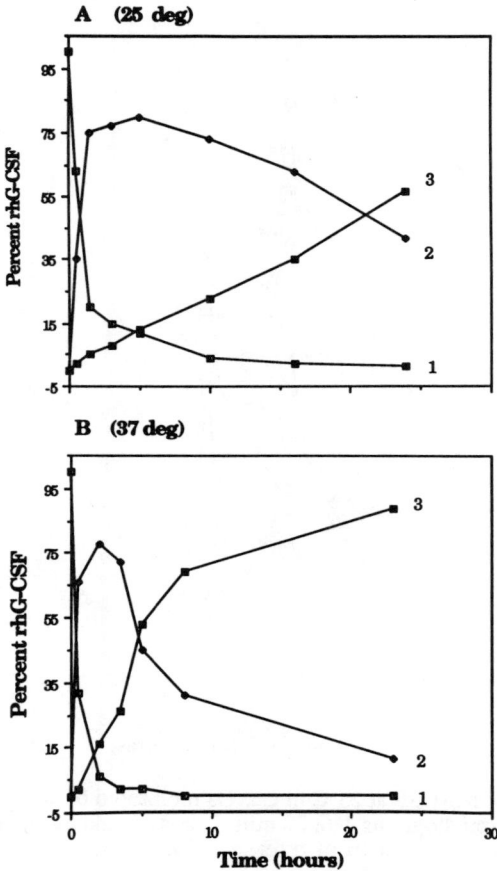

Figure 3. Folding kinetics of hG-CSF at 25°C (A) and 37°C (B) without $CuSO_4$. Curves 1-3, I_1, I_2, and N, respectively.

Table 1. Initial rate constant for disappearance of reduced forms and generation of oxidized form[1]

Folding Products	Conditions		
	$CuSO_4 + 25°C$	No $CuSO_4$, 25°C	No $CuSO_4$, 37°C
	(Percent/sec)		
$I_1 \rightarrow I_2$[2]	1.9×10^{-2}	1.0×10^{-2}	1.0×10^{-2}
$I_2 \rightarrow N$[3]	3.1×10^{-3}	6.6×10^{-4}	1.8×10^{-3}

[1] The initial rate was estimated from the first phase folding kinetics. The folding rate of the second folding phase is slower in all cases and not described here. The error is approximately ±5%.
[2] Estimated from disappearance of reduced form I_1 that corresponds to the formation of intermediate I_2.
[3] Generation of form N, the final oxidized form, corresponds to the disappearance of intermediate I_2.
SOURCE: Reprinted with permission from ref. 19. Copyright 1992 The American Society for Biochemistry & Molecular Biology.

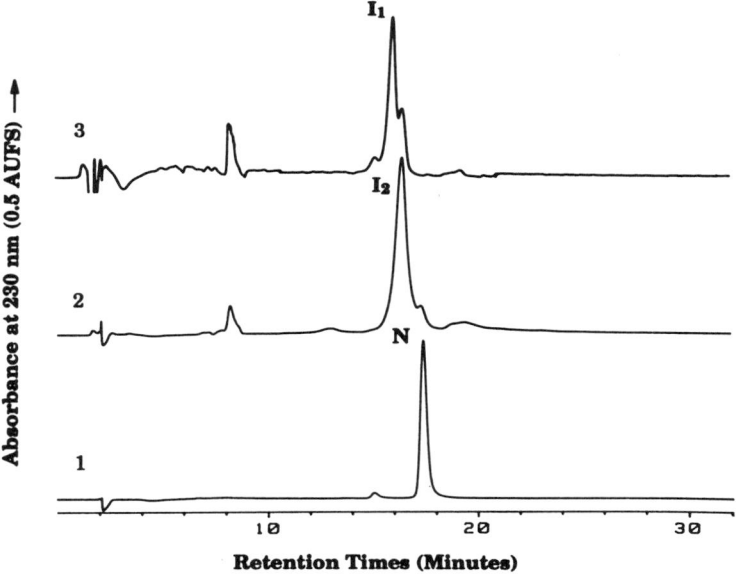

Figure 4. Sulfoethyl cationic exchange HPLC of folding intermediates obtained at 0, 1, and 12 h folding samples after a gel filtration desalting. The peak at 8 min is residual sarkosyl.

Table 2. Some properties of recombinant hG-CSF, folding intermediates, and analogs

Proteins	Iodoacetate labeling[1]	Number of disulfide bonds[6]	Biological activity (× 10⁶ units/mg)	$\Delta G_D^{H_2O}$ [3]	α-Helix (%)	Acid-induced Tyr fluorescence[5]
1. rhG-CSF	0.03 (0) 0.95 (1)[2]	2	100 (±20)	5.4	70	yes
2. I_1	3.74 (4)	0	3 (±1)	2.2	38	no
3. I_2	1.90 (2)	1	5 (±2)	2.6	35	no
4. Ser¹⁷ analog	0.05 (0)	2	100 (±20)	5.4	70	yes
5. Ser³⁶,⁴² analog	0.1 (0)	1	1 (±0.5)	2.6	70	no
6. Ser⁷⁴ analog	N.D.	1	3 (±1)	2.8	34	no

[1] Tritiated iodoacetate in 0.3 M Tris HCl, pH 8.4 was used.
[2] One mole of label was detected when the labeling was performed in 6M GdnHCl.
[3] Thermodynamic constant determined by GdnHCl denaturation, see ref. 15.
[4] Determined by circular dichroic spectropolarimetry at pH 3.5.
[5] hG-CSF exhibits a typical Tyr and Trp fluorescence spectrum at 304 nm and 344 nm at pH 3.0. The Tyr fluorescence disappears at neutral pH. Alkylated intermediates I1 and I2 were used for biological assay, CD, and fluorescence studies.
[6] The assignment of disulfide bonds were separately determined by peptide mapping of the native rhG-CSF after protease digestion (14).

The labeled derivatives were further reduced by dithiothreitol and alkylated with non-radioactive iodoacetate and subsequently subject to *Staphylococcal aureus* protease (V8 strain) digestion and HPLC peptide mapping as described (*19*). Structural analysis of the peptide fractions recovered from the HPLC confirmed that I_1 contains all five free cysteines suggesting that no disulfide formation in this intermediate form. I_2 has a Cys^{36}-Cys^{42} bond, but with both Cys^{64} and Cys^{74} remaining reduced. N has two disulfide bonds, Cys^{36}-Cys^{42} and Cys^{64}-Cys^{74}. Cys^{17} in I_1, I_2, or N is inaccessible to iodoacetate labeling unless the modification is performed in the presence of a denaturant.

G-CSF Analogs Missing Disulfide Bonds. For structural characterization, the iodoacetate-trapped intermediates do not represent the final folded forms. We then prepared and characterized hG-CSF analogs lacking a single disulfide bond. The analysis of the analogs can corroborate the findings obtained from physicochemical characterization of the isolated intermediates since analogs can be folded and oxidized to the respective stable oxidized states and isolated by hG-CSF purification procedures.

Both $Ser^{36,42}$ and Ser^{74} analogs were prepared by site-directed mutagenesis. The replacement of Cys at these positions produces the analogs missing the Cys^{36}-Cys^{42} or Cys^{64}-Cys^{74} disulfide bond (Table 2). It is interesting to note that both analogs had correct disulfide bond formation but exhibited very slow folding in the absence of copper sulfate (several days) and were recovered. The Ser^{74} analog folded more rapidly than the $Ser^{36,42}$ analog, but folded at least four times slower than formation of I_2 from I_1 in hG-CSF in the absence of Cu^{2+} at 25°C. RP-HPLC retention time of $Ser^{36,42}$ analog is similar to the wild type hG-CSF. The Ser^{74} analog coeluted with I_2, suggesting that both molecules lacking the same Cys^{64}-Cys^{74} disulfide bond have similar hydrophobicities.

For comparative analysis, an analog with Ser replacing Cys^{17} was also prepared. The Ser^{17} analog exhibited oxidation and folding rate similar to those of the wild type hG-CSF and could be isolated with equivalent recovery. As seen in Table 2, in iodoacetate labeling experiments, Ser^{17} and $Ser^{36,42}$ analogs give 0.05 and 0.1 mol of label per mol of protein, suggesting the formation of disulfide bonds in G-CSF analogs. The correct disulfide linkages for each isolated analog were further confirmed by peptide map analysis (not shown).

Properties of hG-CSF, Intermediates, and Analogs. Table 2 also summarizes the physicochemical and biological properties of hG-CSF, the intermediates, and analogs. Stimulation of bone marrow cell proliferation by G-CSF was performed by a specific *in vitro* [^{3}H]-thymidine uptake assay (*19*). Wild type hG-CSF usually exhibits a specific biological activity of 1 x10^8 units/mg. We found that intermediates I_1 and I_2, $Ser^{36,42}$ analog, and Ser^{74} analog, all missing either one or two disulfide bonds, have biological activity less than 5% of the wild type molecule. In contrast, Ser^{17} analog is as active as the wild type hG-CSF. These results thus suggest that elimination of a single disulfide bond structure appreciably decreases hG-CSF biological activity.

Conformational stability of hG-CSF, the intermediates, and analogs were evaluated by GdnHCl denaturation studies (*19*). The denaturation profile of wild type

G-CSF approximates a simple two-state denaturation transition when the absorption changes at 290 nm were measured (the GdnHCl induced denaturation was irreversible). The midpoint denaturation transition occurred at approximately 3 M GdnHCl. The denaturation of hG-CSF detected by the change of absorbance at 290 nm is also coincident with the GdnHCl-induced accessibility of Cys[17] by DTNB titration. The thermodynamic constant of hG-CSF ($\Delta G_D^{H_2O}$) determined by GdnHCl denaturation is about 5.4 Kcal/mole. As shown in Table 1, the folding intermediates and analogs (except Ser[17] analog) have much lower denaturation energy (2.2-2.8 Kcal/mole), suggesting that these molecular species are easily denatured and thus conformationally less stable than the wild type molecule and Ser[17] analog.

In CD analysis, all of the molecules tested have α-helical structures. However, the native molecule and Ser[36,42] analog contains substantially higher helical content than the other species examined (Table 2). I_1, I_2, and Ser[74] contains approximately one half of the helical structures when compared to the native hG-CSF, suggesting that the absence of Cys[64]-Cys[74] disulfide bond results in partial recovery of G-CSF native structure.

In fluorescence studies, no significant differences were observed among different G-CSF species analyzed at neutral pH. For the native hG-CSF, the fluorescence emission spectrum is characterized by a single peak with a maximum at 344 nm, typical of somewhat solvent-exposed Trp. There is no detectable Tyr spectrum around 300 nm. The spectra of the two analogs and intermediates are similar to the native hG-CSF, but exhibit higher intensities, reflecting differences in the environment of the two Trp residues.

As shown in Fig 5, the fluorescence spectrum of native G-CSF molecule or Ser[17] analog at lower pH (~3.0) is clearly distinguished from other species tested. The Trp peak of native hG-CSF at a similar wavelength maximum (340 nm) is still evident, but is greatly decreased in intensity, and a peak at 304 nm, attributable to Tyr, is also present. This suggests that the molecule has undergone a reversible change in conformation so that energy transfer from Tyr to Trp no longer occurs. However, such change only occurs for the native molecule or Ser[17] analog. The other species shows a decrease in the intensity of Trp fluorescence, but no change in the Tyr fluorescence. It should be noted that there is no shift in emission wavelength maximum observed for any of the trapped intermediates and analogs at both neutral and acidic pH's. The above data thus conclude that tertiary structure of intermediates I_1 and I_2 as well the two analogs, lacking either disulfide bond, is different from that of the native molecule.

DISCUSSION AND CONCLUSION

Folding Pathway of hG-CSF Is Sequential. Recombinant hG-CSF solubilized from the inclusion bodies is disulfide-reduced and can refold spontaneously in the presence of the detergent, Sarkosyl, by air oxidation. Folding and oxidation of hG-CSF at room temperature by the catalysis of copper sulfate seems to be optimal. The copper ion concentration used for G-CSF oxidation is in the range of 20-40 μM (20). Acceleration of thiol oxidation by copper ion in protein refolding has been reported earlier (21-24). The kinetic data shown here indicates that copper ion increases the folding rate of I_2 from I_1 approximately twofold and the folding rate of N from I_2 5-fold. Copper also promotes oxidation to greater than 95% completion at 25° C. Increasing the temperature can also accelerate folding and oxidation of hG-CSF, but does not appear to increase the rate of the second phase of folding. The kinetic studies and the detection of intermediate forms suggest that the mechanistic pathway for hG-CSF folding under the described conditions is sequential: $I_1 \longrightarrow I_2 \longrightarrow N$. By comparing the folding rates described in Table 1, folding of I_2 from I_1, involving the formation of Cys^{36}-Cys^{42} bond, is very fast. The folding of N from I_2, involving the formation of Cys^{64}-Cys^{74} bond, is rate-limiting.

Similar to the model system of the folding of bovine pancreatic trypsin inhibitor as described by Creighton (1-4), the disulfide-reduced intermediates during hG-CSF folding can be isolated and studied in detail. The intermediate I_1 is fully reduced but contains considerable amount of secondary and tertiary structure. I_2 has a single correct disulfide (Cys^{36}-Cys^{42}) and also contains a moderate amount of native-like structure. However, both intermediates are structurally different from the completely folded form and less stable. Since by HPLC analysis folding intermediates that contain non-native disulfide bonds were not evident, random disulfide formation among the free cysteines is less likely to occur. Intermolecular disulfide formation that can cause molecular aggregation was also not detected during these folding studies.

Role of Disulfide Bonds. The role of disulfide bonds on the structure of G-CSF molecule can be assessed by characterization of the G-CSF analogs. Since the folding, as well as Cys^{64}-Cys^{74} disulfide formation of the $Ser^{36,42}$ analog is slower than the native hG-CSF, the formation of Cys^{36}-Cys^{42} disulfide bond during folding of the wild type G-CSF seems to be important for providing the initial stabilization force for the conformational integrity of the molecule. As detected by far UV-CD analysis, the spectrum of the $Ser^{36,42}$ analog resembles that obtained for the native molecule (~70% helix). The high helical content for $Cys^{36,42}$ analog suggests that the formation of G-CSF secondary structure may be independent of Cys^{36}-Cys^{42} disulfide formation. In contrast, the Ser^{74} analog and I_2, missing Cys^{64}-Cys^{74} disulfide bond, have similar hydrophobicities (by HPLC) and have similar secondary (by CD analysis) and tertiary (by fluorescence analysis) structure. The lower α-helical content observed for both the Ser^{74} analog and I_2 (see Table 1) may suggest that the formation of Cys^{64}-Cys^{74} bond is critical for complete folding of hG-CSF to its oxidized form.

As evidenced from the biological activity of intermediates and analogs, it is apparent that loss of one of the two disulfide bonds in G-CSF molecule dramatically decreases G-CSF stimulatory effect on the proliferation of the isolated bone marrow cells. Since the absence of disulfide bonds in the intermediates or analogs causes conformational change in secondary and/or tertiary structures of hG-CSF, the perturbed structures may lead to these molecules to not tightly bind to the G-CSF receptor and to elicit lower biological activity. A recent report has identified that the functional domain of hG-CSF molecule, responsible for binding to its receptor, is at residues 20-46 and the COOH terminus, as determined by specific neutralizing antibodies (25).

Cys^{17} Is at the N-Terminal Helix. Recently, crystals of recombinant hG-CSF molecule have been successfully prepared. The initial X-ray crystallographic data confirms that G-CSF is abundant in α-helical structures forming an

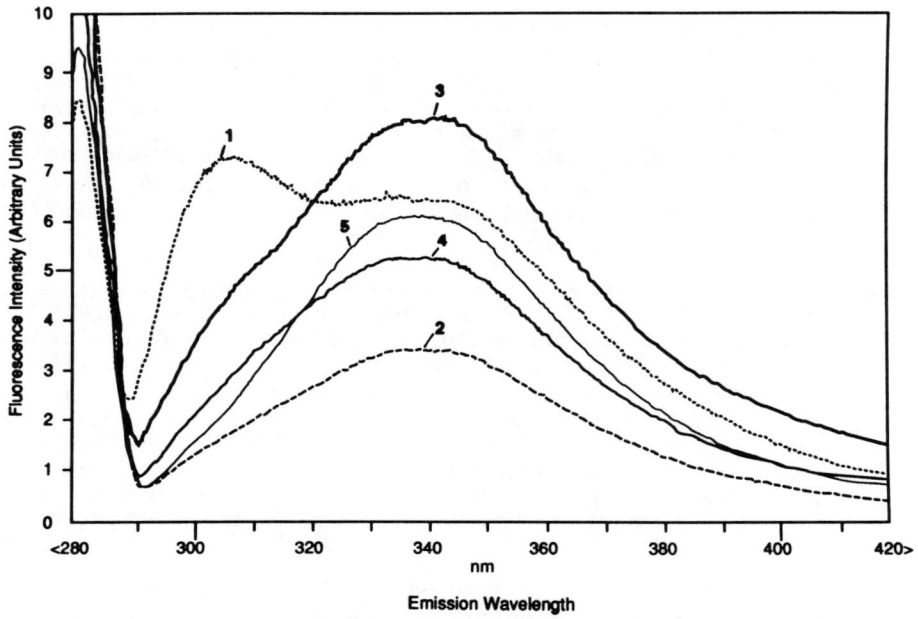

Figure 5. Fluorescence spectra of hG-CSF species at pH 3.2 spectra 1-5: hG-CSF Ser[74] analog, Ser[36,42] analog, I_2 derivative, and I_1 derivative, respectively. Reproduced with permission from reference 19. Copyright 1992 The American Society for Biochemistry & Molecular Biology.)

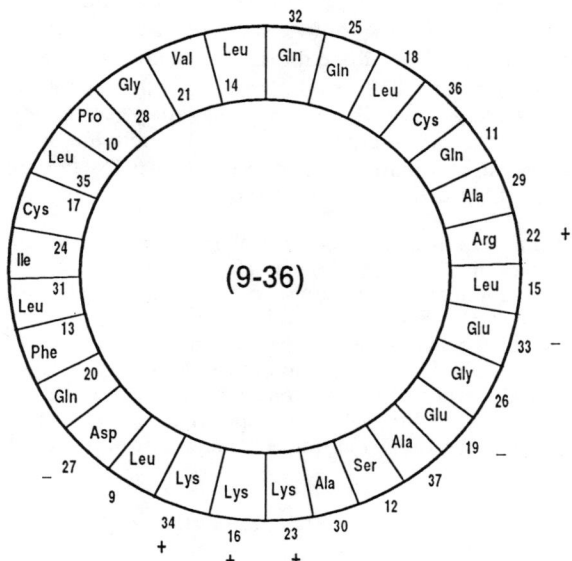

Figure 6. Helical wheel for the predicted N-terminal helix of hG-CF at residues 9-36. Cys[17] resides on the hydrophobic area (upper left).

antiparallel four helical bundle (Tim Osslund and David Eisenberg, submitted for publication), typical of other known proteins such as human growth hormone (26), and granulocyte-macrophage colony stimulating factor (27). Predicted secondary structure of G-CSF revealed that the N-terminal peptide segment, Leu^9 to Ala^{36}, has a high propensity to form an α-helix (14), which is consistent with the deduced X-ray structure. Upon formation of this helical structure, Cys^{17} may reside at the end of the second helical turn, and become oriented in the hydrophobic environment (Fig. 6), which is in close contact with other helices via hydrophobic interaction. The fact that Cys^{17} in I_1 is not available for iodoacetate labeling suggests that the folding of this N-terminal helix occurs earlier than formation of the first disulfide bond. Folding of the secondary structure near the N-terminus together with some tertiary folding observed for I_1 provides some evidence that a correlated, partially folded conformation in I_1 is required for the subsequent folding and for the formation of intramolecular disulfide bonds.

Conclusion. The folding process of hG-CSF was followed in detail by kinetic studies. Unique intermediate forms can be identified and analogs that mimic the intermediates were prepared for further characterization. We have observed that the Ser^{74} analog mimics I_2; however, none of the produced analogs mimic I_1. Detailed folding studies of hG-CSF has greatly impacted the development of hG-CSF manufacturing. In the production of large quantities of rhG-CSF, every process step has to be maximally optimized to obtain a desirable purity and satisfactory recovery yield. Moreover, consistency and reproducibility of the manufacturing processes such as folding and chromatographic separations also have to be carefully validated and characterized. The identification of optimal folding conditions increases the overall yield for the fully oxidized and biologically active hG-CSF form. The efficient folding is also critical to eliminate undesired, partially oxidized intermediate forms in hG-CSF preparation, as large-scale chromatographic separations usually fail to resolve such molecular species with high structural similarity.

ACKNOWLEDGMENT

We are indebted to Joan Bennett for her help in typing this manuscript.

REFERENCES

1. Creighton, T.E. *Methods Enzymol.* **1986**, *131*, 83-106.
2. Creighton, T.E. *J. Mol. Biol.* **1977**, *113*, 273-293.
3. Creighton, T.C. *J. Mol. Biol.* **1977**, *113*, 295-312.
4. Creighton, T.C.; Goldenberg, D.P. *J. Mol. Biol.* **1984**, *179*, 497-526.
5. Metcalf, D.; *The Hematopoietic Colony Stimulating Factors* **1984**, Elsevier, Amsterdam.
6. Morstyn, G.; Burgess, A.W. *Cancer Res.* **1988**, *48*, 5624-5637.
7. Zsebo, K.M.; Cohen, A.M.; Murdock, D.C.; Boone, T.C.; Inoue, H.; Chazin, V.R.; Hines, D.; Souza, L.M. *Immunobiology* **1986**, *172*, 175-184.
8. Cohen, A.M.; Zsebo, K.M.; Inoue, H.; Hines, D.; Boone, T.C.; Chazin, V.R.; Tsai, L.; Ritch, T.; Souza, L.M. *Proc. Natl. Acad. Sci. USA* **1987**, *84*, 2484-2488.

9. Tamura, M.; Hattori, K.; Nomura, H.; Oheda, M.; Kubota, N.; Imazeki, I.; Ono, M.; Ueyama, Y.; Nagata, S.; Shirafuji, N.; Asano, S. *Biochem. Biophys. Res. Commun.* **1987**, *142*, 454-460.
10. Welte, K.; Bonilla, M.A.; Gillio, A.P.; Boone, T.C.; Potter, G.K.; Gabrilove, J.L.; Moore, M.A.S.; O'Reilly, R.J.; Souza, L.M. *J. Exp. Med.* **1987**, *165*, 941-948.
11. Souza, L.M.; Boone, T.C.; Gabrilove, J.; Lai, P.H.; Zsebo, K.M.; Murdock, D.C.; Chazin, V.R.; Bruszewski, J.; Lu, H.; Chen, K.K.; Barendt, J.; Platzer, E.; Moore, M.A.S.; Mertelsmann, R.; Welte, K. *Science* **1986**, *232*, 61-65.
12. Nagata, S.; Tsuchiya, M.; Asano, S.; Kaziro, Y.; Yamazaki, T.; Yamamoto, O.; Hirata, Y.; Kubota, N.; Oheda, M.; Nomura, H.; Ono, M. *Nature* **1986** *319*, 415-418.
13. Bronchud, M.H.; Scarffe, J.H.; Thatcher, N.; Crowther, D.; Souza, L.M.; Alton, N.K.; Testa, N.G.; Dexter, T.M. *Brit. J. Cancer* **1987**, *56*, 809-813.
14. Morstyn, G.; Souza, L.M.; Keech, J.; Sheridan, W.; Campbell, L.; Alton, N.K.; Green, M.; Metcalf, D.; Fox, R. *Lancet* **1988**, *1*, 667-672.
15. Gabrilove, J.L.; Jakubowski, A.; Scher, H.; Sternberg, C.; Wong, G.; Grous, J.; Yagoda, A.; Fain, K.; Moore, M.A.S.; Clarkson, B.; Oettgen, H.; Alton, N.K.; Welte, K.; Souza, L.M. *New Engl. J. Med.* **1988**, *318*, 1414-1422.
16. Bonilla, M.A.; Gillio, A.P.; Ruggeiro, M.; Kernan, N.A.; Brochstein, J.A.; Abboud, M.; Fumagalli, L.; Vincent, M.; Gabrilove, J.L.; Welte, K.; Souza, L.M.; O'Reilly, R.J. *N. Engl. J. Med.* **1989**, *320*, 1574-1580.
17. Sheridan, W.P.; Morstyn, G.; Wolf, M.; Dodds, A.; Lusk, J.; Maher, D.; Layton, J.E.; Green, M.D.; Souza, L.; Fox, R.M. *Lancet* **1989**, *2*, 891-895.
18. Lu, H.S.; Boone, T.C., Souza, L.M.; Lai, P.H. *Arch. Biochem. Biophys.* **1989**, *268*, 81-92.
19. Lu, H.S.; Clogston, C.L.; Merewether, L.M.; Pearl, W.R.; Boone, T.C. *J. Biol. Chem.* **1992**, *267*, 8770-8777.
20. Souza, L.M. **1989**, United States Patent No. 4, 810, 643.
21. Cecil, R.; McPhee, J.R. *Adv. Protein Chem.* **1959**, *14*, 255-389.
22. Takagi, T.; Isemura, T. *Biochem. Biophys. Res. Commun.* **1963**, *13*, 353-359.
23. Yutani, K.; Yutani, A.; Imanishi, A.; Isemura, T. *J. Biochem. (Tokyo)* **1968**, *64*, 449-455.
24. Saxena, P.; Wetlaufer, D.B. *Biochemistry* **1970**, *9*, 5015-5022.
25. Layton, J.E.; Morstyn, G.; Fabri, L.J.; Reid, G.E.; Burgess, A.W.; Simpson, R.J.; Nice, E.C. *J. Biol. Chem.* **1991**, *266*, 23815-23823.
26. Cunningham, B.C.; Ultsch, M.; DeVos, A.M.; Mulkerrin, M.G.; Clauser, K.R.; Wells, J.A. *Science* **1991**, *254*, 821-825.
27. Diederichs, K.; Jacques, S.; Boone, T.; Karplus, P.A. *J. Mol. Biol.* **1991**, *221*, 55-60.

RECEIVED October 26, 1992

Chapter 16

In Vivo Expression of Correctly Folded Antibody Fragments from Microorganisms

Marc Better and Arnold H. Horwitz

XOMA Corporation, 1545 17th Street, Santa Monica, CA 90404

Antibody fragments, such as Fab, Fab', and F(ab')$_2$ are useful targeting agents for treatment and diagnosis of human disease. Co-expression of antibody light chain and truncated heavy chain genes in the bacterium *Escherichia coli* or the yeast *Saccharomyces cerevisiae* allows the production of assembled, correctly folded and active antibody fragments. Linkage of these peptide chains to a signal sequence directs transport through the cytoplasmic membrane, and antibody fragment accumulates in the culture medium. Fermentation technology results in high-level expression of antibody fragments making applications that require large quantities of antibody fragments feasible.

Expression of heterologous proteins in microorganisms provides the opportunity to synthesize large quantities of recombinant material outside of the native organism. In *Escherichia coli*, there are many well characterized expression systems available, and it is possible to localize foreign proteins into a variety of cellular compartments. For example, a protein can be expressed intracellularly where it may remain soluble or aggregate into inclusion bodies. Alternatively, a foreign protein may be linked to a signal sequence and become translocated across the cytoplasmic membrane where it may accumulate in the periplasmic space or, in some cases, in the cell culture supernatant. Under some conditions, proteins may also aggregate into insoluble particles in the periplasm *(1)*. Each of these cellular locations may offer potential production advantages.

Protein expressed intracellularly in an active form can be recovered directly from disrupted cells, but may be difficult to purify from the large number of other cellular proteins or be produced in low yields. Recombinant proteins expressed at high levels (often 10 to 30% of the total cell protein) frequently form insoluble inclusion bodies which may be purified from other cellular components easily, but must then be refolded to an active form. Secreted proteins often fold properly and can sometimes be produced at high yield.

The earliest demonstration of antibody fragment Fab production from *E. coli* relied on intracellular production of the each antibody chain separately as insoluble inclusion bodies with subsequent refolding *in vitro* to an active form *(2)*. This approach had a yield of about 1.4% at 25 µg/ml and was impractical for preparation of large quantities of recombinant material. By the mid-1980's, however, it became clear that disulfide bond formation of heterologous proteins could occur in the environment of the *E. coli* periplasm, and that bacterial signal sequences could direct some heterologous proteins through the cytoplasmic membrane with proper signal peptidase processing *(3, 4)*. We demonstrated in 1988 that assembled, active antibody fragment (Fab) could be secreted from *E. coli (5)*. Skerra and Plückthun *(6)* made a similar observation with the smaller Fv fragment. Successful production of active Fab by secretion from *E. coli* was particularly striking, as it required the proper formation of four intrachain and one interchain disulfide bonds. This expression of Fab was the first demonstration that a complex heterodimeric eukaryotic protein could be properly folded by *E. coli*. We subsequently demonstrated that Fab could be secreted in an active form from the yeast *Saccharomyces cerevisiae* as well *(7)*.

Antibody fragments produced at high yield in an active form from microorganisms have a variety of potential uses, especially as human pharmaceuticals. Antibody fragments are expected to penetrate tissue more easily than whole antibodies due to their smaller size, and may make ideal targeting agents for reagents such as drugs, toxins or radio nuclides. Antibody fragments linked to radio nuclides are also attractive as *in vivo* imaging agents for diagnosis of occult tumors *(8)*. Specific targeting with rapid blood clearance is a particularly attractive feature for this application, since it can result in a high tumor-to-blood ratio.

Microbially-produced antibody fragments can be designed in a wide variety of forms, each with their own particular set of characteristics. Those forms successfully produced in *E. coli* include Fv, Fab, Fab', F(ab')$_2$ and single chain antibodies. The Fv fragment is the smallest unit containing a complete binding site, and is a noncovalently-associated heterodimer of heavy and light chain variable domains. A single chain antibody is an Fv with a flexible peptide linker between heavy and light variable domains. A Fab contains the entire light chain plus the variable and first constant domain of a heavy chain. The Fab' is a Fab with the heavy chain extended to include one or more hinge region cysteine residues. The larger F(ab')$_2$ contains two antibody Fab' arms linked via hinge cysteine residues. Some of these forms have been expressed as gene fusions to cytotoxic proteins as well *(9, 10)*.

Here we demonstrate that correctly folded, active antibody fragments can be secreted from either *E. coli* or yeast at high yield, and that Fab, Fab' or F(ab')$_2$ can be purified easily from fermentation supernatants. *E. coli* components such as DNA and endotoxin can be removed from purified antibody fragments, and purified material is suitable for pharmaceutical development. We have also examined some of the physiological aspects of Fab expression in *E. coli* and yeast, and find that expression in *E. coli* can be limited by translation, while expression in yeast can be limited by secretion. Finally, we demonstrate that specific immunotoxins with the 30 kD subunit of the ricin A chain can be generated with Fab' and F(ab')$_2$, and that each of these has unique and desirable cytotoxic properties.

Construction of Fd and Fd' Modules

Production of Fab and Fab', (or F(ab')$_2$) molecules requires the co-expression of genes encoding the light chain along with heavy chain fragments designated as Fd and Fd', respectively. For Fab expression, we previously described an Fd module that incorporates a variable region gene sequence fused to a human C_H1 and a truncated hinge region *(5)*. A termination codon was introduced at the position of the first inter-heavy chain cysteine residue in the hinge. The only hinge cysteine available for inter-chain disulfide bonding is the residue which forms a disulfide pair with the ultimate cysteine in the kappa chain. For Fab', and/or F(ab')$_2$ expression, we have constructed Fd' modules which incorporate additional hinge-region residues and include either one (Fd'-1C) or both (Fd'-2C) of the human IgG1 cysteine residues involved in inter-heavy chain disulfide formation. The Fd'-1C module incorporates a termination codon at the position of the second inter-heavy chain cysteine residue, while the Fd'-2C module includes the entire hinge-region and the first eight amino acids of C_H2. The amino acid sequences at the 3' end of these three Fd modules are shown in Figure 1.

The constant region segments of the Fd or Fd' genes were constructed in a convenient module so that linkage to a variety of V-regions was practical. Restriction sites were introduced into the J-region of the Fd modules (*Bst*EII), the J-region of kappa (*Hind*III) and at the 3'-end of each constant region gene. Using these sites, different variable regions (mouse, human, "humanized", etc.) are readily fused to the human light chain or Fd/Fd' constant region modules. This can generate a chimeric antibody where the variable and constant regions which fold into structurally distinct domains are from two different species.

Secretion of the chimeric Fab and Fab' from microorganisms is most efficiently accomplished by fusing a species-homologous signal sequence to the mature light chain and Fd/Fd' chain protein sequences. For this purpose, the 5' end of DNA encoding the mature sequences for engineered light chain and Fd or Fd' chains was modified to allow an in-frame fusion with the desired signal sequence. This was accomplished either by incorporating a restriction site at the 5' end of these genes to prepare a blunt end, or by use of appropriate PCR primers that directly generate a blunt 5' end.

Expression of Fab, Fab' and F(ab')$_2$ in *E. coli*

For expression of antibody fragments in *E. coli,* we have positioned the kappa and truncated heavy chain (Fd) genes in a single transcription unit under the control of a strong inducible promoter. Each of the genes has been fused to the *pel*B *(11)* leader sequence (from the pectate lyase gene from *Erwinia carotovora*) to direct protein translocation through the *E. coli* cytoplasmic membrane. The *pel*B leader was chosen since it was known that pectate lyase could be expressed at a high level in *E. coli* and that the protein accumulated in the periplasmic space. The *ara*BAD promoter from *Salmonella typhimurium (12)* was used to control gene expression. This promoter is tightly repressed in uninduced *E. coli* cultures and is expressed at high levels upon induction with L-arabinose.

The general features of this expression system for Fab are outlined in Figure 2. Upon L-arabinose induction of an *E. coli* culture containing the expression plasmid, antibody genes are transcribed in a single dicistronic message, and Fab accumulates in the culture supernatant. The exact mechanism by which Fab escapes from the bacterial periplasmic space into the medium is not known, but this phenomenon allows facile purification of antibody fragment from the culture medium.

Fab purified from bacterial culture medium is active and has the same antigen binding activity as Fab generated from whole antibody by papain digestion *(5)*. This can be demonstrated by direct or competition binding experiments. In all cases tested (greater than 15 antigen specificities), bacterial Fab bound to the appropriate target antigen and competed with intact antibody for antigen binding. When analyzed on SDS polyacrylamide gels in the absence of reducing agent, bacterial Fab migrated as a single band of about 45,000 kDa. When reduced with dithiothreitol, Fab migrated as a doublet of about 23,000 kDa. These observations indicated that Fab released from bacteria is structurally and functionally equivalent to the Fab arm of a whole antibody.

The amount of Fab that accumulates in the culture medium can vary from one antibody specificity to another. Typically upon induction of Fab in a shake flask culture, product yield can vary between 0.1 and 2.0 µg/ml. An example of this variability for two bacterially-produced Fab fragments is shown in Figure 3. For the purposes of this illustration, expression of 2 µg/ml is referred to as "good", while expression at 0.1 µg/ml is referred to as "poor." Accumulation of both light chain and heavy chain RNA and protein were followed after culture induction with L-arabinose. RNA could be detected very soon after induction in both a "good" and "poor" Fab producer and accumulated to the same steady state levels *in vivo*, suggesting that expression by the "poor" Fab producer is not limited by transcription. The Fab both associated with intact cells (cell cytoplasm plus periplasm) and free in the culture medium was detected by ELISA and immunostaining of protein slot blots. Interestingly, the total amount of Fab protein present varied widely in the two cultures, while the distribution of Fab between cells and culture medium (percent of total associated with each compartment) was the same in both a "good" and "poor" producer. This suggests that secretion is not limiting and that translation of protein is the limiting step. A pulse-chase experiment with ^{35}S cysteine demonstrated that the rate of Fab protein turnover is the same in a "good" and "poor" producer (data not shown), thus "poor" expression is not due to increased turnover of Fab. In shake flask cultures, only a small percentage of Fab was released from the cells (as we have previously shown *(13)*) in both the "good" and "poor" producer. Failure of the majority of Fab to be released from the cells is less apparent, however, when cells are grown in a fermenter where oxygen and nutrient feed can be more carefully controlled. Strategies for increasing the expression of Fab might be best aimed at increasing the level of translation.

E. coli transformed with plasmids containing the light chain and Fd or Fd'-1C produced only Fab or Fab', respectively, while those containing light chain and

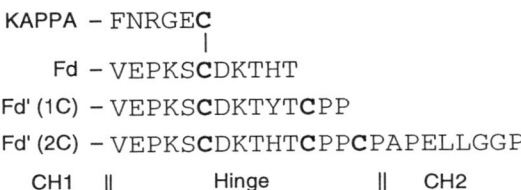

Figure 1. Amino acid structure at carboxy terminal end of Fd and Fd' modules. Module Fd'-1C encodes a single inter-heavy chain cysteine while Fd'-2C encodes both inter-heavy chain cysteines plus an additional nine amino acids. The 3' end of the light chain is shown, as is the location of the Fd'-kappa inter-chain bond.

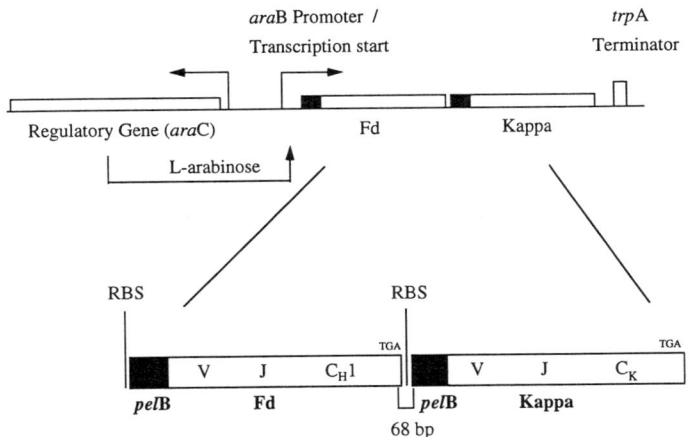

Figure 2. *E. coli* Fab expression vector features. Shown is a schematic view of the Fab transcription unit. The relative positions of the *ara*C regulatory gene, *ara*B promoter, Fd and kappa genes, *trp*A transcriptional terminator, *pel*B leader sequences, and the *pel*B ribosome binding sites (RBS) are shown.

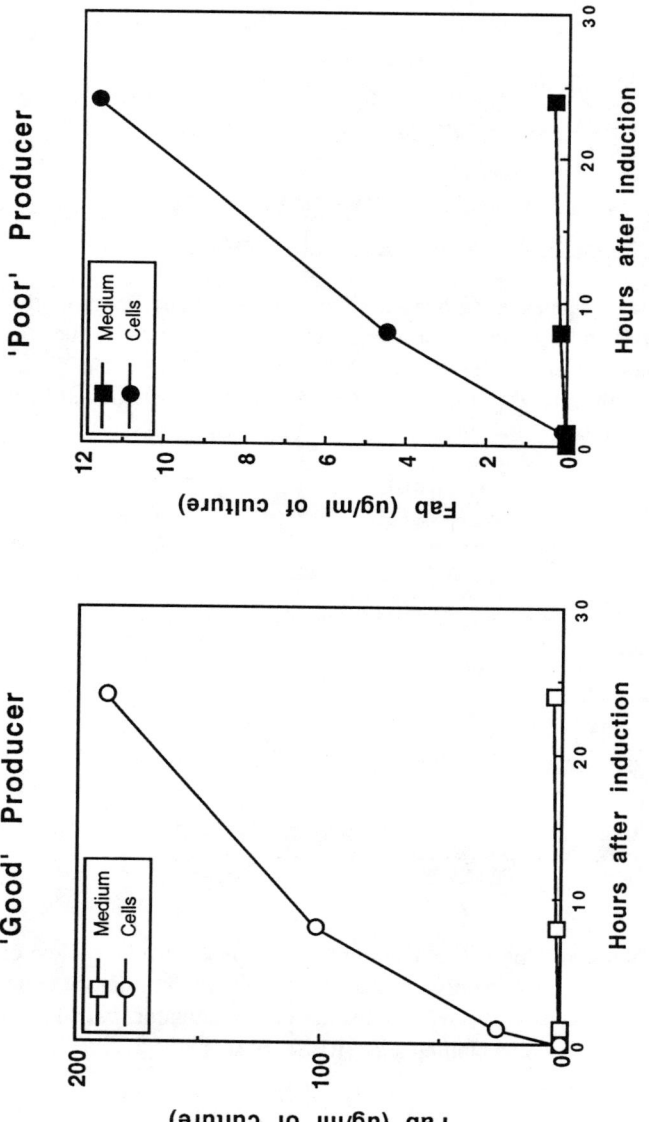

Figure 3. Comparison of Fab protein in E. coli cultures transformed with a Fab vector which secretes 2 μg/ml ("good" producer) and 0.1 μg/ml ("poor" producer). After induction, cells and culture supernatants were separated and analyzed separately for Fab proteins by ELISA, and by immunodetection with ^{125}I-protein A of samples applied to nitrocellulose in a slot-blot manifold. Note that the vertical axes of the two panels have different scales.

Fd'-2C produce a mixture of Fab' and F(ab')$_2$. From 10 to 30% of the antibody fragment present in induced κ, Fd'-2C culture supernatants was F(ab')$_2$. These monovalent and bivalent forms were separated by size exclusion chromatography. F(ab')$_2$ produced by bacteria competed for antigen binding sites in an identical manner to whole antibody while monovalent Fab' competed several fold less well, as expected *(14)*. Monovalent Fab' purified from the culture supernatant could be oxidized to F(ab')$_2$ *in vitro* (Figure 4), as has Fab' obtained from the *E. coli* periplasm *(15)*.

Since F(ab')$_2$ accumulates with the Fd'-2C but not with the Fd'-1C module, we examined hinge variants to determine if both inter-heavy chain cysteine residues are required, or if other inter-hinge interactions may be important for disulfide formation or stabilization. An Fd' module identical to Fd'-2C except that the first cysteine residue involved in inter-heavy chain bonding was mutagenized to an alanine also directed a small amount of F(ab')$_2$ into the *E. coli* culture medium, but the amount was about 10 times less than that formed from the Fd'-2C module. It appears, therefore, that interchain association may be favored when the entire hinge region is present and may then be stabilized by both interchain disulfide bonds. This also seemed to occur with expression of F(ab')$_2$ in yeast, as will be described below.

To prepare large quantities of Fab, Fab' or F(ab')$_2$ for analysis, bacterial cultures were grown in a 14-liter Chemap fermenter. High level expression in the fermenter was accomplished by a series of optimization steps as shown in Table I. Factors which affect fermentation yield include culture temperature, rate of inducer addition and time of culture growth after induction. By this process, cultures were grown to an OD_{600} of 50 to 100 and induced with L-arabinose. Figure 5 demonstrates that Fab accumulated in the culture supernatant over time after induction with L-arabinose. Fab yields of greater than 100 mg/l are typical *(18)*. As noted previously, the final fermentation yield can vary from one antibody fragment to another as would be expected for expression of a variety of recombinant proteins in *E. coli*.

Table I. Fermentation Optimization [a]

	Titer (mg/l)
Shake flask	0.5 - 1
Fermentation - shot induction	
Batch	3
Fed batch	40
Temperature optimization	81
Fermentation - gradient induction	454
Induction time optimization	561

[a] Optimization was for ING2 Fab *(16, 17)*. Each run was done in a 10-liter fermenter in similar medium.

Fab can be purified by a series of ion exchange chromatography steps from fermentation broth. A general purification scheme has been developed which can be

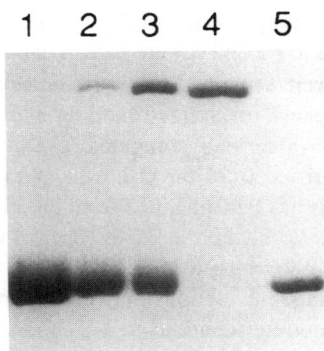

Figure 4. Oxidation of Fab'-2C to F(ab')$_2$ *in vitro*. Hinge cysteine residues were selectively reduced with 0.5 mM dithiothreitol, and Fab' was incubated in the presence or absence of 2 mM cysteine. Lane 1: reduced Fab'; Lane 2: oxidation without cysteine; Lane 3: oxidation with cysteine; Lane 4: purified F(ab')$_2$; and Lane 5: Fab'.

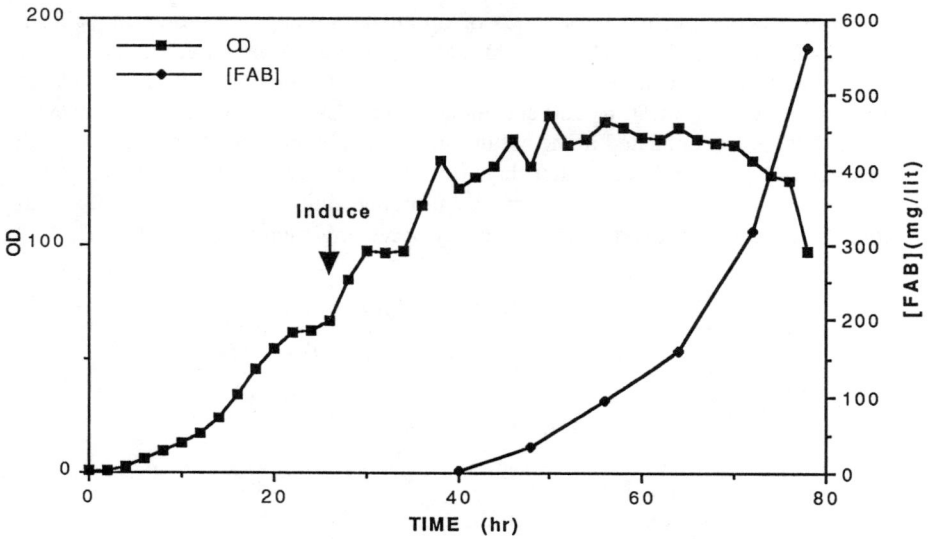

Figure 5. Fermentation time course. An *E. coli* strain which can express ING2 Fab *(16, 17)* was cultured in a Chemap 14L fermenter in 10L of minimal medium supplemented with 0.7% glycerol and induced with L-arabinose when the culture reached an OD$_{600}$≈70. Fab accumulated in the culture supernatant after induction.

used for any Fab purification and is also useful for purification of Fab from yeast cultures *(16)*. The fermentation broth is concentrated and passed over a DEAE-cellulose column. This step removes most contaminating *E. coli* proteins, DNA and endotoxin. The DEAE-cellulose flow-through can be applied to a cation exchange resin, such as CM-cellulose, where the bound antibody fragment can be eluted with a salt gradient. Additional ion exchange steps can be added to remove trace contaminating proteins, DNA and endotoxin. Overall product yield is greater than 80%. As we have shown *(16)*, *E. coli* endotoxin can be reduced to less than 0.1 EU/mg Fab, representing an 8-log decrease from starting material, and contaminating DNA can be removed to less than 1 pg/mg of Fab protein. This purification scheme is suitable for pharmaceutical development.

Expression of Fab, Fab' and F(ab')$_2$ in Yeast

For expression of Fab, Fab' or F(ab')$_2$ in yeast, the light and heavy chain variable regions were fused to the human light chain and Fd chain modules to form light and Fd or Fd' genes. Assembled antibody genes were fused in-frame to the yeast invertase signal sequence *(19)* and cloned between the yeast PGK promoter and 3' untranslated region *(20)*. Both of these transcription units were then inserted into an expression vector containing the yeast Ura3 and Leu2d genes for transformant selection and the 2 micron origin of replication for autonomous replication in yeast *(7)*, Figure 6. Ura$^+$ colonies arising from transformation into Ura$^-$ *S. cerevisiae* were screened for secretion of Fab. Protein of the size expected for Fab was identified in concentrated yeast culture supernatants by Western blot analysis with antibody to the human kappa chain. Cross-reactive bands were observed at the position expected for Fab and Fab' for transformants containing the light chain plus Fd or the Fd'-1C and at the expected positions of Fab' and F(ab')$_2$ for the transformants containing the light chain plus Fd'-2C. A smaller amount of F(ab')$_2$ could be detected in the culture supernatant of yeast expressing kappa plus an Fd module which ended after the proline following the second inter-heavy chain cysteine residue (data not shown). This again suggests that human IgG1 F(ab')$_2$ formation is most efficient when both inter-heavy chain cysteine residues are present along with an intact hinge region.

Yeast cultures were grown in 10-liter fermentors to prepare large quantities of Fab, Fab' and F(ab')$_2$. Expression vectors containing kappa and Fd or Fd'-1C produced primarily yielded Fab or Fab', respectively, while a mixture of Fab' and F(ab')$_2$ (approximate ratio, 3:1) was purified from the culture supernatants of the transformants containing the Fd'-2C. The Fab' and F(ab')$_2$ were separated by gel filtration and ran at ~100 kD and ~50 kD for F(ab')$_2$ and Fab', respectively, on a non-reducing SDS gel, Figure 7.

Antibody fragments purified from yeast cultures were tested for antigen binding by direct and competition binding assays. The results, shown in Figure 8, are typical of these experiments and demonstrate that the F(ab')$_2$ molecules behaved in a manner similar to that of the whole antibody, while the monovalent Fab and Fab' forms bound with a ~3-4 fold lower affinity. We have not observed any differences in the activity of antibody fragments produced in yeast and *E. coli*. The yeast

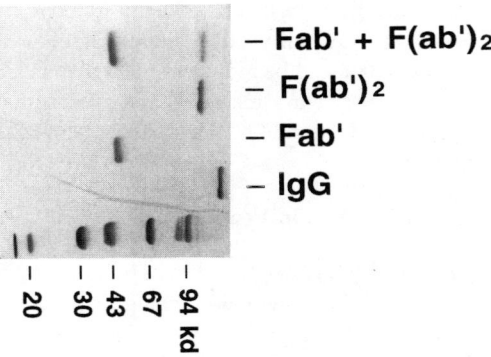

Figure 7. Coomassie blue-stained nonreducing SDS polyacrylamide gel showing Fab', F(ab')$_2$, and chimeric IgG. Size marker molecular masses (kD) are also shown.

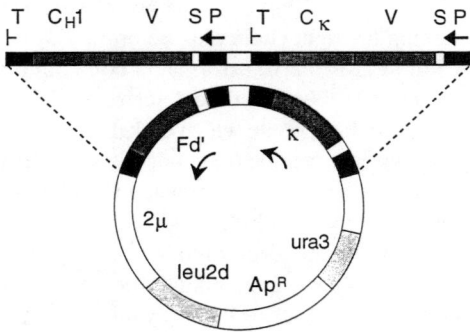

Figure 6. Yeast expression vector features. Shown is a schematic view of the Fd (V-C$_H$1) and kappa (V-Cκ) genes, the PGK promoter (P) and transcription termination signal (T), and the invertase signal sequence (S). Also shown are the relative positions of the Ura3 and Leu2d marker genes, the 2µ circle origin, and the ampicillin resistance gene (Ap^R) for selection in E. coli.

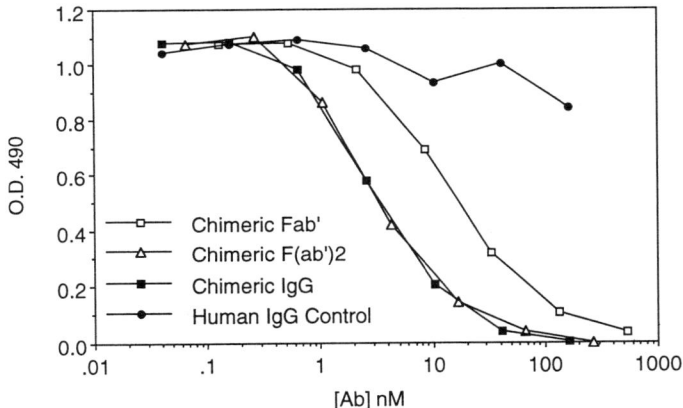

Figure 8. Competition binding assay. Fab', F(ab')$_2$ and IgG were incubated with HT29 cells and biotinylated chimeric IgG (ING4 – recognizing a melanoma-associated cell surface antigen) at 4°C for 3 hours. Bound biotinylated antibody was detected with peroxidase-labeled avidin. Human IgG is included as a negative control.

expression system offers the advantage that culture supernatants contain relatively few proteins other than Fab, and that expression is constitutive. Antibody yield from yeast fermentation is, however, on-average about three-fold lower than that observed with *E. coli*.

In the course of expressing a number of different specificities as Fab in yeast, we noted that, as with *E. coli*, not all fragments were secreted with equal efficiency. In order to determine where a potential block might occur, the levels of intracellular and secreted Fab was determined for "good" and "poor" producing strains. While the secreted Fab levels for the "good" producer were about 10-fold higher than for the "poor" producer (0.13 vs. 0.01 µg/ml), the steady state intracellular levels were similar for both (75 µg/ml, "good" vs. 115 µg/ml, "poor"). Unlike our observation with Fab expression in *E. coli*, the primary block for the "poor" yeast secretor appears to be with secretion rather than protein synthesis. Interestingly, there is not a direct correlation between a highly secreted Fab in yeast and *E. coli*.

Construction of Fab, Fab', and F(ab')$_2$ Immunotoxins

Antibody fragments produced in *E. coli* (or yeast) can be linked to ribosome inactivating proteins to direct potent toxins to target cells. We have expressed chimeric (mouse V-region, human C-region) and human-engineered Fab, Fab', and F(ab')$_2$ fragments in *E. coli* which were derived from the murine H65 antibody *(21)* and recognize the human T-cell antigen CD5. Antibodies to CD5 are attractive candidates for treatment of human T cell-mediated disease since this antigen is found on mature T cells but not on precursor cells. The H65 antibody has previously been conjugated to the ricin A chain *(21)* and was cytotoxic to human T cell lines and peripheral blood mononuclear cells (primarily composed of human T cells) *in vitro*. The H65-ricin A chain immunoconjugate has also proven clinically effective in humans for resolution of steroid-resistant graft versus host disease, a complication of bone marrow transplantation *(22)*.

Both Fab' and F(ab')$_2$ have been linked to the 30 kD form of the ricin A chain (RTA$_{30}$) and evaluated for cytotoxicity against CD5-positive cells *(14)*. It was anticipated that these immunoconjugates might be particularly effective since immunoconjugates of proteolytically-produced anti-CD5 antibody fragments were as cytotoxic to target cells as whole antibody immunoconjugates *(23, 24)*. RTA$_{30}$ contains a unique free cysteine residue which can be used for chemical linkage to antibody fragments. Since Fab' also contains an available free cysteine, a specific conjugate between Fab' and RTA$_{30}$ can be generated by direct oxidation of reduced RTA$_{30}$ and Fab'. This immunoconjugate (Fab'-SS-RTA$_{30}$) does not require nonspecific derivatization with chemical crosslinking agents for its formation.

Alternatively, F(ab')$_2$, which does not contain an available free cysteine residue, can be modified with a crosslinking agent to introduce a reactive thiol for conjugation. The crosslinking agent M2IT *(25)* was linked to F(ab')$_2$ lysine residues and then chemically linked to RTA$_{30}$ to generate F(ab')$_2$-M2IT-RTA$_{30}$.

The binding and cytotoxic activity of these antibody fragment immunoconjugates were compared to the immunoconjugate of chimeric H65 and RTA$_{30}$. The antigen binding characteristics of the immunoconjugates were not altered by linkage to ricin A chain. The cytotoxicity of these immunoconjugates is shown in Table II. Several interesting observations can be made. The Fab'-SS-RTA$_{30}$ is the least active (highest IC$_{50}$) immunoconjugate, while the F(ab')$_2$-M2IT-RTA$_{30}$ is the most active (lowest IC$_{50}$). Also, the extent of kill observed with both of the antibody fragment immunoconjugates is greater than that observed with the whole antibody immunoconjugate.

Table II. Activity of Chimeric H65-RTA$_{30}$ Immunoconjugates

Conjugate	HSB2 [a] IC50 (pM*T) [b]	PBMC [a] IC50 (pM*T) [b]
Fab'-SS-RTA$_{30}$	530	1800
F(ab')$_2$-M2IT-RTA$_{30}$	33	57
IgG-M2IT-RTA$_{30}$	60	400

[a] HSB2 cells (ATCC #CCL 120.1) are a human T cell line that expresses CD5. Peripheral blood mononuclear cells (PBMC) were obtained from healthy donors as described (26). Assays were performed as described (14). Each value represents the average of at least two experiments.

[b] For direct comparison among immunoconjugates, the IC$_{50}$ (immunoconjugate concentration resulting in 50% inhibition) is expressed as toxin concentration (pM*T). pM*T is calculated by multiplying the IC$_{50}$ of the conjugate (pM) by the toxin/antibody ratio (T/A) for a particular conjugate. T/A is obtained from densitometry of immunoconjugates by SDS-PAGE.

Conclusions

In this chapter, we have described the development of E. coli and yeast secretion systems for various types of engineered antibody fragments including Fab, Fab' and F(ab')$_2$. These fragments may have a variety of medical uses, especially for delivery of radio nuclides, drugs and toxins to specific sites in the body and as reagents for imaging solid tumors. To illustrate this potential, we demonstrated that very potent and target-specific immunoconjugates with RTA$_{30}$ can be made with both Fab' and F(ab')$_2$. Immunoconjugates with antibody fragments may offer specific clinical efficacy without the interference of Fc receptors since they lack the region (C$_H$2-C$_H$3) responsible for this activity. Likewise, properly assembled antibody fragments from microorganisms may be useful for *in vitro* clinical analysis and as industrial reagents for affinity purification.

The microbial production approaches described in this chapter represent significant improvement over previous methods for producing antibody fragments since microbial fermentation can be an economic manufacturing method. Indeed, a significant technology base already exists for bacterial and yeast fermentation and is being expanded to included production of other recombinant proteins.

Acknowledgments

We thank Susan L. Bernhard, Stephen F. Carroll, Dianne M. Fishwild, Shan-Shan Hwang, Shau-Ping Lei, and Bob Williams for helping to make this work possible.

Literature Cited

1. Bowden, A. G., Paredes, A. M., Georgiou, G. *Bio/Technology* **1991**, *9*, 725-730.
2. Cabilly, S., Riggs, A. D., Pande, H., Shively, J. E., Holmes, W. E., Rey, M., Perry, L. J., Wetzel, R., Heyneker, H. L. *Proc. Natl. Acad. Sci. U.S.A.* **1984**, *81*, 3273-3277.
3. Hsuing, H. M., Mayne, N. G. and Becker, G. W. *Bio/Technology* **1986**, *4*, 991-995.
4. Oka, T., Sakamoto, S., Miyoshi, K.-I., Fuwa, T., Yoda, K., Yamasaki, M., Tamura, G., Miyake, T. *Proc. Natl. Acad. Sci. U.S.A.* **1985**, *82*, 7212-7216.
5. Better, M., Chang, C. P., Robinson, R. R., Horwitz, A. H. *Science* **1988**, *240*, 1041-1043.
6. Skerra, A., Plückthun, A. *Science* **1988**, *240*, 1038-1041.
7. Horwitz, A. H., Chang, C. P., Better, M., Hellstrom, K. E., Robinson, R. R. *Proc. Natl. Acad. Sci. U.S.A.* **1988**, *85*, 8678-8682.
8. Delaloye, B., Bischolf-Delaloye, A., Buchegger, F., von Fleidner, V., Grob, J.-P., Volant, J.-C., Peltavel, J., Mach, J.-P. *J. Clin. Invest.* **1986**, *77*, 301-311.
9. Chovnick, A., Schneider, W. P., Tso, J. Y., Queen, C., Chang, C. N. *Cancer Res.* **1991**, *51*, 465-467.
10. Chaudhary, V. K., Queen, C., Junghans, R. P., Waldman, T. A., FitzGerald, D. J., Pastan, I. *Nature* **1989**, *339*, 394-397.
11. Lei, S-P., Lin, H. C., Wang, S.-S., Callaway, J., Wilcox, G. *J. Bacteriol.* **1987**, *169*, 4379-4383.
12. Johnston, S., Lee, J. H., Ray, D. S. *Gene* **1985**, *34*, 137-145.
13. Better, M., Horwitz, A. H., In *Methods in Enzymology;* Langone, J. J., Ed.; Academic Press, San Diego, CA, **1989**, *Vol. 178*; pp 476-496.
14. Better, M., Bernhard, S. L., Lei, S.-P., Fishwild, D. M., Lane, J. A., Carroll, S. F., Horwitz, A. H. *Proc. Natl. Acad. Sci. U.S.A.*, in press.
15. Carter, P., Kelley, R. F., Rodrigues, M. L., Snedecor, B., Covarrubias, M., Velligan, M. D., Wong, W.-L., Rowland, A. M., Kotts, C. E., Carver, M. E., Yang, M., Bourell, J. H., Shepard, H. M., Henner, D. *Bio/Technology* **1992**, *10*, 163-167.
16. Gavit, P., Walker, M., Wheeler, T., Bui, P., Lei, S.-P., Weickmann, J. *Bio/Pharm.* **1992**, *5*, 28-29.
17. Robinson, R. R., Chartier, J., Jr., Chang, C. P., Horwitz, A. H., Better, M. *Hum. Antibod. Hybridomas* **1991**, *2*, 84-93.
18. Better, M., Weickmann, J., Lin, Y.-L. *ICSU Short Reports* **1990**, *10*, 105.

19. Taussig, R., Carlson, M. *Nucleic Acids Res.* **1983**, *11*, 1943-1954.
20. Hitzeman, R. A., Hagie, F. E., Hayflick, J. S., Chen, C. Y., Seeburg, P. H., Derynck, R. *Nucleic Acids Res.* **1982**, *10*, 7791-7808.
21. Kernan, N. A., Knowles, R. W., Burns, M. J., Broxmeyer, H. E., Lu, L., Lee, H. M., Kawahata, R. T., Scannon, P. J., Dupont, B. *J. Immunology* **1984**, *133*, 137-146.
22. Byers, V. S., Henslee, P. J., Kernan, N. A., Blazer, B. R., Gingrich, R., Phillips, G. L., LeMaistre, C. F., Gilliland, G., Antin, J. H., Martin, P., Tutscha, P. J., Trown, P., Ackerman, S. K., O'Reilly, R. J., Scannon, P. J. *Blood* **1990**, *75*, 1426-1432.
23. Derocq, J. M, Casellas, P., Laurent, G., Ravel, S., Vidal, H., Jansen, F. *J. Immunol.* **1988**, *141*, 2837-2843.
24. Rostaing-Capaillon, O., Casellas, P. *Cancer Immunol. Immunother.* **1991**, *34*, 24-30.
25. Goff, D. A. and Carroll, S. F. *Bioconjugate Chem.* **1990**, *1*, 381-386
26. Fishwild, D. M., Staskawicz, M. O., Wu, H.-M., Carroll, S. F. *Clin. Exp. Immunol.* **1991**, *86*, 506-513.

RECEIVED October 26, 1992

Chapter 17

Characterization of Humanized Anti-p185[HER2] Antibody Fab Fragments Produced in *Escherichia coli*

R. F. Kelley[1], M. P. O'Connell[1], P. Carter[1], L. Presta[1], C. Eigenbrot[1], M. Covarrubias[2], B. Snedecor[2], R. Speckart[3], G. Blank[3], D. Vetterlein[3], and C. Kotts[4]

[1]Protein Engineering, [2]Fermentation, [3]Process Sciences, and [4]Medicinal and Analytical Chemistry Departments, Genentech, Inc., 460 Point San Bruno Boulevard, South San Francisco, CA 94080

> We have been using biochemical and biophysical methods to characterize chimeric and humanized variants of the murine monoclonal antibody 4D5, directed against human epidermal growth factor receptor 2 (p185HER2). These studies were performed on antibody Fab fragments produced by secretion from E. coli. Humanized Fab fragment (hu4D5-8 Fab) was expressed at very high levels (1-2 g/L), whereas chimeric Fab (ch4D5 Fab) was expressed at much lower titers (5-20 mg/L), as determined by antigen-binding ELISA of supernatants from 10 L fermentations. Hu4D5-8 Fab and ch4D5 Fab purified by using affinity chromatography on immobilized bacterial IgG-binding proteins gave identical far UV-CD spectra characteristic of the immunoglobulin fold. Thermodynamic studies of antigen binding show comparable affinities (ΔG) for ch and hu4D5-8 Fab, but different ΔH values suggesting slight differences in the mechanism of binding. This difference is reflected in the anti-proliferative activity of these fragments on human breast tumor cells.

Monoclonal antibodies (*1*) offer great promise for the treatment of human disease because of their specificity and affinity in binding antigen, effector functions and long serum half-lives. This potential, however, has often been thwarted by a human immune response against the therapeutic antibody (*2*). Current technology for eliciting a monoclonal antibody (MAb) requires immunization of a rodent with the target antigen. These rodent MAbs are often recognized as foreign by the human immune system. Although much work is in progress on developing methods for producing human MAbs (reviewed in *3*), these methods have not yet yielded an antibody with affinity and specificity equal to that obtained via rodent immunization. A process called "humanization" (*4*), however, offers an alternative route to generating a therapeutic MAb. In this procedure, protein engineering methods are used to convert a rodent MAb into a human antibody, which should be less immunogenic, while retaining the antigen affinity and specificity of the original rodent MAb. As outlined in Figure 1, humanization

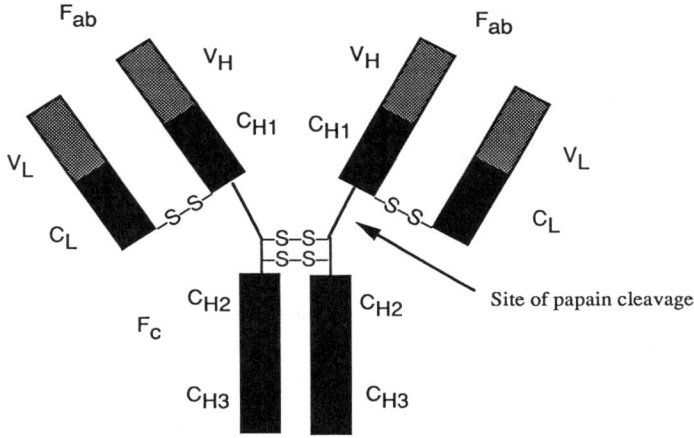
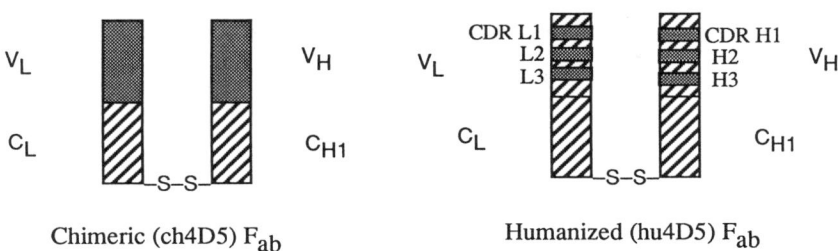

Figure 1. Schematic diagram of antibody humanization. Murine variable domain sequences are shown *stipled* and constant domains in *black*. Human consensus sequence is shown *cross-hatched*.

involves grafting the complementarity determining regions (CDR) of the murine variable domains into a human antibody. However, since a few framework residues may contact antigen or are important for determining the conformation of the antigen binding segments, molecular modeling must be used to identify which murine framework residues should also be included in the humanized form. We summarize here work on humanizing a murine MAb (4D5) which is directed against the extracellular domain (ECD) of human epidermal growth factor receptor 2 (p185^{HER2}).

Role of p185^{HER2} in Breast Cancer. p185^{HER2} is overexpressed in about 30 % of breast and ovarian cancers and the level of expression is correlated with poor prognosis for survival of the disease (5). p185^{HER2} is homologous to the epidermal growth factor (EGF) receptor and thus is a receptor tyrosine kinase. This class of receptor has an extracellular ligand binding domain, a single transmembrane sequence, and a cytoplasmic tyrosine kinase domain. The ligand for p185^{HER2}, called "heregulin", has only recently been identified (6). Heregulin stimulates phosphorylation of p185^{HER2} as well as the growth of cells which express p185^{HER2}.

Properties of MAb4D5. muMAb4D5 was originally elicited by immunizing mice with a fibroblast cell line transformed by overexpression of p185^{HER2} (7). This antibody (IgG$_1$-k) binds p185^{HER2}-ECD with high affinity (8; K$_D$ = 60 pM) and also displays an anti-proliferative effect on tumor cell lines which overexpress p185^{HER2}. The anti-proliferative effect is specific for tumor cell lines which express p185^{HER2}. These results suggest that muMAb4D5 may have therapeutic potential for treatment of breast and ovarian cancer patients whose tumors show amplified p185^{HER2} expression (5). In order to test 4D5 as an anti-cancer agent in humans, this MAb has been humanized (9).

Humanization of 4D5 was successful in generating the antigen affinity of the murine MAb but did not fully regenerate the anti-proliferative activity (9). In order to further investigate the linkage between antigen binding and growth inhibition, we have used biochemical and biophysical techniques to characterize Fab fragments of 4D5. An expression system in which Fab fragments are secreted from *E. coli* (10) was employed to generate quantities of material sufficient for biophysical characterization. The role of framework residues implicated as being important for antigen binding has been addressed by characterizing Fab variants differing at these sites (Table I). For several of these variants, a complete set of thermodynamic parameters describing antigen binding has been determined, including ΔC_p, which should enable assignment of the contribution to binding from the hydrophobic effect. A few of these Fab variants have been successfully crystallized and high resolution x-ray structures determined (C. Eigenbrot, unpublished results).

Antibody Structure. For the purposes of discussing humanization of 4D5, the salient features of IgG structure are briefly reviewed here. A schematic diagram of a typical IgG antibody is shown in Figure 1. Antibodies have two Fab arms, which are responsible for antigen binding, and an Fc portion which is required for the effector functions of the immune system, i.e. complement activation. Intact antibodies are composed of two "heavy" and two "light" chains. The two heavy chains are linked together through disulfide bonds in the hinge region and each light chain is disulfide bonded to a heavy chain. Each chain has a

Table I. Sequence differences between 4D5 F$_{ab}$ variants[a]

| F$_{ab}$ | V$_H$ residue | | | | | | V$_L$ residue | | Expected Mass | Observed Mass[b] | T$_m$ °C[c] |
	71 FR3	73 FR3	78 FR3	93 FR3	102 CDR3		55 CDR2	66 FR3			
ch4D5	A	T	A	S	Y		Y	R	48,120	48,121±5	72.4
hu4D5-1[d]	R	D	L	A	V		E	G	47,666	47,666±3	ND
hu4D5-2	A	D	L	A	V		E	G	47,581	47,583±3	ND
hu4D5-3	A	T	A	S	V		E	G	47,541	47,528±4	84.2
hu4D5-4	A	T	L	S	V		E	R	47,682	47,672±4	79.6
hu4D5-5	A	T	A	S	V		E	R	47,640	47,641±3	82.2
hu4D5-6	A	T	A	S	V		Y	R	47,674	47,662±5	84.6
hu4D5-7	A	T	A	S	Y		E	R	47,704	47,693±7	ND
hu4D5-8	A	T	A	S	Y		Y	R	47,738	47,738±3	82.5

[a]This set of residues, in addition to CDR residues, appeared to be important for antigen binding based on molecular modeling and functional analysis of variant MAbs (9). Residues are numbered according to Kabat et al. (15). In addition to the differences shown, ch4D5 differs from hu4D5 at 21 framework sites in V$_L$ and 32 framework sites in V$_H$.
[b]Molecular weights were determined by electrospray-ionization mass spectrometry.
[c]Melting temperature (T$_m$) at pH 5 was determined by differential scanning calorimetry. ND = not determined.
[d]human consensus sequence

domain architecture, four immunoglobulin domains in the heavy chain and two in the light chain. Immunoglobulin domains are about 110 amino acids in length and have a β-sheet sandwich structure linked by a conserved, intradomain disulfide bond (*11-12*). Fab fragments are traditionally generated from MAbs by proteolytic cleavage with papain (*13*).

X-ray structures of complexes formed between antigen and antibody Fab fragments (*14*) indicate that the majority of contacts with antigen are contributed by the complementarity determining regions (CDR) of the variable domains. These segments, 3 in the light chain and 3 in the heavy chain, have hypervariable sequences (*15*) thus generating the diversity of the immune response. The CDR segments are located on loop regions whereas the "framework" regions between the CDRs form the β-sheet of the domain (Figure 2). The folding of the immunoglobulin domain brings the discontinuous CDR segments together to make a contiguous antigen binding surface. As shown in Figure 2, interdomain contacts are much more extensive between chains as compared with contacts between domains on the same chain. Thus, folding and association of the two chains is an intricately coupled process.

Although most of the contacts with antigen involve CDR residues, a few contacts are also observed with framework residues (*14*). In addition, some framework residues contact buried CDR residues and thus are important for determining the conformation of the CDR loops (*16*).

Materials and Methods

Humanization of muMAb4D5. Humanization of 4D5 has been described in detail elsewhere (*9*) and is only briefly summarized here. Models of the V_L and V_H domains of muMAb4D5 were constructed by using the coordinates of seven Fab crystal structures from the Brookhaven Protein Data Bank. A similar procedure was used to generate a model for a human antibody by using consensus sequences from the most abundant human Ig subclasses – V_L k_I and V_{HIII}. A model of humanized 4D5 was then generated by transferring the 6 CDR segments from the model of the murine antibody into the human consensus model. Inspection of the models identified a set of framework residues (Table I) which appeared important for antigen binding either because the side chain could lie near the CDRs and thus contact antigen or because the side chain might help determine the conformation of a CDR loop. For example, V_L residue 66, which is normally a Glycine residue in human and murine κ chains, is an Arginine in mu4D5. Modeling suggested that the Arginine side chain could influence the conformation of CDR segments L1 and L2 or might contact antigen.

The muMAb4D5 variable domain genes were isolated by PCR amplification of the mRNA from the hybridoma. These genes were fused with human constant domains to generate a plasmid encoding a chimeric version of the 4D5 Fab. (In this and subsequent plasmids the heavy chain sequence terminates 4 residues after the cysteine which makes a disulfide bond with a cysteine from the light chain). The framework regions were then mutagenized in one step to generate the prototype humanized sequence (hu4D5-5) as previously described (*9*). Variants of hu4D5-5 having the substitutions shown in Table I were constructed by site-directed mutagenesis.

Expression and Purification of 4D5 F_{ab} Fragments. Both the chimeric and humanized sequences were cloned into an *E. coli* expression plasmid designed for co-secretion of light chain and heavy chain Fd (*10*). This plasmid contains an expression cassette encoded on an EcoRI/SphI fragment that has light chain and heavy chain Fd under the control of a single alkaline phosphatase (*phoA*) promoter. Both the light chain and heavy chain Fd are preceded by bacterial signal sequences. Humanized (hu) and chimeric (ch) 4D5 Fab fragments were expressed in *E. coli* by growth of the transformed strain in a 10 L fermentor. Fab was found in the supernatant fraction obtained upon removal of the cells by centrifugation. Antigen-binding ELISA indicated very high expression levels of humanized Fab (1–2 g/L) and much lower titers of chimeric Fab (5–20 mg/L). Fab was purified from the supernatant as described previously (*8*) and briefly summarized here. The fermentation supernatant (ca. 8 L) was microfiltered by tangential flow filtration (TFF) using a 0.16 µm membrane. The clarified supernatant was then concentrated to 200 mLs by TFF using a 30 kD cellulose membrane and diafiltered with 3 x 5 volumes of PBS. This fraction was subjected to affinity chromatography either on immobilized Protein A (huFab) or Protein G (chFab). Peak fractions eluted with 0.1 M Glycine pH 3 were pooled and further purified by using hydrophobic interaction chromatography. Concentrations of 4D5 Fab were determined by absorbance measurements using an ε_{280} of 67 mM^{-1} cm^{-1}.

Purification of p185^{HER2}-ECD. The extracellular domain of p185^{HER2} was purified from the cell culture medium of a CHO cell line expressing a truncated form of the receptor using the protocol of Fendly et al. (*17*). Concentrations of solutions of p185^{HER2} were determined by absorbance measurements using an of ε_{280} 76 mM^{-1} cm^{-1}.

Circular Dichroic Spectroscopy. CD spectra were recorded on a Cary/Aviv spectropolarimeter using cylindrical cells having a pathlength of 1 cm (near ultraviolet region) or 0.05 cm (far ultraviolet region). A series of 10 spectra were collected using a time constant of 1.0 s, a bandwidth of 1.0 nm and a wavelength interval of 0.5 nm, and then averaged to obtain the spectra shown in Figure 4.

Differential Scanning Calorimetry. Thermal denaturation curves were obtained by using a Microcal, Inc. MC-2 calorimeter operated at a scan rate of 1.0 °C min^{-1}. Protein solutions were prepared for measurements by extensive dialysis versus 50 mM sodium acetate pH 5.0. Concentrations were adjusted to 2 mg mL^{-1} for measurements.

Determination of Thermodynamics of Antigen Binding. K_D values for binding of Fab to p185^{HER2}-ECD were determined by using a radioimmunoassay (RIA) as previously described (*8*). For some of the variants, K_D values were also determined from the kinetics of binding to immobilized antigen by using the method of Surface Plasmon Resonance (SPR) (*18*). A Microcal, Inc. OMEGA titration calorimeter was used to measure the enthalpy of binding. Solutions were prepared by dialysis versus 20 mM sodium phosphate pH 7.5, 100 mM sodium chloride. In a typical experiment, 1.3943 mLs of a solution of 7–10 µM p185^{HER2}-ECD was titrated by addition of 6, 12 µL aliquots of a concentrated (0.2 – 0.5 mM) solution of Fab. For most of the Fab variants examined here, the "c" value (*19*) is greater than 10^4 thus precluding a calorimetric measurement of the binding constant. The observed heat pulses were integrated using the software supplied by the manufacturer. Heats of Fab dilution were measured in separate experiments by injection of Fab into a solution of buffer. These blank heats were subtracted from the heats measured for antibody–antigen interaction and the result divided by the moles of receptor in the calorimeter cell to obtain ΔH. ΔC_p was determined from the temperature dependence of ΔH in the range of 25 – 45 °C. In this range ΔH shows a linear dependence on temperature suggesting that ΔC_p is independent of temperature.

Results and Discussion

Expression and Purification of 4D5 Fab. A secretion strategy was chosen for 4D5 Fab production based on the previous success of Better et al. (20) and Skerra and Plückthun (21) in *E. coli* expression of antibody Fab and Fv fragments, respectively. Secretion of the chains is necessary because the periplasmic space of *E. coli* has a more oxidative environment than the intracellular compartment, thus fostering disulfide bond formation. Both chains must be expressed together because isolated antibody Fd chains or V_H domains generally have poor solubility properties. All of the hu variants were expressed at high levels (0.5 – 2.0 g/L) as judged by quantitative measurements of antigen binding by using an ELISA. ch4D5 Fab was expressed at much lower levels and required growth at 30 °C rather that 37 °C to obtain optimal expression (0.02 g/L).

Both hu and ch 4D5 Fab could be rapidly purified using the protocol outlined in the *Methods* section. Due to the tight binding of hu4D5 Fab to *S. aureus* Protein A and high expression levels, 1-2 g of this material could be purified in a single day from 8 L of fermentation supernatant. Although data indicating binding of *S. aureus* Protein A to the Fab portion of some IgG molecules has been previously reported (22-23), this is the first demonstration, to our knowledge, of the use of Protein A in the purification of human IgG Fab fragments. Protein A is widely believed to bind only to the Fc region of IgG and a crystal structure of a fragment of Protein A in complex with Fc has been determined (24). ch4D5 Fab does not bind to Protein A suggesting that the Protein A binding site of hu Fab is located on the variable domains. ch4D5 Fab could be purified by using immobilized *Streptococcal* Protein G. The yield of this form was considerably lower because of the decreased expression levels and weak affinity of Fab binding to Protein G.

As shown by SDS-PAGE (Figure 3), Fabs purified from *E. coli* supernatants are fairly homogenous and also gave molecular weights upon electrospray-ionization mass spectrometry (Table I) consistent with their amino acid sequence. These data show that the Fab produced by secretion from *E. coli* have an interchain disulfide bond and are not proteolytically clipped or otherwise modified.

Conformation and Stability of E. coli Produced Fab. The circular dichroic spectra of ch and hu 4D5-5 Fab are compared in Figure 4. Both proteins gave similar far UV-CD spectra in having a weak negative CD band at 217 nm and a stronger positive CD band at 202 nm. These spectra are consistent with CD spectra observed for Fab fragments generated by proteolytic digestion of IgG (25) and indicate that Fabs purified from *E. coli* supernatants have the immunoglobulin fold.

The thermal stabilities of ch and hu 4D5 were compared by using differential scanning calorimetry as shown in Figure 5A. Thermal denaturation of 4D5 Fabs is not completely reversible as judged by cooling and reheating the sample. Both

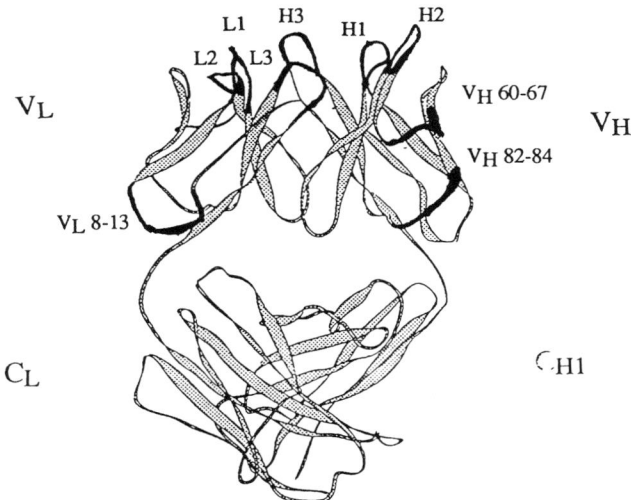

Figure 2. X-ray structure (backbone only) determined for hu4D5 Fab (C. Eigenbrot, unpublished results). The light chain is on the left with the variable domain at top. CDR loops, 3 from V_L and 3 from V_H, are designated L1, L2, L3, H1, H2, and H3 and are highlighted in black. Loop segments swapped to test role in anti-proliferative activity are also highlighted.

Figure 3. SDS-PAGE on Fab purified from *E. coli* fermentation supernatant. Lanes a–f, h: purifed Fab; Lane g: molecular weight standards. (Reproduced with permission from ref. 8. Copy right 1992 American Chemical Society.)

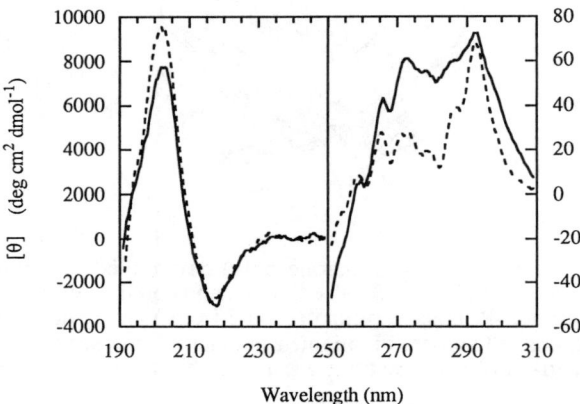

Figure 4. Circular dichroic spectra of hu4D5-5 (dashed line) and ch4D5 (solid line) Fab. Spectra were recorded on solutions containing 4 mM sodium phosphate pH 7.5, 20 mM NaCl at 25 °C. (Reproduced with permission from ref. 8. Copy right 1992 American Chemical Society.)

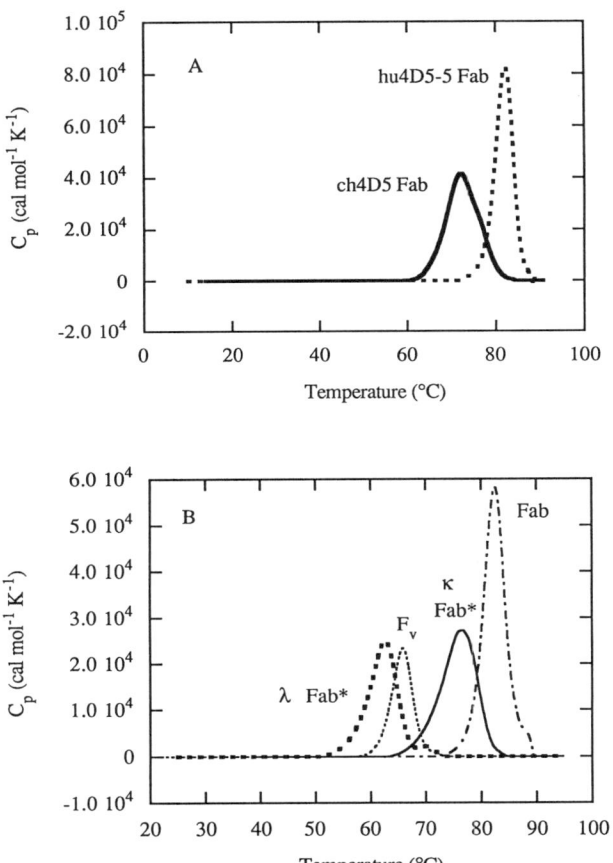

Figure 5. Thermal denaturation of 4D5 fragments. Baselines were subtracted using the splines rountine of the Microcal DA-2 software and then the curves were normalized to the moles of protein in the calorimeter cell. A) Comparison of thermal stability of hu4D5-5 and ch4D5 Fab. B) Thermal stability of non-covalently linked versions of hu4D5-8. k Fab* differs from Fab in that the cysteines which make an interchain disulfide bond have been mutated to serines. λ Fab* has an additional change in that the C_k domain has been replaced with a heterologous $C\lambda$ domain. Fv is a fragment composed of non-covalently associated V_H and V_L domains.

proteins display high melting temperatures (T_m) with ch4D5 Fab having a T_m of 72 °C and hu4D5-5 Fab having a T_m of 82 °C. The thermal transitions cannot be precisely described by a single two-state process; $\Delta H_{cal}/\Delta H_{VH}$ ratios are 3.7 and 2.3 for ch4D5 and hu4D5-5 Fab, respectively. These data suggest that the 4 immunoglobulin domains of chimeric Fab unfold more independently than the 4 domains of the humanized Fab. Since ch4D5 and hu4D5 Fab have identical constant domains, these changes must reflect differences in the stability and cooperativity of the V_L/V_H unit. This conclusion is consistent with thermal denaturation studies (data not shown) on a hybrid Fab in which the murine V_L domain of the chimeric Fab has been replaced with the humanized V_L. The T_m for this Fab is 76 °C and the $\Delta H_{cal}/\Delta H_{VH}$ ratio is reduced relative to the chimeric Fab. We speculate that the differences in *E. coli* expression level between chimeric and humanized 4D5 Fab may be related to the observed differences in thermal stability and interchain cooperativity.

We have begun denaturation studies aimed at quantitating interdomain interactions. Our approach is to attempt to dissect the free energy of chain association from the folding free energy by examining the concentration dependence of denaturation for non-covalently linked antibody fragments. For these experiments a non-covalently linked Fab fragment (Fab*) has been produced by mutating the cysteines which join the two chains to serines (P. Carter, unpublished results). In addition, the hu4D5-8 Fv fragment, a molecule consisting of only V_L and V_H domains and thus is non-covalently linked, has been expressed by secretion from *E. coli* and purified (26). Preliminary thermal denaturation experiments with these variants are shown in Figure 5B. Removal of the interchain disulfide bond results in an approximately 6 °C decrease in the T_m. hu4D5-8 Fab* displays a small temperature dependence of unfolding with the T_m increasing 0.7 °C upon increasing the concentration from 23 μM to 328 μM (data not shown). This result suggests that the interaction between the two chains is quite strong such that the two chains are mostly associated even at the lowest concentration. As shown in Figure 5B, replacement of the C_κ domain with a heterologous C_λ domain generates a molecule with a much decreased T_m. This result emphasizes the importance of interdomain interactions in the stability of Fab fragments.

hu4D5-8 Fv retains high thermal stability as indicated by a T_m of 65 °C for a 2 mg mL^{-1} protein solution. Unfolding of this variant shows a greater concentration dependence than observed for Fab* in that the T_m varies by 5°C in the concentration range of 20 – 200 μM (J. Livingstone, unpublished results). The Fv fragment has a higher T_m than the C_λ Fab* suggesting that changes in the light chain constant domain may decrease the stability of both the V_H/V_L and C_{H1}/C_L units.

Antigen Binding Thermodynamics. A rigorous test of the humanized antibody can be made by determining the complete thermodynamics of antigen binding since the information present in ΔG, ΔH, ΔS and ΔC_p can help illuminate the binding mechanism. For example, the magnitude of ΔC_p should be related to the contribution to binding from the hydrophobic effect (27) and thus correlate with the amount of non-polar surface that becomes buried in the complex (28). We have used a radioimmunoassay (RIA) to determine the K_D for binding of Fab to the extracellular domain (ECD) of p185^{HER2} and an isothermal calorimeter to measure the ΔH of binding (8). A Scatchard plot of a typical set of RIA data is shown in Figure 6 and a calorimetry experiment is shown in Figure 7. ΔG and ΔS were calculated using equations 1 and 2 and ΔCp was calculated from the temperature dependence of ΔH using equation 3.

$$\Delta G = -RT \ln(1/K_D) \quad (1)$$
$$\Delta G = \Delta H - T\Delta S \quad (2)$$
$$\Delta H(T_2) = \Delta H(T_1) + \Delta C_p(T_2 - T_1) \quad (3)$$
$$\Delta S(T_2) = \Delta S(T_1) + \Delta C_p \ln(T_2/T_1) \quad (4)$$

As indicated in Table II, the prototype humanized Fab (hu4D5-5) binds antigen about 3-fold more weakly than the chimeric Fab. hu4D5-5 has a much less exothermic ΔH of binding and also a less negative ΔCp than measured for chimeric F_{ab}. Incorporation of two additional murine residues, V_L-Y55 and V_H-Y102, results in a variant (hu4D5-8) which has antigen affinity equal to or perhaps slightly greater than that measured for ch4D5 Fab. These changes also make ΔH and ΔCp more negative; nonetheless, the ΔH of binding is still 4 kcal mol^{-1} less exothermic than measured for ch4D5 Fab. Although the antigen affinity of 4D5 can be reproduced in a humanized form, the thermodynamic data suggest differences in the mechanism of antigen binding between ch and hu 4D5. This difference may be important for anti-proliferative activity as described in the next section.

A value of 960 ± 100 Å^2 for the amount of solvent-accessible nonpolar surface area which becomes buried upon complex formation is calculated from the magnitude of ΔCp (8). Since typical protein-protein interaction surfaces are about 55 % nonpolar (29), this suggests that about 1750 Å^2 of total surface is buried in the interface with antigen. This surface area includes contributions from both antibody and antigen and is consistent with the mean value of 1600 ± 285 Å^2 of contact area observed in x-ray crystal structures of antibody-antigen complexes (29). As shown by comparing the thermodynamic data of Table II, the side chain of V_L-Y55 appears to become buried in the complex with antigen and contributes to binding through the hydrophobic effect. In contrast, V_H-Y102 does not appear to contribute to binding through the hydrophobic effect suggesting that this residue contributes a side chain hydrogen bond to the interaction. The effect of this mutation is dependent on the presence of V_L-Y55 since variant hu4D5-7 binds antigen with the same affinity as hu4D5-5. In the x-ray structure determined for hu4D5-8 Fv, the V_L-Y55 and V_H-Y102 side chains neighbor each other consistent with the non-additive effects on antigen binding of changes at these sites.

Role of Framework Residues in Antigen Binding. The role of framework residues in antigen binding was determined by measuring antigen binding thermodynamics for hu4D5 variants having the changes from hu4D5-5 shown in Table III. Retention of human residues at the 5 positions identified as critical for antigen binding from modeling studies produces a variant (hu4D5-1) which binds antigen with much reduced affinity. This result highlights the importance of modeling studies in reproducing the affinity of a murine antibody in humanized form. About half of the reduction in ΔG can be attributed to the V_H-A71R substitution, shown by comparing variants hu4D5-1 and hu4D5-2. V_H residue 71 has been previously suggested as important for maintaining CDR H2 conformation (30). The substitution V_L-R66G decreases the affinity for antigen by about 3-fold primarily due to an unfavorable effect on the ΔH of binding. The

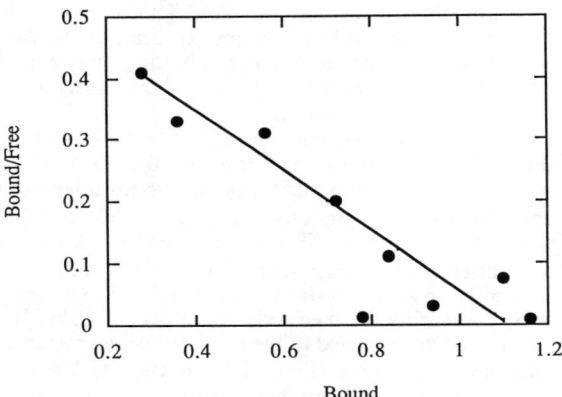

Figure 6. Scatchard plot of RIA data for binding of ch4D5 Fab to p185^{HER2}-ECD. (Reproduced with permission from ref. 8. Copy right 1992 American Chemical Society.)

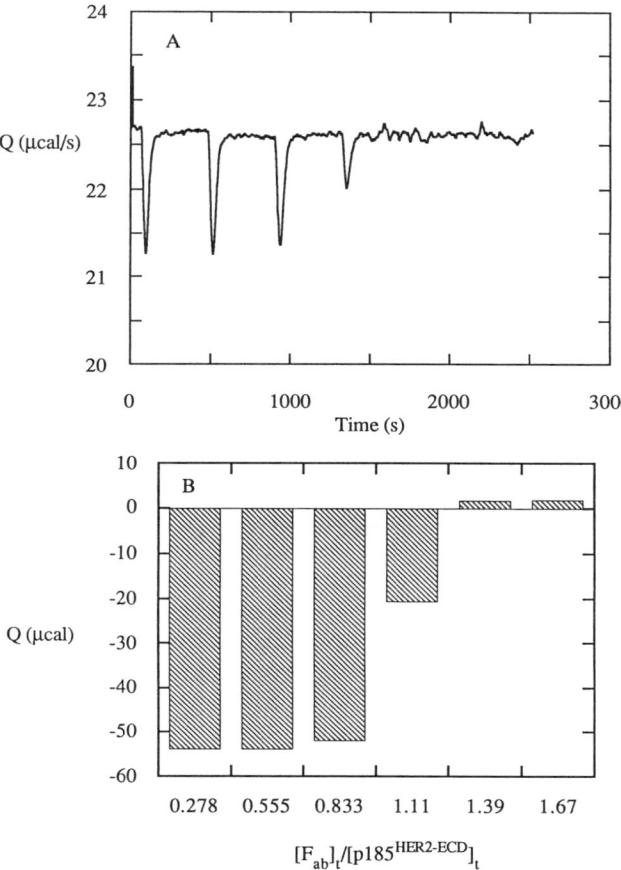

Figure 7. Determination of the ΔH of binding by using isothermal titration calorimetry. A) Titration of 8.5 μM p185^{HER2}-ECD by 6 injections of 12 μL each of a solution of 270 μM ch4D5 Fab. B) Integrated heats as a function of final antibody/antigen ratio. (Adapted from ref. 8.)

Table II. Antigen binding thermodynamics for 4D5 Fab fragments[a]

4D5 Fab[b]	K_D (pM)	$\Delta G°$ (kcal mol^{-1})	ΔH (kcal mol^{-1})	$\Delta S°$ (cal mol^{-1} °K^{-1})	ΔC_p[c] (cal mol^{-1} °K^{-1})
ch4D5	150 ± 90	-13.5 ± 0.4	-17.2 ± 1.5	-12	-400 ± 40
hu4D5-5 (V$_H$-V102, V$_L$-E55)	430 ± 120	-12.8 ± 0.2	-10.0 ± 0.6	+9	-320 ± 20
hu4D5-6 (V$_H$-V102, V$_L$-Y55)	159	-13.4	-11.4	+7	-400 ± 30
hu4D5-8 (V$_H$-Y102, V$_L$-Y55)	90 ± 30	-13.7 ± 0.1	-12.9 ± 0.4	+3	-370 ± 30

[a]Standard state is unit molarity. $\Delta G°$ was calculated using the standard equation, $\Delta G° = -RT \ln K$, where K=1/K$_D$, T = 298.55 °K, R = 1.99 cal mol^{-1} °K^{-1}. ND = not determined. K$_D$ and ΔH values shown with standard deviations are the mean of 3 or more determinations, other values are from single measurements.

[b]Residues at sites V$_L$-55 and V$_H$-102 are shown in parentheses for humanized variants.

[c]ΔC_p was determined from the slope of the temperature dependence of ΔH between 25 and 45 °C. In this range ΔH displays a linear dependence on temperature suggesting that ΔC_p is independent of temperature.

Table III. Effect of hu4D5 variable domain substitutions on the thermodynamics of antigen binding[a]

4D5 Fab	Residue changes[b]	$\Delta\Delta G°$ (kcal mol^{-1})	$\Delta\Delta H$ (kcal mol^{-1})	$\Delta\Delta S°$ (cal mol^{-1} °K^{-1})
hu4D5-1	V_H-A71R, T73D, A78L, S93A V_L-R66G	2.4	1.5	-3
hu4D5-2	V_H-T73D, A78L, S93A V_L-R66G	1.2	-0.4	-5
hu4D5-3	V_L-R66G	0.6	1.4	3
hu4D5-4	V_H-A78L	0	-1.5	-5
hu4D5-6	V_L-E55Y	-0.6	-1.4	-2
hu4D5-7	V_H-V102Y	0	0.1	1
hu4D5-8	V_H-V102Y V_L-E55Y	-0.9	-2.9	-6

[a]$\Delta\Delta J$ calculated by subtracting ΔJ value for hu4D5-5 from that measured for variant. J = G, H or S.

[b]Amino acid replacements in going from hu4D5-5 to specified variant.

nonpolar portion of the V_L-R66 side chain is buried in the x-ray structure determined for hu4D5-8 F_v (C. Eigenbrot, unpublished results), NH1 is involved in a salt bridge with V_L-D28 whereas NH2 is free and could hydrogen bond with antigen. The thermodynamic parameters attributed to the V_L-R66 side chain are consistent with that expected for peptide hydrogen bond formation in water (27); nonetheless, a role for V_L-R66 in determining CDR L1 conformation or neutralizing the charge on V_L-D28 cannot be excluded.

Modeling studies had suggested that V_H-A78 might be important for maintaining the conformation of CDR H1 and thus was retained in the prototype humanized molecule. Mutation of this residue to the human consensus (Leu) does not result in an effect on affinity for antigen but does result in a less favorable ΔH of binding. A comparison of x-ray structures of two variants differing at this site indicates that the Leu substitution is accommodated by movement of a β-strand without significant change in CDR H1 (C. Eigenbrot, unpublished results).

Anti-Proliferative Activity of 4D5 Fab Fragments. The effect of addition of 4D5 Fab fragments on the growth of the human breast tumor cell line SK-BR-3 was assayed as previously described (9). As shown in Figure 8A, the bivalent muMAb4D5 inhibits 50–60 % of the growth of SK-BR-3 cells relative to a control of no antibody addition. Monovalent, ch4D5 Fab displays about half the anti-proliferative effect observed for muMAb4D5. In contrast, hu4D5-8 Fab is inactive even though it binds antigen with affinity comparable to that measured for ch4D5 Fab. hu4D5-8 Fab will block the anti-proliferative effect of ch4D5 Fab suggesting that these two variants bind to the same target on SK-BR-3 cells. hu4D5-8 is active, however, if produced as a bivalent F(ab')2 (10) or full length MAb (9). These results indicate that high affinity binding is required but not sufficient for anti-proliferative activity. Since ch4D5 Fab is active, antibody-mediated receptor dimerization is not obligatory for activity but may enhance the effect. The difference in antigen binding thermodynamics between ch and hu 4D5 Fab (Table II) suggests that the difference in biological activity has its origins in interactions with the extracellular domain of p185^{HER2}. Since ch and hu 4D5 Fab have identical CDR regions and constant domains, residues responsible for anti-proliferative activity must reside in the framework regions of the variable domains.

Recruitment of Anti-Proliferative Effect into Humanized 4D5 Fab. We attempted to recruit anti-proliferative activity into hu4D5-8 Fab by replacing solvent accessible, non-CDR loops with murine sequence. Examination of the models of murine and humanized 4D5 Fv identified 3 framework regions that were solvent accessible and considerably different in sequence between the two forms. These segments are V_L8-13, V_H60-67 and V_H82-84 (Table IV and Figure 2). Single replacements of loops V_L8-13 or V_H82-84 with murine sequence gave a variant with a more exothermic ΔH of binding without significant effect on ΔG whereas the replacement of the V_H60-67 loop resulted in weaker binding. Both the V_L8-13 and V_H82-84 changes resulted in small increases in anti-proliferative effect and combining the two loop changes gave a variant with anti-proliferative activity similar to the chimeric Fab (Figure 8B). These two loops are far from each other and far from the CDR loops (Figure 2) suggesting that they do not directly contact antigen. A conformational change in the Fab:antigen complex that is obligatory for the anti-proliferative effect is consistent with the unfavorable ΔS for antigen binding.

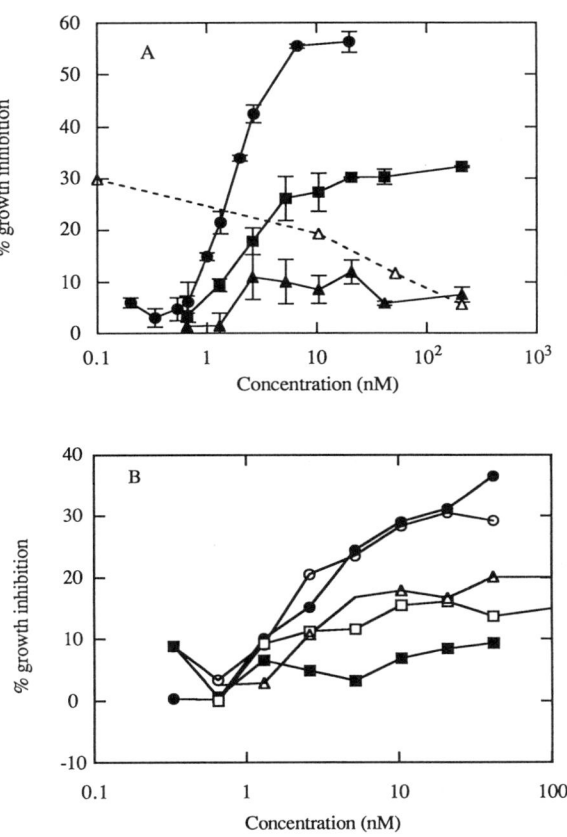

Figure 8. Anti-proliferative effects of 4D5 variants. Inhibition of the growth of SK-BR-3 cells during a 4-day assay was determined as described previously (9). A) Comparison of activities of MAb and Fab variants. Effects of increasing concentrations of muMAb4D5 (filled circles), ch4D5 Fab (filled squares) and hu4D5-8 Fab (filled triangles) are shown. Inhibition of anti-proliferative effect of 20 nM ch4D5 Fab by increasing concentrations of hu4D5-8 Fab is shown as open triangles. B) Effect of loop replacements on anti-proliferative activity of hu4D5-8 Fab. hu4D5 (filled squares), hu4D5-8.A (open squares), hu4D5-8.C (open triangles), hu4D5-8.D (open circles), and ch4D5 (filled circles).

Table IV. Antigen binding thermodynamics of hu4D5 Fab loop variants[a]

hu4D5 Fab	Loop Swap	K_D (pM)	ΔG (kcal mol^{-1})	ΔH (kcal mol^{-1})
hu4D5–8	None	90 ± 30	–13.7 ± 0.1	–12.9 ± 0.4
hu4D5–8.A	V_L8–13	120	–13.6	–15.8
hu4D5–8.B	V_H60–67	280	–13.1	–10.4
hu4D5–8.C	V_H82–84	140	–13.5	–16.7
hu4D5–8.D	V_L8–13/V_H82–84	110	–13.6	ND

[a]Variants of hu4D5–8 were constructed in which the human sequence of the indicated segment was replaced with the corresponding murine sequence.

Antigen Binding Determinants of hu4D5 Fab. We have mapped the functional epitope of hu4D5-5 Fab by determining the effect of individual alanine substitutions in the CDR loops on antigen binding affinity. The effect of these mutations on the K_D for antigen binding was determined by using SPR (*18*). As shown in Figure 9, we find that four residues, H91 in V_L and R50, W95 and Y100a in V_H, make large contributions to antigen binding. These residues are found in three CDR loops: H91 in CDRL3, R50 in CDRH2, and W95 and Y100a in CDRH3. Most of the other CDR residues make only modest contributions to antigen binding ($\Delta\Delta G < 1$ kcal mol^{-1}). These results are consistent with free energy calculations (*31*) which suggest that the set of residues that actively contribute to antigen binding is much smaller than the set of residues observed to contact antigen in structures determined by x-ray crystallography (*14*). Residues which make a large contribution to antigen binding are clustered in a shallow depression on the surface of hu4D5 (*32*) and thus define a putative antigen binding pocket. Less functionally important residues are arranged about the periphery of this pocket. A few mutations, mostly in V_L, result in small but significant (ΔG Standard error = –0.15 kcal mol^{-1}) increases in affinity. This result suggests that the antigen affinity of the 4D5 antibody could be improved by combining mutations in V_L or by combinatorial sorting of a library of human V_L domains.

Summary

Studies on 4D5 Fab fragments indicate that hu4D5-8 binds antigen with affinity comparable to that measured for ch4D5. Variants of hu4D5 with human residues at critical framework sites bind antigen with decreased affinity confirming the importance of molecular modeling in antibody humanization. ch4D5 Fab but not hu4D5-8 Fab has anti-proliferative activity suggesting that other framework

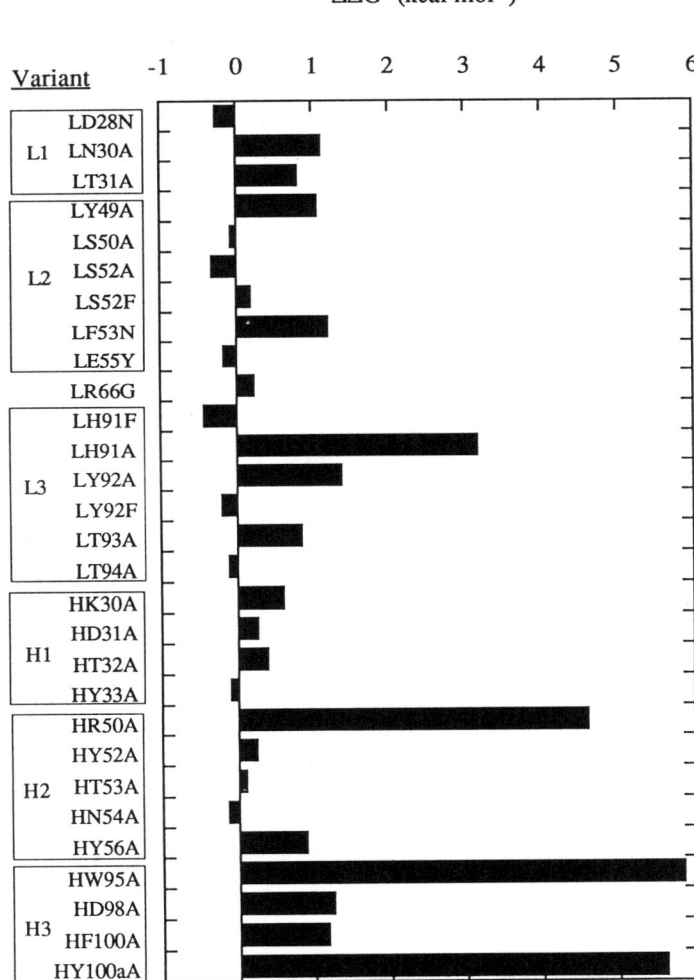

Figure 9. Contribution of CDR residues to antigen binding. Effect of alanine substitutions on antigen affinity is shown as a $\Delta\Delta G$ value where $\Delta G = -RT\ln(1/K_D)$ and $\Delta\Delta G = \Delta G_{mutant} - \Delta G_{wild-type}$. K_D values were determined by using SPR (18) in which p185^{HER2}-ECD was immobilized on the sensor chip. Mutants are named with the first letter denoting light (L) or heavy (H) chain, the second letter denoting the wild-type residue followed by the sequence position and the residue in the mutant.

residues are important for the biological effects of the antibody. These residues appear to reside in the loops V_L8-13 and V_H82-84 which are far from the putative antigen binding surface. Biologically active fragments also have a more exothermic ΔH of binding suggesting that the differences in biological activity are inherent in the interactions with the receptor extracellular domain. These results suggest that a conformational change occurs in the antibody/antigen complex for the active but not inactive fragments. Analysis of a complete set of CDR mutations indicates that the energetics of antigen binding are dominated by a few residues which are clustered in a pocket on the surface of the antibody.

Literature Cited

1. Köhler, G.; Milstein,C. *Nature* **1975**, *256*, 52-53.
2. Miller, R.A.; Oseroff, A.R.; Stratte, P.T.; Levy, R. *Blood* **1983**, *62*, 988-995.
3. Winter, G.; Milstein, C. *Nature* **1991**, *349*, 293-299.
4. Riechmann, L.; Clark, M.; Waldmann, H.; Winter, G. *Nature* **1988**, *332*, 323-327.
5. Shepard, H.M.; Lewis, G.D.; Sarup, J.C.; Fendly, B.M.; Maneval, D.; Mordenti, J.; Figari, I.; Kotts, C.E.; Palladino, M.A.; Ullrich, A.; Slamon, D. *J. Clin. Immunol.* **1991**, *11*, 117-127.
6. Holmes, W.E.; Sliwkowski, M.X.; Akita, R.W.; Henzel, W.J.; Lee, J.; Park, J.W.; Yansura, D.; Abadi, N.; Raab, H.; Lewis, G.D.; Shepard, H.M.; Kuang, W.-J.; Wood, W.I.; Goeddel, D.V.; Vandlen, R.L. *Science* **1992**, *256*, 1205-1210.
7. Hudziak, R.M.; Lewis, G.D.; Winget, M.; Fendly, B.M.; Shepard, H.M.; Ullrich, A. *Mol. Cell. Biol.* **1989**, *9*, 1165-1172.
8. Kelley, R.F.; O'Connell, M.P.; Carter, P.; Presta, L.; Eigenbrot, C.; Covarrubias, M.; Snedecor, B.; Bourell, J.H.; Vetterlein, D. *Biochemistry* **1992**, *31*, 5434-5441.
9. Carter, P.; Presta, L.; Gorman, C.M.; Ridgway, J.B.B.; Henner, D.; Wong, W.-L.; Rowland, A.M.; Kotts, C.E.; Carver, M.E.; Shepard, H.M. *Proc. Natl. Acad. Sci. USA* **1992**, *89*, 4285-4289.
10. Carter, P.; Kelley, R.F.; Rodrigues, M.L.; Snedecor, B.; Covarrubias, M.; Velligan, M.D.; Wong, W.-L.; Rowland, A.M.; Kotts, C.E.; Carver, M.E.; Yang, M.; Bourell, J.H.; Shepard, H.M.; Henner, D. *Bio/Technology* **1992**, *10*, 163-167.
11. Amzel, L.M.; Poljak, R.J. *Ann. Rev. Biochem.* **1979**, *48*, 961-967.
12. Davies, D.R.; Metzger, H. *Ann. Rev. Immunol.* **1983**, *1*, 87-117.
13. Parham, P. In *Cellular Immunology*; Weir, E.M., Ed.; Blackwell Scientific Press: Oxford, UK, 1, 1401-1423.
14. Davies, D.R.; Padlan, E.A.; Sheriff, S. *Ann. Rev. Biochem.* **1990**, *59*, 439-473.
15. Kabat, E.A.; Wu, T.T.; Perry, H.M.; Gottesman, K.S.; Foeller, C. *Sequences of Proteins of Immunological Interest* **1991**, Natl. Inst. Health, Bethesda, MD.
16. Chothia, C.; Lesk, A.M.; *J. Mol. Biol.* **1987**, *196*, 901-917.
17. Fendly, B.M.; Kotts, C.; Vetterlein, D.; Lewis, G.D.; Winget, M.; Carver, M.E.; Watson, S.R.; Sarup, J.; Saks, S.; Ullrich, A.; Shepard, H.M. *J. Biol. Response Modif.* **1990**, *9*, 449-455.
18. Karlsson, R.; Michaelsson, A.; Mattson, L. *J. Immunol. Meth.* **1991**, *145*, 229-240.
19. Wiseman, T.; Williston, S.; Brandts, J.F.; Lin, L.-N. *Anal. Biochem.* **1989**, *179*, 131-137.
20. Better, M.; Chang, C.P.; Robinson, R.R.; Horwitz, A.H. *Science* **1988**, *240*, 1041-1043.

21. Skerra, A.; Plückthun, A. *Science* **1988**, *240*, 1038-1041.
22. Inganäs, M.; Johansson, S.G.O.; Bennich, H.H. *Scad. J. Immunol.* **1980**, *12*, 23-31.
23. Sasso, E.H.; Silverman, G.J.; Mannik, M. *J. Immunol.* **1989**, *142*, 2778-2783.
24. Deisenhofer, J. *Biochemistry* **1981**, *20*, 2362-2396.
25. Doi, E.; Jirgensons, B. *Biochemistry* **1970**, *9*, 1066-1073.
26. Rodrigues, M.L.; Randal, M.; Eigenbrot, C.; Carter, P. *J. Mol. Biol.* **1992**, submitted.
27. Kauzmann, W. *Adv. Protein Chem.* **1959**, *14*, 1-63.
28. Livingstone, J.R.; Spolar, R.S.; Record, M.T. *Biochemistry* **1991**, *30*, 4237-4244.
29. Janin, J.; Chothia, C. *J. Biol. Chem.* **1990**, *265*, 16027-16030.
30. Tramontano, A.; Chothia, C.; Lesk, A.M. *J. Mol. Biol.* **1990**, *215*, 175-182.
31. Novotny, J.; Bruccoleri, R.E.; Saul, F.A. *Biochemistry* **1989**, *28*, 4735-4749.
32. Kelley, R.F.; O'Connell, M.P. *Biochemistry* **1992**, submitted.

RECEIVED October 26, 1992

Chapter 18

Spectral Analysis of Site-Directed Mutants of Human Growth Hormone

Michael G. Mulkerrin[1] and Brian C. Cunningham[2]

[1]Medicinal and Analytical Chemistry and [2] Protein Engineering Departments, Genentech, Inc., 460 Point San Bruno Boulevard, South San Francisco, CA 94080

Site directed mutagenesis has been used to identify receptor binding determinants on (hGH) by the substitution of alanine for residues in the hGH molecule (Cunningham and Wells, Science 244:1081-1985 (1989)). A number of these engineered molecules show greater than a 10-fold reduction in binding affinity to the human growth hormone binding protein (hGHbp) ($\Delta\Delta G$ >1.3 kcal/mole). Presently, although a crystal structure is available for hGH, there is no structure available for the mutants that would allow an accurate assessment of the perturbations occurring in hGH mutant molecular structure. In this report we use the spectroscopic tools, circular dichroism, second derivative absorption spectroscopy, and fluorescence spectroscopy to assess the structural integrity of the mutant proteins. Overall we detect little change in the structure of these engineered molecules. The structural changes we do observe in these mutants appears to be largely confined to alterations in the packing of the interior of the four helix bundle but in no case is there any significant change in the helix content of these molecules. For one of the mutant proteins (hGH D169N) though, there is a change in the permeability of the collisional quenching reagent, acrylamide, to the interior of the four helix bundle. The single tryptophan, Trp86, is a major spectral determinant of the hydrophobic core of hGH and is sensitive to subtle structural changes induced by site directed mutagenesis.

The growth hormone family of proteins consists of the growth hormones, prolactins, and placental lactogens. This family of proteins are single chain polypeptides of 191 to 199 amino acids and approximately 22,000 Daltons. Based on sequence homology this entire family is believed to be comprised of helical coiled coil 4 helix bundles with a topological arrangement similar to human growth hormone hGH (1) and to porcine growth hormone (pGH) (2), for which there are crystal structures. This family of proteins are potent cytokines involved in growth and reproduction in many vertebrate species (3).

Several mammalian growth hormones have been the subject of intense study including human growth hormone (hGH), bovine growth hormone (bGH), and porcine growth hormone (pGH). Site directed mutagenesis has been used to discern which residues on both the hormone (4,5), and the receptor (6) are involved in binding in the human system. Similar mutagenesis strategies have shown which residues are involved in binding hGH to the growth hormone receptor when using a human prolactin scaffold (7).

Alanine scanning mutagenesis has proven to be a powerful mutagenesis strategy for determining the role of specific residues in protein-protein interactions. Alanine scanning mutagenesis removes the side chain beyond the β-carbon to test the functionality of a specific side chain. However, assigning a functional effect to a specific side chain requires that the alanine substitution causes only localized effects and does not disrupt the global structure of the protein. Ideally, mutational analysis of the structure or function of a protein would be followed by an analysis of the perturbation on the protein structure. Examples of such analysis include T4 lysozyme (8), and tyrosyl tRNA synthetase (9) in which the mutational analysis of enzymatic function has been complemented with detailed crystallographic structural analysis. However, such a comprehensive analysis of structure/function relationships simply is not possible for most proteins because techniques are not yet available to crystallize any protein of choice and solve the structure quickly to a resolution sufficient for a comprehensive structure/function analysis. In light of this limitation, a number of techniques have been used in an attempt to access the impact of site directed mutagenesis on protein structure. These include the use of monoclonal antibodies (10), or biophysical techniques, such as fluorescence spectroscopy (11), circular dichroism (12,13), and NMR spectroscopy (14). In this report we show for human growth hormone a combination of spectral techniques that allow us to focus on structural changes in the variant hormones so that we can identify the side chain substitutions that cause specific localized effects rather than global structural disruptions. Such an analysis allows a more accurate interpretation of mutational analysis data.

MATERIALS AND METHODS

Protein samples were dialyzed in 0.01 M Tris pH 7.5 overnight. Protein concentration was determined from the absorption spectrum using $\varepsilon_{277\,nm}^{0.1\%} = 0.82$ cm^{-1} for hGH and the mutants. Circular dichroism spectra were measured on an Aviv 60DS spectropolarimeter. Near UV CD spectra were measured over the wavelength range 320 nm to 250 nm at a 0.5 nm interval. The far UV CD spectra were obtained in the wavelength range of 250 nm to 190 nm at a 0.2 nm interval. Each spectrum is the sum of 5 scans with a two second integration time for each datum.

Absorption spectra were measured from 350 nm to 250 nm at an interval of 0.1 nm with a 0.1 nm band pass on an Aviv 14DS absorption spectrophotometer. Light scattering was corrected for by calculating the log OD vs. log wavelength line from 325 nm to 350 nm and subtracting the extrapolated curve from the spectrum (15). From the absorption spectrum the second derivative spectrum is calculated using the vendor supplied software. Each spectrum is the sum of 5 scans with a 5 second integration time for each datum.

Mutant proteins were expressed in *E. coli*, purified, and binding activity was assayed as previously described by Cunningham et al. (5).

Fluorescence quenching experiments are carried out using additions of 4 to 10 μl of either a 1 M or 5 M acrylamide solution into a 1.0 ml quartz cuvette. Protein solutions are 50 μg of hGH or hGH mutant in 1.0 ml of 0.01 M Tris, pH 7.5. Results are corrected for dilution and the acrylamide used did not increase the total optical density to 0.1 OD at 295 nm in these experiments.

Fluorescence quenching is described by the Stern Volmer equation (*16*)

$$\frac{F_o}{F} = 1 + K_{SV}[Q] \qquad (1)$$

where F_o is the fluorescence in the absence of quenching, F is the fluorescence of the solution with quenching reagent added, K_{SV} is the Stern Volmer quenching constant, and [Q] is the concentration of the quenching reagent in molarity. From a plot of $\frac{F_o}{F}$ versus [Q] a linear plot is obtained, and the Stern Volmer quenching constant (K_{SV}) is the slope of the line.

Lehrer modified Stern Volmer analysis (*16*) of the data is described by the equation,

$$\frac{F_o}{\Delta F} = \frac{1}{[f_a]K[Q]} + \frac{1}{f_a} \qquad (2)$$

where f_a is the fraction of the total fluorescence accessible to the quenching reagent.

RESULTS AND DISCUSSION

Each of the site directed mutant proteins to be discussed displays a greater than 10-fold decrease in binding to the growth hormone binding protein except E174A, which displays a 4-fold increase in binding. These mutant proteins were identified as part of a screen of 65 mutants in a systematic mutational analysis strategy and were proposed to constitute the receptor binding determinants on hGH (4). After identifying mutants of hGH which reduce receptor binding it becomes important to verify that each of these mutations specifically disrupt binding rather than indirectly affect receptor affinity through global perturbations in hormone structure. In this systematic mutational analysis the mutant protein D169A could not be expressed. Therefore, the mutant hGH protein D169N was designed as a more conservative mutation and was used to probe the proposed hydrogen bond between the tryptophan (Trp86) and the aspartate (Asp169) (Bewley, personal communication).

As a spectral probe the far UV CD (190 to 250 nm) is sensitive to the content of the secondary structural elements in a protein. Human growth hormone is 60% α-helical (1) and the contribution of the α-helical elements dominate the far UV CD spectrum. Therefore, the far UV CD would be sensitive to the disruption of the helical elements should this occur. In contrast, the near UV portion of the spectrum is sensitive to the local environment of the aromatic chromophores tryptophan, tyrosine, phenylalanine, and the disulfides. Although these chromophores are not evenly distributed throughout the hGH molecule, these residues do occur at many different positions and are shown along with the sites of the mutations in Figure 1. These chromophores are sensitive to their immediate environment, and the greatest effect will be observed from those chromophores nearest the site of mutation.

Human growth hormone contains 22 aromatic residues: 1 tryptophan, 8 tyrosines, 13 phenylalanines, and 2 disulfide bonds. Though the chromophores are distributed throughout the molecule, there is a significant difference between the sensitivity of each of the aromatic residues to changes in their environment, both in the absorption spectrum and in the near UV CD spectrum. Tryptophan is the most sensitive to changes in the polarity of the environment with potential shifts in the wavelength maximum of ~3.0 nanometers for an absorption spectrum. Again tryptophan is the most sensitive residue contributing to the near UV CD with the

18. MULKERRIN & CUNNINGHAM *Mutants of Human Growth Hormone* 243

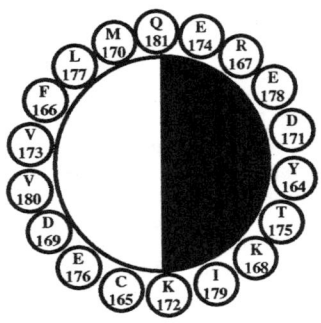

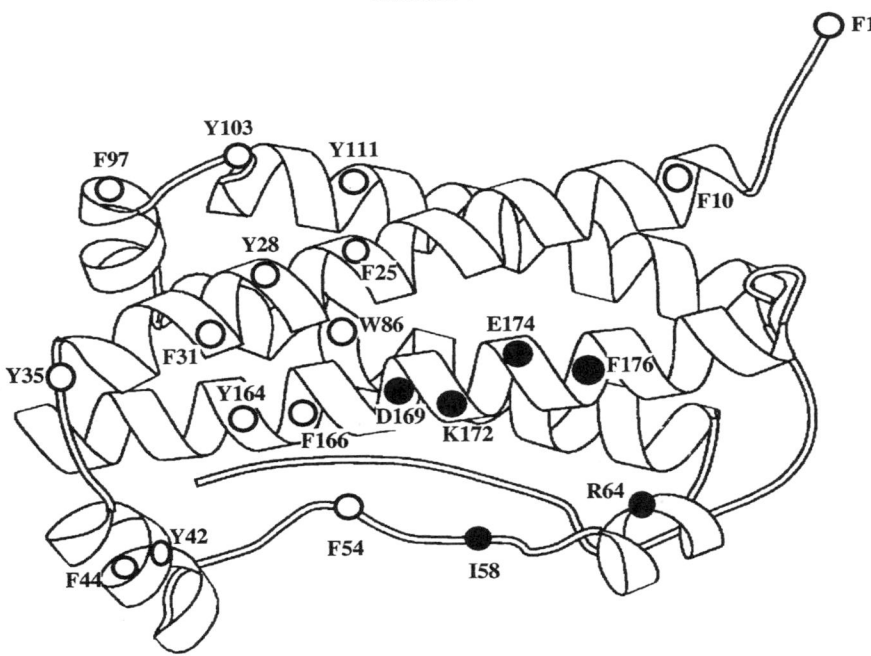

Figure 1: Distribution of the aromatic amino acids and the mutants over the hGH molecule. This ribbon structure of hGH is based upon the X-ray structure (1) and the ribbon diagram was created using the program Molscript (27). Positions of the aromatic residues are shown (O) along with the residues changed in the protein described in the text (●). The helix wheel projection for helix 4 shows the amphipathic character of the helix with polar and charged residues on the hydrophilic face of the helix and the apolar residues on the hydrophobic face of the helix. Residue Phe191 is not observed in the X-ray crystal structure and is not shown on the figure. Residue Phe92 is on helix 2 hidden by helix 1.

largest potential change in rotatory strength (*17*). For both absorption spectroscopy and for the near UV CD spectrum tyrosine is less sensitive to its environment and phenylalanine is the least sensitive (*17*).

The near UV CD spectrum is the sum of the intensity for each of the chromophores, but for each of the chromophores there is a region of the spectrum where that chromophore will dominate. The near UV CD spectrum of hGH is dominated by the tryptophan band centered at 292 nanometers while the tyrosine and phenylalanine residues comprise most of the spectrum from 250 nm to 285 nm with the two disulfides also contributing in this region (*19*). Each of the bands in the hGH spectrum, both the near UV CD and the far UV CD have been previously assigned by Holladay et al. (*18*), although there is some disagreement between these assignments and those made by Bewley and Li (*19*) from their interpretation of the second derivative absorption spectrum of hGH.

The CD spectrum of the two mutants, K172A and E174A, are shown in Figure 2A, B, and are compared with the spectrum of the wild type (wt) protein. In these spectra there is no change in the CD spectra either in the near or far UV CD, although these mutants reduce binding by 15-fold or increase binding by 4-fold, respectively (*5*). In the helix wheel projection in Figure 1 these residues are shown to be on the hydrophilic face. Therefore, these residues are exposed to solvent and do not contribute to the internal packing of the four helix bundle. In neither the far UV CD spectrum nor in the near UV CD spectrum is there any perturbation observed.

A second series of mutants I58A, D169N, F176A with similar levels of decreased binding are shown in Figure 3A, B. These all show a reduction in the intensity of the far UV CD bands and marked alteration in the near UV CD spectra. Difference spectra (not shown) show a change in the intensity of the far UV CD spectrum increased from long to short wavelength. This phenomenon has been observed by Manning et al. (*20*), Manning and Woody (*21*) and Cooper and Woody (*22*). Their data show that a change in the alignment of the helices in a pair of helical peptides causes similar variations in the intensity of the CD spectrum with no change in the total number of residues in α-helical structure. Therefore a decrease in the intensity of the far UV CD does not necessarily imply there is a change in the total number of residues in the α-helix conformation in the protein but may result from a change helix packing.

In the near UV CD spectra of these molecules (Figure 3B) the major change in the spectrum is in the region of 275 nm to 305 nm. In this region the chromophore that is responsible for much of the intensity is the single tryptophan (Trp86). The vibronic bands of Trp86 as assigned by Holladay et al. (*18*) are all positive in hGH and are as follows [0 - 0] ^1L_a Trp at 304 nm although our assignment is at 303.5 nm and Bewley and Li (*19*) assigned the peak to 304.2 nm. The peak at 292 nm is the [0 - 0] Trp ^1L_b as discussed by Bewley and Li (*19*) as opposed to the assignment made by Holladay et al. (*18*).

The position in the structure (*1*) of each of the mutants F176A, D169N, the double mutant F176A/K172A, are shown in Figure 1 and these residues are on the hydrophobic side of helix 4, except K172A. Whereas, I58A is in the loop between helix 1 and 2 and as determined from the X-ray structure (*1*) this residue interacts with helix 2. Taken together these spectra show the mutants in helix 4 on the hydrophobic side of this amphipathic helix and the hydrophobic residue Ile58 appears to alter the packing of the helices in the four helix bundle resulting in a diminution of the intensity of the far UV CD signal and a decrease in the intensity of the tryptophan CD signal.

We use the second derivative (*24*) of the absorption spectrum to determine the position of individual peaks under the envelope of the all of the contributing chromophores in the absorption spectrum of hGH. Figure 4 is an overlay of the

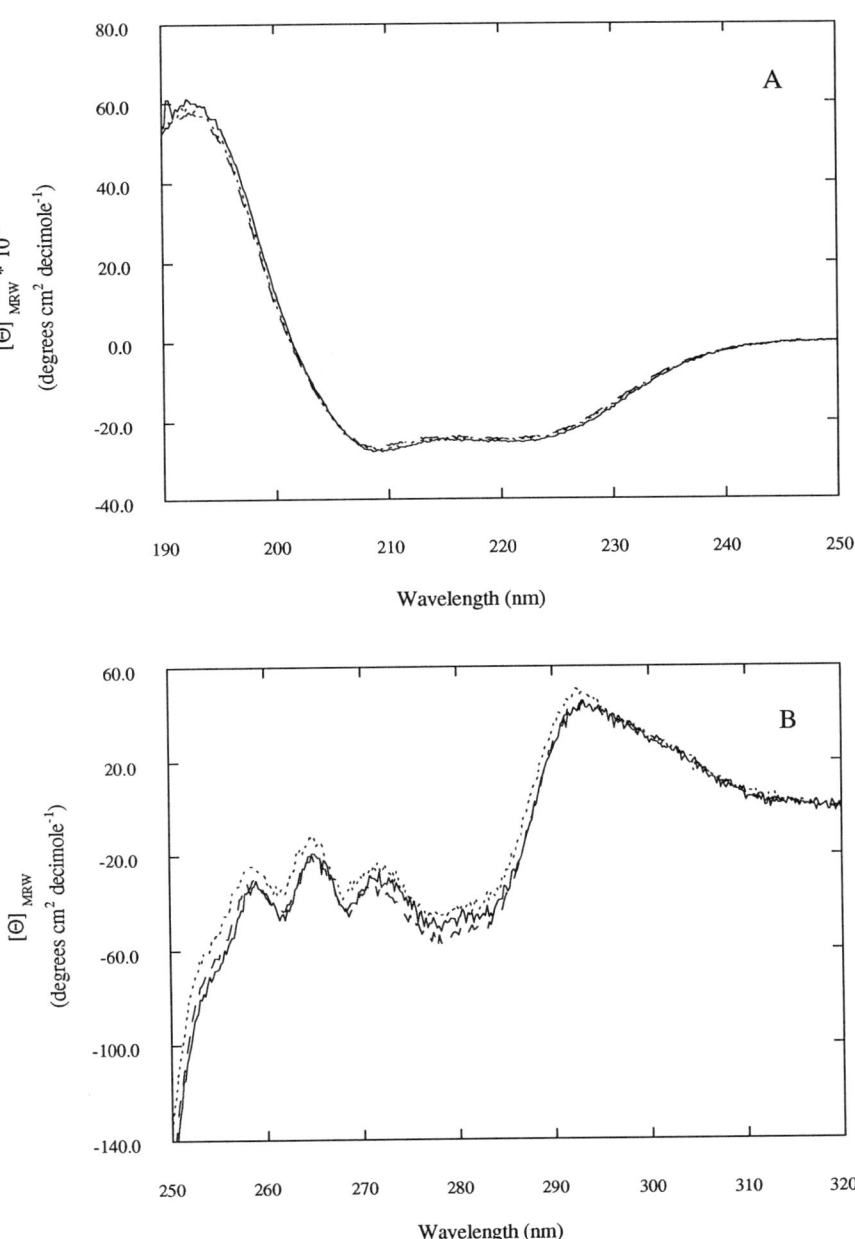

Figure 2: Circular dichroism spectra of wthGH (———) and the proteins K172A (-----) and E174A (——). Panel A is the far UV CD spectrum and panel B is the near UV CD spectrum.

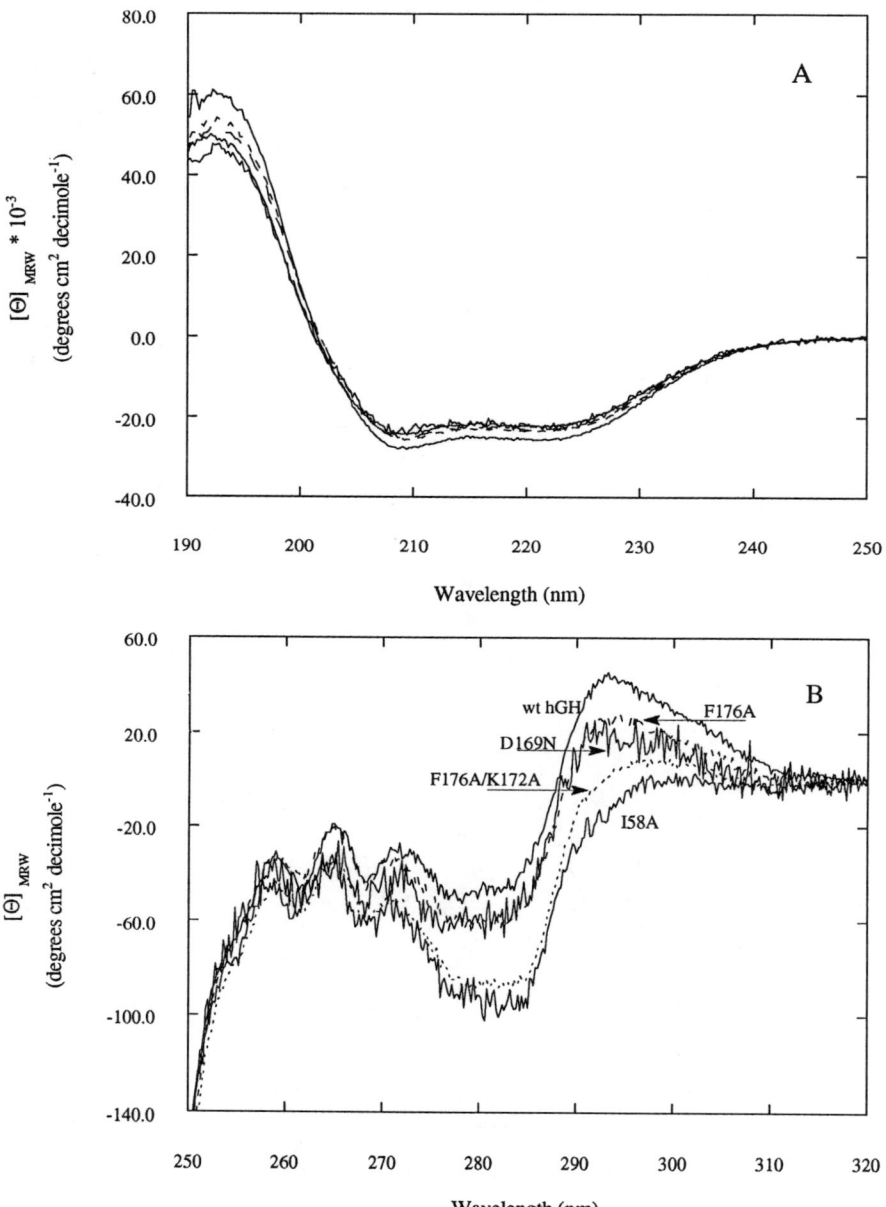

Figure 3: Circular dichroism spectra of wt hGH (———) and the mutants I58A (———), D169N (–––––), F176A (– – –), and the double mutant F176A/K172A (-------). Panel A is the far UV CD spectrum and panel B is the near UV CD spectrum.

absorption spectrum of hGH and the calculated second derivative of the absorption spectrum. This figure shows the correspondence of the minima in the second derivative of the absorption spectrum with the peaks in the zero order absorption spectrum, first determined by Bewley and Li (*19*). The tryptophan ^1L_a band is one of two overlapping $\pi \Rightarrow \pi^*$ electronic transitions in the plane of the indole ring. The indole chromophore is shown in Figure 5 with the transition dipole moments of the ^1L_a and the ^1L_b indicated (*25*). The ^1L_a transition dipole results in a greater charge density on the indolyl nitrogen and the reason for the red shift is that the hydrogen bond can help to delocalize the charge density, the result of which is to lower the energy of the transition. Changes in the wavelength position of this minimum would therefore reflect alterations in the hydrogen bonding of the tryptophan to the hydrogen bond acceptor.

Demonstration that the changes in the structure are caused by slightly altered packing is seen in the second derivative spectra shown in Figure 6A, B. The spectra of wild type hGH along with the proteins E174A, F176A, K172A, is shown in Figure 6A. In these spectra the wavelength minima are coincident indicating no change in the polarity of the environment for these chromophores. In Figure 6B is the spectra of D169N, I58A, and the double mutant F176A/K172A. In this series of spectra a single chromophoric band at 303 nm is shifted toward shorter wavelengths (summarized in Table 1). This band is on the red edge of the hGH absorption spectrum, at 303 nm, has been assigned by Bewley and Li (*19*) to the tryptophan $0-0$ ^1L_a. The 12 nm red shift of this band is due to the hydrogen bond of the proton on the tryptophan (W86) indole nitrogen to a strong hydrogen bond acceptor, the aspartic acid (D169) in helix 2.

Therefore, this perturbation in hGH appears to be highly localized in that it is a perturbation of the hydrogen bond between the tryptophan at 86 and the aspartic acid at 169. There is also a 1.0 nm shift of the band at 291 nm, which is consistent with the shift in the ^1L_b band observed by Strickland (*19*). The tryptophan ^1L_b band is also sensitive to hydrogen bonding at the indole nitrogen, but the maximum shift is only about 2 nm (*19*). Otherwise, the minima in the second derivative spectra are coincident indicating that the effect is localized to the tryptophan and there is not a major change in the conformation of the molecule affecting the other chromophores.

These data also suggest that Ile58 perturbs Trp86 or influences the packing between helices so as to alter either the length or the angle of the hydrogen bond between Asp169 and Trp86. In an analysis of the structure (*1*) we find that the side chain of Ile58 packs between the side chains of Leu81 and Leu82 in helix 2. These two residues are 1.5 turns away from the tryptophan at position 86 and the distance between the side chains of Trp86 and Ile58 is approximately 10 angstroms. It is not obvious as to why the mutation I58A has such a dramatic effect on the circular dichroic and absorption spectra but packing against the residues Leu81 and Leu82 in helix 2 may restrict local mobility. Alternatively, Ile58 may force helix 2 closer to helix IV providing for a more optimal hydrogen bond between Asp169 and Trp86. The double mutant F176A/K172A appears to be folded correctly from the far UV circular dichroic spectrum but there appears to be an alteration in the packing between helix II and IV manifest as a blue shift of the 303 nm band which could change only if there is a change in the hydrogen bond between the indole nitrogen proton and the hydrogen bond acceptor.

Bewley and Li (*19*) originally proposed that W86 is buried in the interior of the hGH molecule, from an analysis of the absorption spectra, corroborated by Havel et al. (*23*) in an analysis of (I⁻), acrylamide, and trichloroethanol quenching of the homologous protein, bovine growth hormone (bGH), and this observation has now been confirmed by the X-ray crystal structure of this protein (*1*). The limited

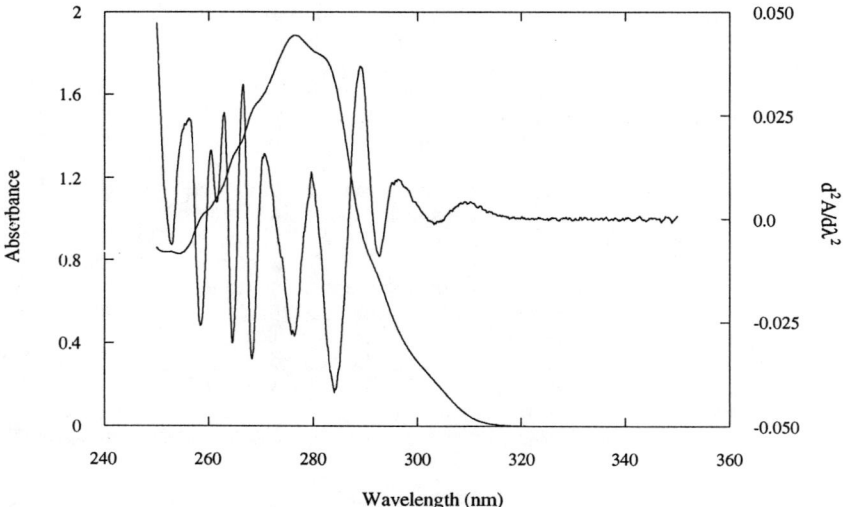

Figure 4: The absorption spectrum of hGH obtained in 0.01 M TRIS pH 7.5 at 20°C, left ordinate, and the calculated second derivative of that spectrum, right ordinate.

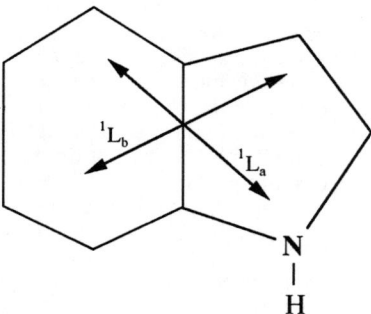

Figure 5: Schematic of the tryptophan indole structure with the ^1L_a and ^1L_b transition moments shown for tryptophan.

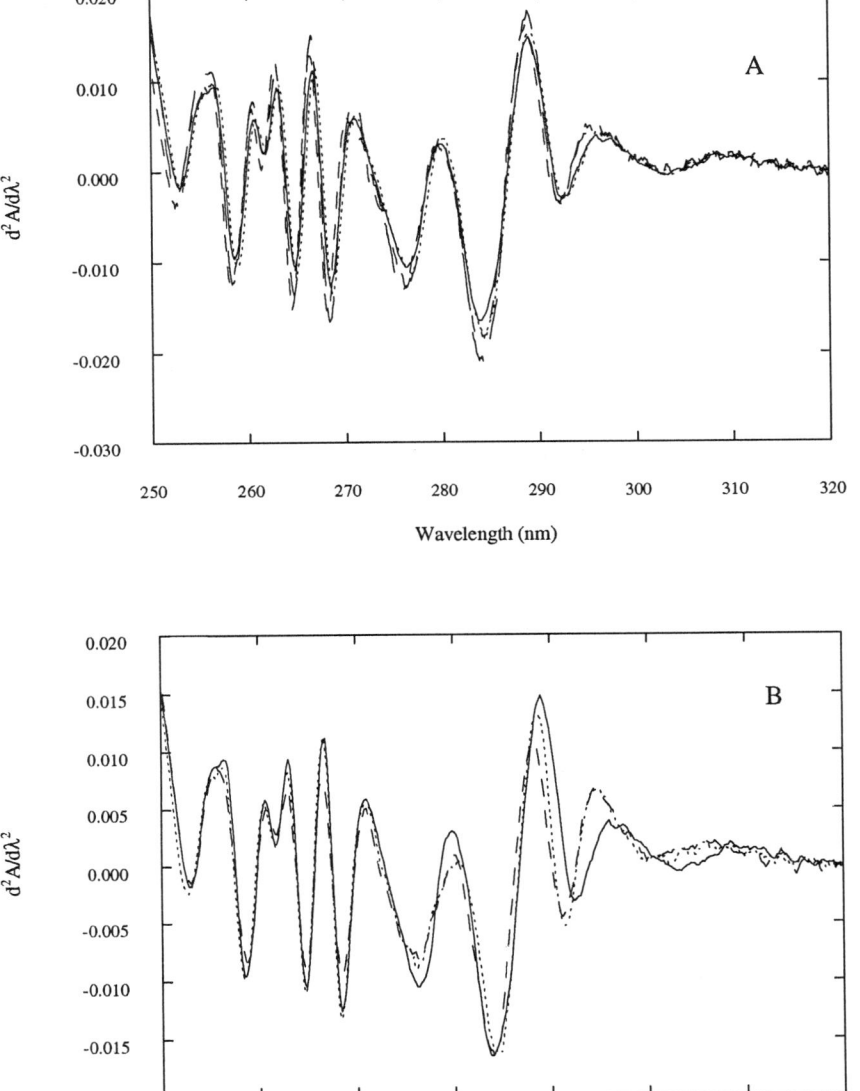

Figure 6: Second derivative spectra of hGH and the hGH mutants. Panel A is the second derivative of the spectra of wt hGH (———), F176A (– – –), K172A (-------), E174A (···········). Panel B is the second derivative of the spectra of wthGH (———), I58A (-----), and D169N (– – –).

accessibility of such buried tryptophans to collisional quenching reagents is recognized as a probe of protein dynamics, because changes in quenching can be related to changes in accessibility of the probe to the interior of the protein (26) and can be measured by Stern-Volmer analysis. Since our data show that some hGH mutants perturb the internal packing of the 4 helix bundle we may be able to detect alterations in the dynamics of these proteins that affect the solvent accessibility of W86. For these experiments we chose acrylamide as the quenching reagent since it is a more strongly penetrating quencher than iodide ion. The Stern Volmer quenching constant (K_{sv}) for bGH using acrylamide is 1.4 times higher than for iodide ion (23) and should permit sensitive measurements. If a given mutant does alter the dynamic properties of the protein such that the accessibility of W86 to acrylamide is increased we can detect it through an associated increase in the Stern Volmer quenching constant, (K_{sv}).

Figure 7 shows the Stern Volmer plot of each mutant protein discussed thus far. The data for proteins wt-hGH, K172A, F176A, E174A, K172A/F176A have been averaged and the linear least squares fit represents the average of this data, we calculate K_{sv} =2.2 M^{-1} from this average. This is lower than K_{sv} for bGH (2.7 M^{-1}) but not substantially different considering these two molecules are 67% homologous. The only outlier in this data is the variant hGH D169N. These data indicate there is an increase in the accessibility of the tryptophan, also these data are nonlinear. Using the method of Lehrer (16) we can calculate an apparent exposure (f_a, equation 2) for the single tryptophan in these molecules; for growth hormone and the mutants I58A, K172A, E174A, F176A, F176A/K172A, f_a is 0.11 and for D169N, f_a is 0.5. The residue Asp169 (Bewley, personal communication) was suggested as the most likely proton acceptor for the hydrogen bond with the single tryptophan (Trp86). This hydrogen bond has been shown in the X-ray crystal structure recently reported (1). Bewley's identification of Asp169 was originally based on the high degree of conservation of this tryptophan and aspartic acid within the growth hormone/prolactin family (3) and the apparent proximity of these residues in the pGH X-ray structure (2). Only for the mutant, D169N, is there an alteration in the permeability to the interior of the protein for acrylamide. Iodide ion on the other hand which has reduced permeability to the protein interior compared to acrylamide shows no changes in quenching for any of the proteins discussed (data not shown).

In summary, although most of these hGH molecules were chosen for their significant effect on the free energy of binding to the growth hormone binding protein. The far UV CD spectra indicates they retain the overall structure of the wt-hGH molecule. There are however, subtle differences in the structure of the interior of these molecules that are differentially reflected in the far UV CD, the near UV CD and in the quenching of tryptophan fluorescence. In the far UV CD, mutations made on the hydrophilic face of helix 4 are not expected to alter helix packing and do not alter the far UV CD signal. These mutations also have no effect on the near UV CD, nor would they be expected to. There is apparently no effect of these mutations on helix packing and aromatic residues exposed to solvent are expected to have little if any CD signal.

When the mutated residues are in the hydrophobic face of helix 4, changes are observed both in the far UV CD and in the near UV CD. These residues make up part of the hydrophobic core of the protein and when changed to alanine they can alter the packing of the four helix bundle. The changes observed in the far UV CD of these proteins are consistent with alterations in the packing of the hydrophobic core and do not necessarily imply a change in the content of secondary structure in the protein. Mutations made on the hydrophobic face of helix 4 consistently alter the near UV CD spectrum and in large part this is observed as a decrease in the rotational strength of the tryptophan CD signal.

Table 1. Wavelength position of the red shifted TRP^1L_a peak

Hormone	Tryptophan ^1L_a (nm)
hGH	303.5
I58A	302
K172A/F176A	301
D169N	300

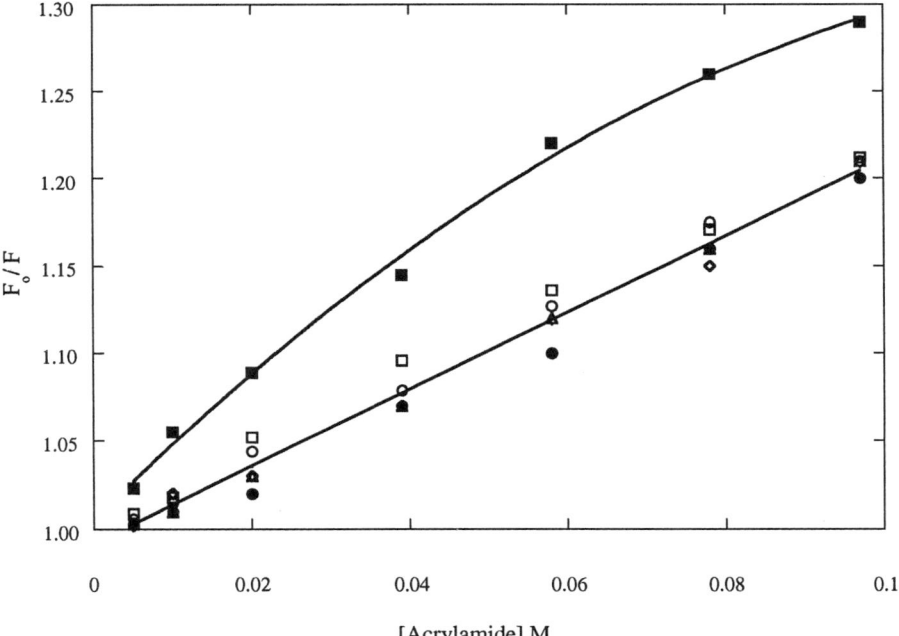

Figure 7: Stern Volmer analysis of tryptophan fluorescence in hGH and the hGH mutants. Data is plotted according to equation 1. Shown is the data for wt hGH (●), D169N (■), E174A (◊), I58A (○), K172A (Δ), and F176A/K172A (□). The lower solid line represents a linear least squares fit to the data from all of the mutants except the D169N. The line through the data for D169N has no theoretical significance.

Although changes seen in the near UV CD for these mutant proteins implies there may be significant changes in the internal packing of the protein. Stern-Volmer analysis coupled with the near and far UV CD spectra demonstrate that any changes are highly localized. The only exception to this is the protein hGH D169N. This protein has increased permeability to acrylamide. For hGH D169N, this experiment indicates an increase in the dynamics of the four helix bundle that allows greater permeability of acrylamide to the protein interior and implies the hydrogen bond between Trp86 and Asp169 may provide an important degree of structural stability to hGH.

In combination, these techniques have provided significant detailed information about the structure of the mutants of hGH that affect the binding of the hormone to the receptor. Mutants in this region of the protein which reside in the hydrophobic core preferentially effect the environment of the tryptophan (Trp86). For hGH the tryptophan provides a particularly useful handle to probe the mutagenesis of hGH. Moreover, as is shown here and shown by Ogasahara et al. (12) in their analysis of tryptophan synthetase, having a large number of mutants for a single protein allows one to observe and analyze structural perturbations in some detail. Analysis of a number of mutants allows one to see trends in changes of structure. Saturation mutagenesis at position 49 in the tryptophan synthetase α subunit have shown a correlation between stability of the protein, the hydrophobicity of the residues substituted at position 49 and the rotatory strength of tyrosine in the near UV CD (12). Analysis such as these, particularly when they can be correlated with the X-ray crystal structure should help to provide a detailed analysis of the CD signal which can be used for protein structural analysis.

Acknowledgments

We want to thank Thomas Bewley, Abraham de Vos, and Jim Wells and for helpful discussions and encouragement; Abraham de Vos for help with Figure 1, Mike Brochier for help with fermentation; and M. Vasser, P. Ng, P. Jhurani for synthetic oligonucleotides.

LITERATURE CITED

1. DeVos, A.M.; Ultch, M.; Kossiakoff, AA. *Science* **1992**, 255, 306-312.
2. Abdel-Meguid, S. S.; Sheih, H. S.; Smith, W. W.; Dayringer, H. E.; Violand, B. N.; Bertle, L. A. *Proc. Natl. Acad. Sci. USA* **1987**, 84, 6434-6437.
3. Nicol, C. S.; Mayer, G. L.; Russel, S. M. *Endocrine Reviews* **1986**, 7, 169-202.
4. Cunningham, B. C.; Jhurani, P.; Ng, P.; Wells, J. A. *Science* **1989**, 243, 1330-1335.
5. Cunningham, B. C.; Wells, J. A. *Science* **1989**, 244, 1081-1085.
6. Bass, S.; Mulkerrin, M. G.; and Wells, J. A. *Proc. Natl. Acad. Sci. USA* **1991**, 88, 4498-4502.
7. Cunningham, B. C.; Henner, D. J.; and Wells, J. A. *Science* **1990**, 247, 1461-1465.
8. Bell, J. A., Becktel, W. J., Sauer, U., Baase, W. A., Matthews, B. W. *Biochem.* **1992**, 31, 3590-3596.
9. Fothergill, M. D.; Fersht, A. R. *Biochem.* **1991**, 30, 5157-5164.

10. Collawn, J. F.; Wallace, C. J. A.; Proudfood, A. E. I.; Patterson, Y. *J. Biol. Chem.* **1988**, 263, 8625-8634.
11. Wright, G.; Freedman, R. B. *Prot. Eng.* **1989**, 2, 583-588.
12. Ogasahara, K.; Sawada, S.; Yutani, K. *Proteins: Structure, Function and Genetics* **1987**, 5, 211-217.
13. Yutani, K.; Hayashi, S.; Sujisaki, Y.; Ogasahara, K. *Proteins: Structure, Function, and Genetics* **1991**, 9, 90-98.
14. Moy, F. J.; Scheraga, H. A.; Kui, J. F.; Wu, R.; Montelione, G. T. *Proc. Natl. Acad. Sci., USA* **1989**, 86, 9836-9840.
15. Schauenstein, E.; Bayzer, H. *J. Polymer Sci.* **1955**, 16, 45-51.
16. Lehrer, S. S. *Biochem.* **1971**, 10, 3254-3263.
17. Strickland, E. H. CRC Crit. Rev. *Biochem.* **1974**, 2, 113-175.
18. Holladay, L. A.; Hammonds, R. G.; Puett, D. *Biochem.* **1974**, 13, 1653-1661.
19. Bewley, T.A.; Li, C.H. *Arch. of Biochem. and Biophys.* **1984**, 233, 219-227.
20. Manning, M. C.; Illangusekare, M.; Woody, R. W. *Biophys. Chem.* **1988**, 31, 71-86.
21. Manning, M. C.; Woody, R. W. *Biopolymers* **1991**, 31, 569-586.
22. Cooper, T. M.; Woody, R. W. *Biopolymers* **1990**, 30, 657-676.
23. Havel, H.A.; Kaufman, E. W.; Elzinga, P. A. *Biochem. Biophys. Acta* **1988**, 955, 154-163.
24. Butler W. L.; Hopkins, D. W.; *Photochemistry and Photobiology* **1970**, 12, 439.
25. Yamomoto, Y.; and Tanaka, J.; *Bulletin of the Chemical Society of Japan* **1972**, 45, 1362-1366.
26. Lakowicz, J. R.; Weber, G. Q. *Biochem.* **1979**, 12, 4171-4179.
27. Kraulis, P. J. *J. Appl. Cryst.* **1991**, 24, 946-950.

RECEIVED October 26, 1992

Chapter 19

Altering the Self-Association and Stability of Insulin by Amino Acid Replacement

David N. Brems, Patricia L. Brown, Christopher Bryant,
Ronald E. Chance, Richard D. DiMarchi, L. Kenney Green,
Daniel C. Howey, Harlan B. Long, Alita A. Miller, Rohn Millican,
Allen H. Pekar, James E. Shields, and Bruce H. Frank

Lilly Research Laboratories, Eli Lilly and Company, Indianapolis, IN 46285

We have manipulated the amino acid sequence of insulin to speed the absorption rate from the subcutaneous injection site and to improve the storage stability. The self-association of insulin was shown to be particularly sensitive to amino acid changes in the C-terminus of the B chain. The self-association of insulin was completely abolished in some analogs, which resulted in the biological effect of faster absorption rates from the subcutaneous injection site. The storage stability of Zn-free insulin was shown to be limited by the lability of the disulfide bonds. The maintenance of the native state of insulin was shown to be important in protecting the disulfides, indicating that the storage stability of insulin is under the thermodynamic control of the conformational equilibria. The storage stability of these insulin analogs differed by 100-fold.

Altering Self-Association

Insulin association is well documented but not fully understood. The insulin association behavior is known to be complex, with the metal-free species exhibiting a pH, ionic strength, and protein concentration-dependent association pattern consisting of monomer, dimer, tetramer, and higher ordered polymers all in dynamic equilibrium *(1-6)*. The association constant for the metal-free porcine monomer-dimer equilibrium at pH 7.0 is 1.4-7.5 x 10^5 M^{-1} *(3-7)*.

The most detailed atomic information concerning insulin association comes from X-ray crystallography. A variety of crystal forms of insulin have been studied. The N- and C-termini of the B chain undergo significant conformational changes between the different crystal forms *(8-12)*. The relationship between conformational flexibility of insulin, determined from X-ray crystallography, and the self-association in solution is of considerable importance. The crystal structure of the 2-Zn hexameric form and thermodynamic studies of insulin dimerization indicate that both hydrophobic interactions and hydrogen bonding are responsible for stabilizing association *(1,2,6)*. X-ray crystallography results demonstrate that most of the non-polar dimer contacts involve the C-terminal end of the B chain, with B^{23-26} and B^{28} being the most predominant residues *(10)* (see Figure 1 for the amino acid sequence of human insulin). Association of the dimer is secured by a small antiparallel ß-sheet of hydrogen

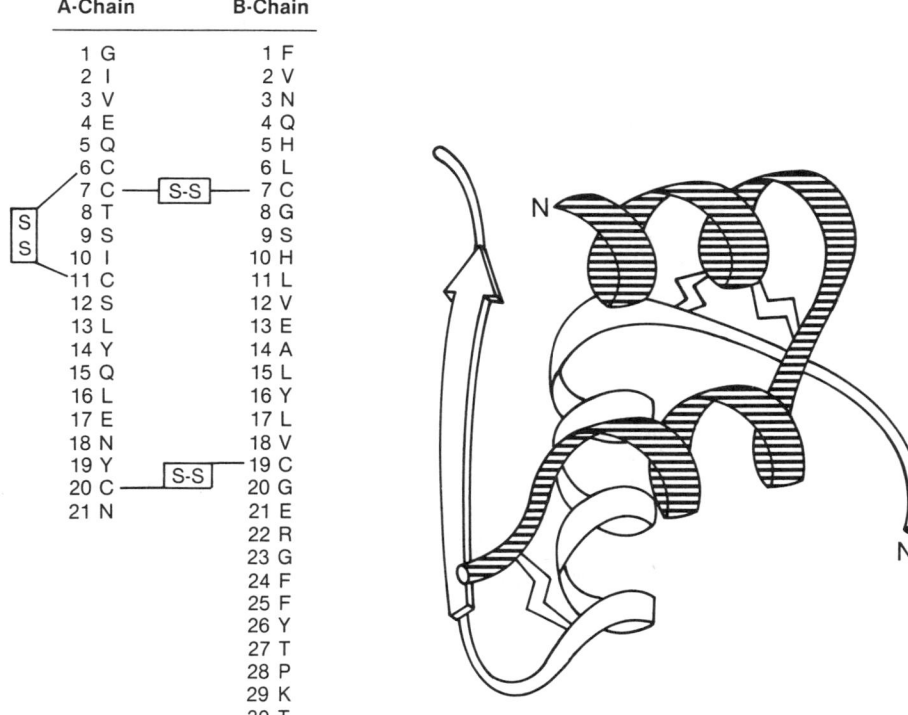

Figure 1. Amino acid sequence of human insulin and ribbon drawing of the monomeric structure (the A chain is striped).

bonds involving residues B^{24-26}. Formation of the dimer packs Pro^{B28} against residues B^{20-23}, which are in a ß-turn *(10)*. The hexamer is mainly stabilized through the Zn coordination, but additional polar and non-polar residues are buried between the dimers as a result of hexamer assembly *(10)*.

The importance of the C-terminus of the B chain is further evidenced by removal of B^{26-30} in despentapeptide insulin, which does not significantly alter the rest of the molecule but abolishes dimerization *(13)*. In this study, the C-terminus of the B chain was systematically truncated and it was established that Pro^{B28} is critical to the stabilization of insulin self-association in solution. Amino acid replacement was used to further investigate the role of B^{28} and B^{29} on self-association.

The ability to design an active insulin with diminished self-association is important for future diabetes therapy. All commercial pharmaceutical formulations contain insulin in the self-associated state and predominantly in the hexamer form *(14)*. Current models propose that the rate-limiting step in absorption of insulin from a subcutaneous injection site is the dissociation to monomer, which is readily absorbed *(15)*. Indeed, researchers at both the Novo Research Institute and the Lilly Research Laboratories have shown that monomeric insulin analogs act more rapidly than current formulations *(16, 17)*.

The Effect on Association of Truncating the C-Terminus of the B Chain. The ultracentrifugation properties of intact insulin, Des^{B30} insulin, Des^{B29-30} insulin, Des^{B28-30} insulin and Des^{B27-30} insulin, in the absence of Zn, are illustrated in Figure 2 *(see ref. 18 & 19 for experimental details)*. Removal of B^{30} or B^{29} has little effect on aggregation, but removal of B^{28-30} or B^{27-30} results in much less self-association over the concentration range of 0-5 mg/ml. The solid lines in Figure 2 represent calculated curves that best fit the experimental data. The actual data for Des^{B30} insulin and Des^{B28-30} insulin are illustrated to demonstrate the typical variation between the data and the calculated curve. We conclude from these results that B^{30} and B^{29} are not critical, but the C-terminal three amino acids are essential for stabilizing association. Removal of Pro^{B28} removes the critical intermolecular contact with B^{21}. These truncated insulins are at least 50% equipotent to insulin, as measured by receptor binding using membrane preparations of human placenta, and are at least 60% equipotent to insulin in lowering blood glucose in rats. Therefore removal of the C-terminal segment of the B chain does not destroy the conformational integrity nor activity.

The Effect on Association of Altering B^{28} and B^{29}. The ultracentrifugation properties of several insulin analogs, altered at B^{28} and/or B^{29} in the absence of Zn, are illustrated in Figure 3 *(see ref. 18 & 19 for experimental details)*. The solid lines represent the calculated curves that best fit the experimental data. The actual data for the insulin sample are illustrated to demonstrate the typical variation between the data and the calculated curve. The calculated curves in Figure 3 were obtained using an even-aggregate model of association *(5)* and the equilibrium constants were varied to achieve the best fit to the experimental data. The data are not sufficiently precise to exclude alternative models. The results show that substitution of Pro^{B28} reduces the extent of aggregation, with the least aggregation observed for $Asp^{B28} < Ala^{B28} \approx Lys^{B28}$. Variation of B^{28} plus replacement of Lys^{B29} with Pro diminishes aggregation further. For the Xaa^{B28} or the $Xaa^{B28}Pro^{B29}$ series of analogs, the least aggregation results when B^{28} is an acid side chain and the most aggregation results when B^{28} is an imino acid side chain. The equilibrium constants $K_{1,2}$ and K were determined for the analogs and are shown in Table I *(see 18 & 19 for experimental details)*. The data in Table I show that for the analogs the dimerization constants are altered more than the constant for higher order association.

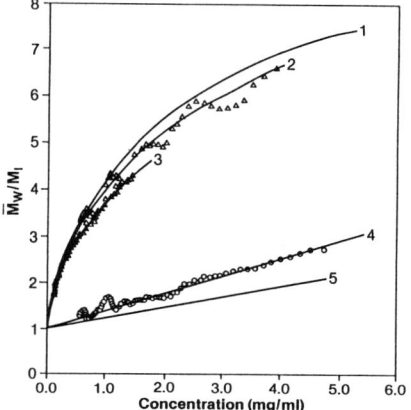

Figure 2. Equilibrium ultracentrifugation of human insulin and analogs. Effect of protein concentration of Zn-free insulin (curve 1) and Zn-free C-terminal truncated analogs: Des^{B30} insulin, curve 2; Des^{B29-30} insulin, curve 3; Des^{B28-30} insulin, curve 4; and Des^{B27-30} insulin, curve 5. (Δ), observed data for Des^{B30} and (O), Des^{B28-30} insulin. Each line represents a calculated curve that best fits the data according to the even-aggregate model of association (see Table I). M_W/M_1 represents the weight average molecular weight obtained from ultracentrifugation divided by the molecular weight of insulin monomer. (Reproduced with permission from ref. 19. Copyright 1992 IRL Press.)

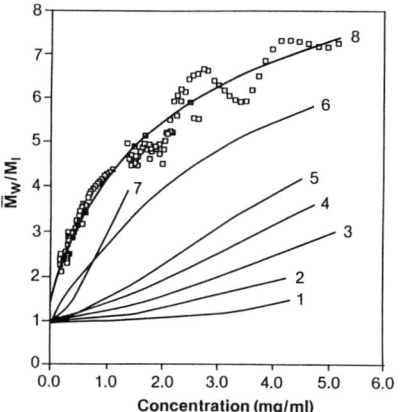

Figure 3. Equilibrium ultracentrifugation of human insulin and analogs. Effect of protein concentration on ultracentrifugation in the absence of Zn of: Asp^{B28}Pro^{B29} insulin, curve 1; Ala^{B28}Pro^{B29} insulin, curve 2; Lys^{B28}Pro^{B29} insulin, curve 3; Pro^{B29} insulin, curve 4; Asp^{B28} insulin, curve 5; Ala^{B28} insulin, curve 6; Lys^{B28} insulin, curve 7; and insulin, curve 8. The symbols represent the observed data for insulin. Each line represents a calculated curve that best fits the data according to the even-aggregated model of association (see Table I). (Reproduced with permission from ref. 19. Copyright 1992 IRL Press.)

Table I. Equilibrium Constants for an Even-Aggregation Model[a] as Determined by Ultracentrifugation

Analog (Zn-free)	$K_{1,2}$ (x 10^3 M^{-1})	K_{2n+2} (x 10^3 M^{-1})
Human insulin (HI)	300	12
Ala^{B28}HI	15	7.5
Lys^{B28}HI	2.3	27
Asp^{B28}HI	1.5	6.5
Pro^{B29}HI	1.1	4.8
Lys^{B28}Pro^{B29} HI	0.9	3.5
Ala^{B28}Pro^{B29} HI	0.5	3.6
Asp^{B28}Pro^{B29} HI	0.1	7.5

[a]Even-aggregate model $I_1 \leftrightarrow I_2 \leftrightarrow I_4 \leftrightarrow I_6 ... I_{12} \leftrightarrow I_{14}$. $K_{1,2} = [I_2]/[I_1]^2$ and $K_{2n+2} = [I_{2n+2}]/[I_2][I_{2n}]$ where n = 1-6 *(see ref. 5 for details)*.

Mechanism for Decreased Self-Association. Insulin in dilute concentrations has an altered conformation compared to insulin in more concentrated solutions *(6)*. Dilute insulin has a 15-20% increase in ß-sheet and a concomitant decrease in random coil *(20.)*. The far-UV circular dichroism data of the predominantly monomeric analogs are each similar and indistinguishable from that of dilute insulin (data not shown). Our results show that the location of a Pro residue at B^{28} is crucial for high-affinity dimerization of insulins. Thus, the absence of Pro at B^{28} in a series of C-terminal truncated insulins or amino acid replacements at B^{28} leads to decreased self-association. Replacing Lys^{B29} with Pro and varying the amino acid at B^{28} caused even greater diminution of self-association. The X-ray structure of 2-Zn insulin hexamer indicates that Pro^{B28} makes a non-bonded intermolecular contact of ≤3.5Å to B^{21}. Removal or replacement of Pro^{B28} with other amino acids destroys this critical self-association contact, and in solution disrupts dimerization to the extent shown in Table I. Lys^{B29} does not make critical intermolecular interactions as evidenced by the self-association of Des^{B29-30} insulin. Charge repulsion at B^{28} is not the cause for the disruption in self-association *(21)*, because neutral, acid, and basic amino acid substitutions at B^{28} all significantly disrupt dimerization (Table I). We suggest that insertion of Pro at B^{29} creates a new surface that interferes with dimerization. The combination of removing the Pro^{B28} intermolecular contact and introducing an interfering contact with Pro at B^{29} is most effective at disrupting insulin self-association without destroying the insulin monomeric conformation.

Time-Action Profile for a Monomeric Insulin Analog. The rate-limiting step of insulin absorption from the subcutaneous injection site is thought to be the dissociation process *(15)*. Insulin analogs that are predominantly monomeric at the normal concentrations used for injection therapy (3.5 mg/ml) are expected to diffuse more quickly from subcutaneous injection to plasma circulation. Lys^{B28}Pro^{B29} insulin was tested for its time-dependent activity profile in normal human volunteers. Lys^{B28}Pro^{B29} insulin was chosen because it was a fully potent insulin agonist by *in vitro* binding assessment relative to insulin in IM-9 lymphocytes, glucose transport in isolated rat adipocytes, *in vivo* activity in rats and rabbits, and that it displayed extremely weak self-association. The time action profile of Lys^{B28}Pro^{B29} insulin (Zn-free) has been compared to Humulin R (Zn-human insulin formulated for commercial sale) *(17)*. The results of a randomized crossover design study in 9 healthy men using the euglycemichyperinsulinemic clamp technique *(22)* is shown in Table II.

Table II. **Human Clinical Study of the Time Action of Lys^{B28}Pro^{B29} insulin compared to Humulin R**

Physiological response[a]	Lys^{B28}Pro^{B29} insulin	Humulin R	pvalue[b]
Cmax[c] (ng/ml)	5.26	1.80	0.0003
Tmax[d] (hr)	0.56	2.36	0.0045
Ka[e] (1/hr)	0.829	0.373	0.0001
Rmax[f] (mg/min)	501	436	ns[g]
TRmax[h] (hr)	1.03	2.17	0.0038
DA[i] (hr)	3.60	7.10	0.0001

[a]10 units (28.57 units/mg) of each insulin injected subcutaneously (the *in vivo* potency of Lys^{B28}Pro^{B29} insulin is equal to that of human insulin).
[b]analysis of variance (i.e. a value of 0.0003 implies that the differences observed in this trial are expected only 3 times in 10,000 similar trials if insulin and the analog did not really promote different time actions).
[c]peak concentration of insulin or analog.
[d]time to peak.
[e]absorption rate constant.
[f]peak glucose infusion rate.
[g]not significant.
[h]time to peak glucose infusion rate.
[i]duration of action.

Table II demonstrates that Lys^{B28}Pro^{B29} insulin displays a more rapid onset and disappearance of action than Humulin R. The time action of Lys^{B28}Pro^{B29} insulin more closely approximates the normal physiological response to a meal. Clinical studies are currently ongoing to assess the degree of improvement in diabetes management that might be achievable through use of this fast-acting insulin analog or other appropriate monomeric analogs.

Altering Stability

In vitro protein stability is important in all investigations regarding proteins. Lack of stability or degradation can prohibit or compromise the understanding of protein structure-function relationships. When proteins are utilized as pharmaceuticals (i.e., insulin), stability is of paramount importance to millions of patients. Formulations of insulin consist of liquid preparations that are administered by injection. Some degradation products are inactive and may cause unnecessary complications *(23)*. In current insulin formulations, every effort is made to keep insulin degradation to a minimum. The current expiration date for insulin formulations ranges from 18 to 24 months and is governed by the intrinsic degradation rate of insulin *(23-24)*. It may be possible to increase the stability of insulin through appropriate replacement or deletions of amino acids, and thereby reduce unwanted degradation and extend the expiration time. The native state of proteins is known to provide protection against degradation, which suggests that the solution-state storage of a protein may be governed by the equilibrium constant of unfolding;

$K_{eq} = U / N$, where: N = native and U = unfolded; for the reaction N <----> U.

Excipients or amino acid replacements that increase the equilibrium constant are predicted to increase chemical degradation, and those that decrease the equilibrium constant are predicted to decrease chemical degradation. To establish that such a relationship exists, the rate of chemical degradation and the conformational stability of numerous insulin analogs were examined. The chemical degradation studies were conducted by incubating solutions of insulin or insulin analogs at 50°C for varying time periods and measuring the formation of large molecular weight polymers by size-exclusion chromatography. Conformational stability was determined by guanidine hydrochloride (GdnHCl)-induced equilibrium denaturation *(25)*.

Chemical Stability. Insulin and analogs were stored at 50°C for varying times. Chemical degradation accelerates at a temperature of 50°C but heat denaturation is not induced (unpublished results). Degradation at 50°C was detected by size-exclusion chromatography and the results are illustrated in Figure 4 *(see ref. 18 & 19 for experimental details)*. The chromatography mobile phase contained 6M GdnHCl to disrupt any noncovalent self-association of insulin. Therefore, peaks eluting earlier than monomer represent covalent adducts of insulin. The primary degradation product is a large molecular weight polymer that elutes near the exclusion limit of the column. When DTT was added to a sample that had been degraded for 6 days, the polymer disappeared (Figure 4) and reduced to authentic insulin A chain and insulin B chain, as determined by RP-HPLC (data not shown). Peaks eluting after 720 s (Figure 4) correspond to the buffers and preservatives that were present in the incubation mixture.

The mechanism of the large molecular weight polymer formation was explored. Table III *(see ref. 18 & 19 for experimental details)* shows that the rate of polymer formation is dependent on pH, temperature, and metal ions.

Table III. **Effect of pH, Metal Ion, and Temperature on the Rate of Polymer Formation** [a,b]

	50°C				40°C	30°C
	pH 9	pH 8	pH 7.4	Cu^{2+}(pH 8)[c]	pH 8	pH 8
Human insulin[d]	12.4	15	139	56	392	4029

[a]Polymer was detected by size-exclusion chromatography using 6M GdnHCl, 0.1M phosphate, pH 3.0, as the eluting solvent.
[b]Values are half-lives, in days, determined from the slope of a plot of the ln of the area of the high molecular weight polymer peak versus time.
[c]100 μM Cu^{2+}.
[d]3.5 mg/ml.

The polymerization rate increases with increasing pH, suggesting that the mechanism is dependent on the concentration of hydroxyl ions. The polymerization rate increases with higher temperatures. Redox metal ions, such as Cu^{2+}, retard the polymerization

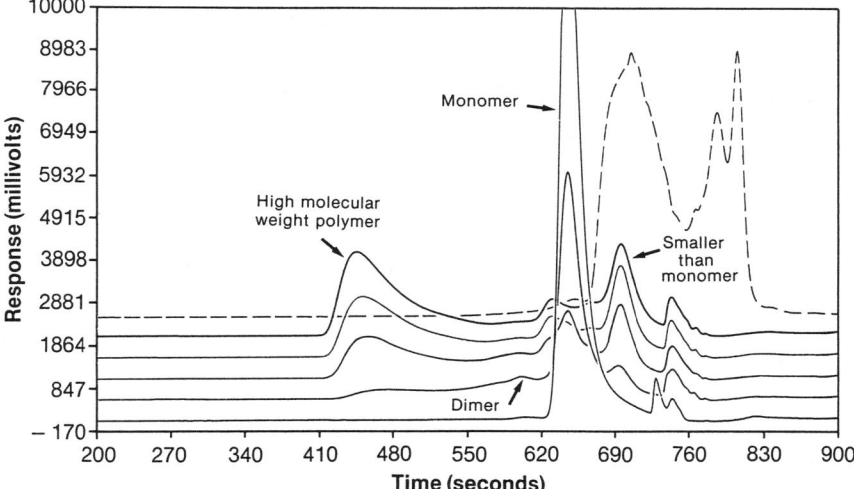

Figure 4. Size-exclusion chromatography of human insulin incubated for varying time periods at 50°C, in 0.1 M sodium phosphate, pH 8.0, 0.25% m-cresol and 1.6% glycerol. The bottom chromatogram sample was not incubated, and in ascending order the chromatograms were incubated for 1, 2, 3, and 6 days, and 6 days with DTT added after incubation but prior to chromatography (dashed line). The chromatography elution solvent was 6M GdnHCl, 0.1M sodium phosphate, pH 3.0. (Reproduced with permission from ref. 18. Copyright 1992 IRL Press.)

reaction (the oxidized form is required for the retardation). These results are consistent with a β-elimination mechanism of the disulfide bonds and a disulfide interchange reaction, as previously established by Zale and Klibanov (26) as a major degradation pathway for insulin at 100°C. The polymerization rate of metal-free insulin at 40 and 30°C was explored and the results are shown in Table III. At all temperatures tested, a high molecular weight polymer is the major degradation product. The rate of polymerization was linear with respect to temperature over the range investigated. For insulin, 50°C is an accelerator for chemical degradation and is representative of the polymerization mechanism that occurs at lower temperatures. The possibility that the OH^- concentration and metal ions affect the rate of polymerization by altering the equilibrium constant for self-association has been considered. However, monomeric analogs that remain monomeric under the variations of Table III still behave as expected for a β-elimination and disulfide interchange mechanism. The rate of polymerization to high molecular weight forms at 50°C was investigated for numerous insulin analogs. The results for the rate of polymerization for analogs and insulin are illustrated in Figures 5 and 7 *(see ref. 18 & 19 for experimental details)*. As can be seen in Figure 5, the polymerization rate of insulin analogs varies greatly, with $Asp^{B10}Asp^{B28}Pro^{B29}$ insulin demonstrating greater stability and $Lys^{B28}Pro^{B29}$ insulin showing lesser stability compared with insulin.

Conformational Stability. Guanidine hydrochloride-induced equilibrium denaturation was utilized to determine the conformational stability of insulin and insulin analogs. Denaturation was measured by circular dichroism at 224 nm. Previous studies of insulin have established that the denaturation transitions are reversible and conform to a two-state mechanism, with the two states being monomeric native and monomeric denatured *(27)*. The denaturation experiments contained 20% ethanol to disrupt insulin self-association and to avoid the complications of intermolecular interactions *(27)*. Figures 6, 7, and Table IV *(see ref. 18 & 19 for experimental details)* show the equilibrium denaturation results for numerous insulin analogs. Figure 6 shows the equilibrium denaturation results for insulin, $Lys^{B28}Pro^{B29}$ insulin (the least stable analog), and $Asp^{B10}Asp^{B28}Pro^{B29}$ insulin (the most stable analog). The free energy of unfolding varies by 2 kcal/mol (Figure 6 & Table IV) for the different analogs. The majority of analogs belong to a related series for which Lys^{B29} was changed to proline and B^{28} was varied. In this series of analogs, the conformational stabilities are comparable or less than that for insulin. Changing His^{B10} to aspartic acid significantly increased the free energy of unfolding. The effect of Asp^{B10} was also evident for analogs in which the C-terminus was altered. We suggest that the stabilizing effect of Asp^{B10} and the $Xaa^{B28}Pro^{B29}$ could be caused by a positive helix dipole effect *(28-30)* or a hydrogen bond acceptor near the N-terminal end of a helix *(31-32)*. Residues B^{9-19} form an α-helix containing a net helix dipole of one positive charge near B^9 and one negative charge near B^{19} (Figure 1). A negative charge at B^{10} from an aspartic side chain might be expected to favorably interact with the positive charge dipole and cause a net stabilization of the helix and an overall increase in the protein's conformational stability. An alternative explanation for the stabilizing effect of Asp^{B10} is that the ß-carboxyl of aspartic acid can form hydrogen bonds with any of the initial four main chain NH groups of the helix B^{9-19} that are otherwise unpaired.

Correlation of Chemical and Conformational Stability. Figure 7 shows the rate of chemical degradation versus the midpoint of denaturation for the different insulin analogs. The straight line in Figure 7 was determined by a least-squares fit to the data and has an R^2 value of 0.7 and $P < 0.0005$. Insulin analogs with the greatest chemical stability have the largest midpoint of denaturation and those analogs with the least chemical stability have the smallest midpoint of denaturation. The positive correlation for the rate of polymerization versus the midpoint of denaturation indicates that the

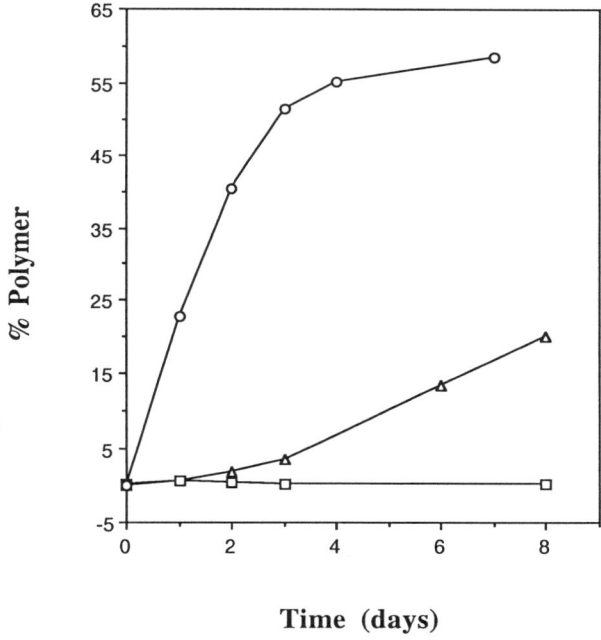

Figure 5. Stability to chemical degradation (tendency to form polymers). Human insulin (△), Asp^{B10}Asp^{B28}Pro^{B29} insulin (□), and Lys^{B28}Pro^{B29} insulin (○) were incubated for various times at 50°C as in Figure 4 except at pH 7.4. The formation of polymer was determined by size-exclusion chromatography with 6M GdnHCl, 0.1M sodium phosphate, pH 3 as the chromatography elution solvent. (Reproduced with permission from ref. 18. Copyright 1992 IRL Press.)

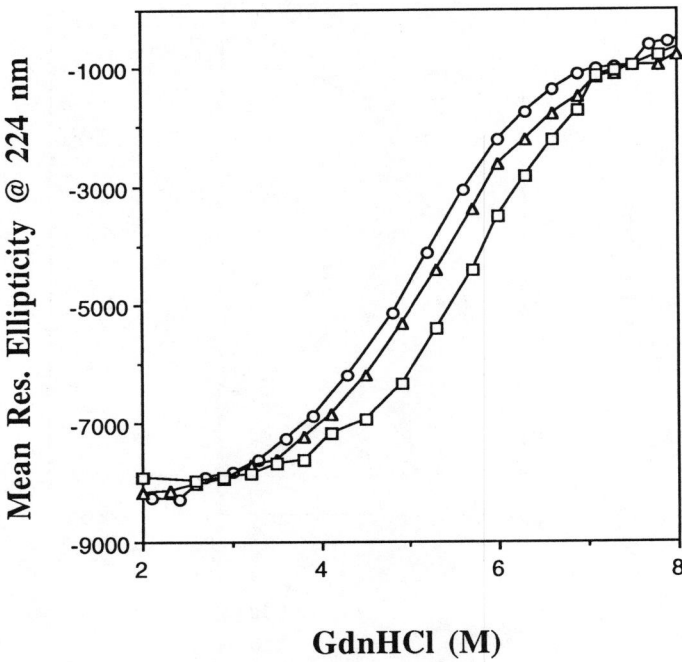

Figure 6. Stability of the conformational equilibria. GdnHCl-induced equilibrium denaturation of insulin (Δ), Asp^{B10}Asp^{B28}Pro^{B29} insulin (□), and Lys^{B28}Pro^{B29} insulin (O), measured by far-UV CD as described in ref. 27. (Reproduced with permission from ref. 18. Copyright 1992 IRL press.)

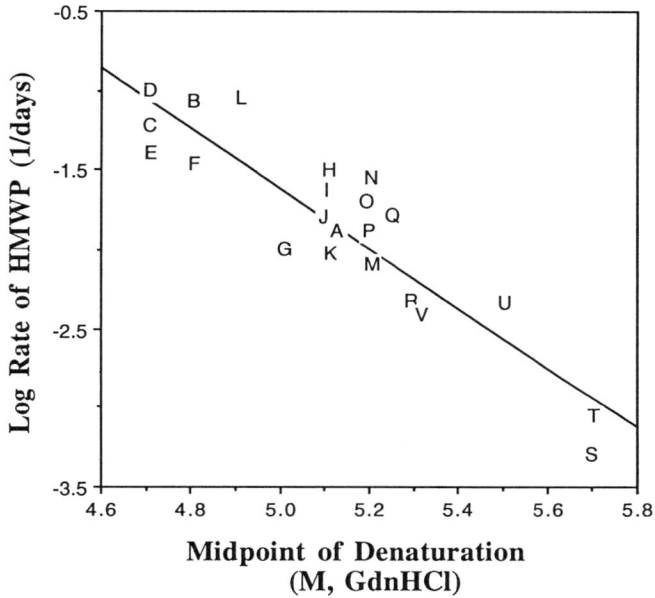

Figure 7. Correlation between chemical stability and conformational stability. Chemical stability was determined by the tendency to form polymer at 50°C as in Figure 4. Conformational stability was estimated by the transition midpoint of the GdnHCl-induced equilibrium denaturation as described in ref. 27. The letter code for each analog is contained in Table IV. (Reproduced with permission from ref. 18. Copyright 1992 IRL press.)

chemical stability of insulin is under the thermodynamic control of the protein conformational equilibria.

Since the main pathway for chemical degradation for Zn-free insulin is through disulfide destruction, we explored the rate of chemically induced reduction of protein disulfides for the analogs that differed the most in chemical and conformational stability. Insulin or insulin analogs were exposed to DTT for varying times and disulfide reduction was determined by reversed-phase chromatography. The results of such an experiment are illustrated in Figure 8 *(see ref. 18 & 19 for experimental details)*. The ordinate in Figure 8 represents the percent insulin that remains native or unreacted. As can be seen in Figure 8, $Asp^{B10}Asp^{B28}Pro^{B29}$ insulin is the most resistant to reduction, insulin showed intermediate susceptiblity, and $Lys^{B28}Pro^{B29}$ insulin was the most labile. To verify that the different susceptibilities observed for insulin and its analogs are conformationally related, the reduction was carried out in denaturing concentrations of GdnHCl. The results of this control experiment are illustrated in Figure 8 and demonstrate that the susceptibility to reduction of insulin and its analogs is indistinguishable in GdnHCl. Guanidine hydrochloride was shown not to alter the reducing potential of the reaction mixture by measuring the rate of DTT-induced reduction of glutathione in the presence and absence of GdnHCl (data not shown). Therefore, different susceptibilities observed for insulin in the presence and absence of denaturant is a measure of the disulfide protection provided by the native structure. In order to develop $lys^{B28}pro^{B29}$ insulin as a commercial product of

Table IV. **Results from Equilibrium Denaturation**

Code	Insulin analog	Midpoint[a] (M GdnHCl)	ΔG (kcal/mol)[b]
A	human insulin (HI)	5.1	4.4
B	A^{B28}-HI	4.8	4.0
C	des^{B23-30}-HI	4.7	3.8
D	$W^{B28}P^{B29}$-HI	4.7	3.9
E	$L^{B28}P^{B29}$-HI	4.7	4.2
F	$S^{B28}P^{B29}$-HI	4.8	4.0
G	des^{B30}-HI	5.0	4.6
H	$F^{B28}P^{B29}$-HI	5.1	4.9
I	G^{B29}-HI	5.1	4.3
J	$G^{B28}P^{B29}$-HI	5.1	4.6
K	$Q^{B28}P^{B29}$-HI	5.1	4.8
L	$K^{B28}P^{B29}$-HI	4.9	3.6
M	$E^{B28}P^{B29}$-HI	5.2	4.6
N	$A^{B28}P^{B29}$-HI	5.2	4.8
O	$Aba^{B28}P^{B29}$-HI	5.2	5.4
P	$V^{B28}P^{B29}$-HI	5.2	4.7
Q	$D^{B10}K^{B28}P^{B29}$-HI	5.3	4.7
R	D^{B10}-HI	5.3	5.1
S	$D^{B10}V^{B28}P^{B29}$-HI	5.7	4.9
T	$D^{B10}D^{B28}P^{B29}$-HI	5.7	5.7
U	$D^{B10}E^{B28}P^{B29}$-HI	5.5	5.2
V	$D^{B10}Q^{B28}P^{B29}$-HI	5.3	5.4

[a]The midpoint of the denaturation transition where: $K_{eq} = U/N = 1$. Repetitive measurements of separate sample preparations show a variation of ±0.1 M GdnHCl.

[b]Gibbs free energy of unfolding in the absence of denaturant, which equals $\Delta G + m$ (GdnHCl), where m is a measure of the dependence of ΔG on [GdnHCl]. Repetitive measurements of separate sample preparations show a variation of ±0.5 kcal/mol.

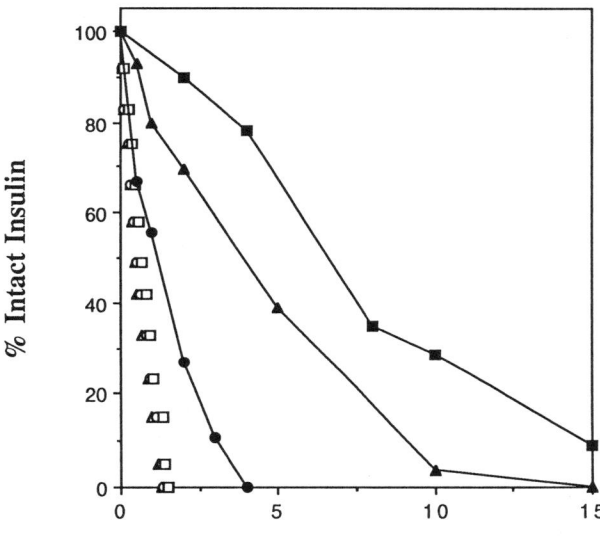

Figure 8. Susceptibility to reduction. Insulin (▲), Lys^{B28}Pro^{B29} insulin (●), and Asp^{B10}Asp^{B28}Pro^{B29} insulin (■) at a concentration of 1.2 mg/ml were incubated in a reducing solution containing 25 mM DTT, pH 7.5, 0.2 M Tris at 23°C, and at various times samples were withdrawn, quenched in acid and analyzed for unreduced (intact) insulin by reversed-phase HPLC *(18)*. The open symbols are the results when the reducing solution contained 7 M GdnHCl. (Reproduced with permission from ref. 18. Copyright 1992 IRL Press.)

adequate stability, an alternative strategy that incorporates stabilizing additives into the formulation will need to be developed. Instability is not inherent to monomeric insulins as demonstrated by the extreme stability of analogs containing asp^{B10}.

We conclude the different rates of polymer formation for the analogs are caused by the differing extents of susceptibility of their disulfide bonds which, in turn, are dependent on the equilibrium constant of unfolding for the conformational equilibria of the insulin analog. These results confirm that the chemical stability of insulin is under the thermodynamic control of the conformational equilibrium constant for unfolding, which indicates that the rate of polymer formation is greater from the unfolded than the folded state.

Literature Cited

1) Fredericq, E. *Arch. Biochem. Biophys.*, **1956**, *65*, 218-228.
2) Jeffrey, P. D.; Coates, J. H. *Biochemistry*, **1966**, *5*, 489-498.
3) Pekar, A. H.; Frank, B. H. *Biochemistry*, **1972**, *11*, 4013-4016.
4) Goldman, J.; Carpenter, F. H. *Biochemistry*, **1974**, *13*, 4566-4574.
5) Jeffrey, P. D.; Milthorpe, B. K.; Nichol, L. W. *Biochemistry*, **1976**, *15*, 4660-4665.
6) Pocker, Y.; Biswas, S. B. *Biochemistry*, **1981**, *20*, 4354-4361.
7) Strazza, S.; Hunter, R.; Walker, E.; Darnall, D. W. *Arch. Biochem. Biophys.*, **1985**, *238*, 30-42.
8) Bentley, G.; Dodson, E.; Dodson, G.; Hodgkin, D.; Mercola, D. *Nature*, **1976**, *261*, 166-168.
9) Chothia, C.; Lesk, A. M.; Dodson, G. G.; Hodgkin, D.C. *Nature*, **1983**, *302*, 500-505.
10) Baker, E. N.; Blundell, T. L.; Cutfield, J. F.; Cutfield, S. M.; Dodson, E. J.; Dodson, G. G.; Crowfoot Hodgkin, D. M.; Hubbard, R. E.; Isaacs, N. W.; Reynolds, C. D.; Sakabe, K.; Sakabe, N.; Vijayan, N. M. *Phil. Trans. R. Soc. Lond., B*, **1988**, *319*, 369-456.
11) Derewenda, U.; Derewenda, Z.; Dodson, E. J.; Dodson, G. G.; Reynolds, C. D.; Smith, G. D.; Sparks, C.; Swenson, D. *Nature*, **1989**, *338*, 594-596.
12) Badger, J.; Harris, M. R.; Reynolds, C. D.; Evans, A. C.; Dodson, E. J.; Dodson, G. G.; North, A. C. T. *Acta Crystallogr.*, **1991**, *B47*, 127-136.
13) Bi, R. C.; Dauter, Z.; Dodson, E. J.; Dodson, G. G.; Giordano, F.; Reynolds, C. D. *Biopolymers*, **1984**, *32*, 391-395.
14) Blundell, T. L.; Dodson, G. G.; Hodgkin, D. C.; Mercola, D. *Adv. Protein Chem.*, **1972**, *26*, 279-402.
15) Binder, C. In *Artificial Systems for Insulin Delivery;* Brunetti, P.; Alberti, K. G. M. M.; Albisser, A. M.; Hepp, K. D.; MassiBendetti, M., Eds.; Raven Press: New York, 1983; pp. 53-57.
16) Brange, J.; Dodson, G. G.; Xiao, B. *Curr. Opin. Struct. Biol.*, **1991**, *1*, 934-940. *8*, 280-286.
17) DiMarchi, R.D.; Mayer, J.P.; Fan, L.; Brems, D.N.; Frank, B.H.; Green, L.K.;, Hoffmann, J.A.; Howey, D.C.; Long, H.B.; Shaw, W.N.; Shields, J.E.; Slieker, L.J.; Su, K.S.E.; Sundell, K.L.; Chance, R.E. In *Peptides: Chemistry and Biology, Proceedings of the Twelfth American Peptide Symposium;* Smith, J.E.; Rivier, J.E., Eds; ESCOM, Leiden, 1992; pp. 26-28.
18) Brems, D. N.; Brown, P. L.; Bryant, C.; Chance, R. E.; Green, L. K.; Long, H. B.; Miller, A. A.; Millican, R.; Shields, J. E.; Frank, B. H. *Protein Engineering*, **1992**, *5*, 519-525.
19) Brems, D. N.; Alter, L. A.; Beckage, M. J.; Chance, R. E.; DiMarchi, R. D.; Green, L. K.; Long, H. B.; Pekar, A. H.; Shields, J. E.; Frank B. H. *Protein Engineering*, **1992**, *5*, 527-533.

20) Melberg, S. G.; Johnson, W. C., Jr. *Proteins: Struct. Funct. Genet.*, **1990**,
21) Brange, J.; Ribel, U.; Hansen, J. F.; Dodson, G.; Hansen, M. T.; Havelund, S.; Melberg, S. G.; Norris, F.; Norris, K.; Snel, I.; Sorensen, A. R.; Voigt, H. O. *Nature*, **1988**, *333*, 679-682.
22) Andres, R.; Swerdloff, R.; Pozefsky, T.; Coleman, D. In *Automation in Analytical Chemistry;* Skeggs, L.T. Jr., Ed.; Medaid: New York, NY, 1966; pp. 486-491.
23) Brange, J. *Galenics of Insulin;* Springer Verlag: New York, NY, 1987; pp. 1-101.
24) In *United States Pharmacopeia*; XXII Mack Printing Co.: Easton, Rockville, PA, 1990; pp. 692-700, 1513-1514..
25) Pace, C. N. *Methods Enzym.*, **1986**, *131*, 266-280.
26) Zale, S. E.; Klibanov, A. M. *Biochemistry,* **1986**, *25*, 5432-5444.
27) Brems, D. N.; Brown, P. L.; Heckenlaible, L. A.; Frank, B. H. *Biochemistry*, **1990**, *29*, 9289-9293.
28) Hol, W. G. J. *Prog. Biophys. Molec. Biol.*, **1985**, *45*, 149-195.
29) Shoemaker, K. R.; Kim, P. S.; York, E.; Stewart, J. M.; Baldwin, R. L. *Nature*, **1987**, *326*, 563-567.
30) Fairman, R.; Shoemaker, K. R.; York, E. J.; Stewart, J. M.; Baldwin, R. L. *Proteins: Struct., Funct., Genet.*, **1989**, *5*, 1-7.
31) Presta, L. G.; Rose, G. D. *Science,* **1988**, *240*, 1632-1641.
32) Richardson, J. S.; Richardson, D. C. *Science*, **1988**, *240*, 1648-1652.

RECEIVED November 11, 1992

INDEXES

Author Index

Aubert, J.-P., 38
Baldwin, R. L., 166
Baneyx, François, 133
Béguin, P., 38
Better, Marc, 203
Blank, G., 218
Boone, Thomas C., 189
Brems, David N., 254
Brown, Patricia L., 254
Bryant, Christopher, 254
Carter, P., 218
Chakrabartty, A., 166
Chance, Ronald E., 254
Chang, Judy Y., 178
Chaudhuri, Bhabatosh, 102
Chrunyk, Boris A., 46
Cleland, Jeffrey L., xi,1
Clogston, Christi L., 189
Covarrubias, M., 218
Cunningham, Brian C., 240
Dhurjati, Prasad, 38,59
DiMarchi, Richard D., 254
Eigenbrot, C., 218
Evans, Judy, 46
Frank, Bruce H., 254
Fujino, Tsuchiyoshi, 38
Gatenby, Anthony A., 72,133
Gordon, Carl, 24
Green, L. Kenney, 254
Haase-Pettingell, Cameron, 24
Horowitz, P. M., 156
Horwitz, Arnold H., 203
Howey, Daniel C., 254
Kelley, R. F., 218
Kiefhaber, Thomas, 142

King, Jonathan, 24
Klein, Jim, 59
Kotts, C., 218
Long, Harlan B., 254
Lorimer, George H., 72
Lu, Hsieng S., 189
Mayr, Lorenz M., 142
Merewether, Lee Ann, 189
Miller, Alita A., 254
Millet, J., 38
Millican, Rohn, 254
Mitraki, Anna, 24
Mulkerrin, Michael G., 240
Narhi, Linda O., 189
O'Connell, M. P., 218
Pekar, Allen H., 254
Presta, L., 218
Robinson, Anne S., 121
Salamitou, Sylvie, 38
Sather, Susan, 24
Schmid, Franz X., 142
Shields, James E., 254
Snedecor, B., 218
Speckart, R., 218
Stephan, Christine, 102
Swartz, James R., 178
Tokatlidis, K., 38
van der Vies, Saskia M., 72
Vetterlein, D., 218
Viitanen, Paul V., 72
Wetzel, Ronald, 46
Wiech, H., 84
Wittrup, K. Dane, 121
Zimmermann, R., 84

Affiliation Index

Abteilung Biochemie II der Georg-August Universität, 84
Amgen Inc., 189
Centre Médical Universitaire, 72
Ciba-Geigy Ltd., 102
E. I. du Pont de Nemours and Company, 72,133
Eli Lilly and Company, 254
Genentech, Inc., xi,1,178,218,240
Massachusetts Institute of Technology, 24

SmithKline Beecham Pharmaceuticals, 46
Stanford University, 166
Unité de Physiologie Cellulaire and URA, 38
Universität Bayreuth, 142
University of Delaware, 38,59

University of Illinois, 121
University of Texas Health Science Center, 156
XOMA Corporation, 203

Subject Index

A

α-Factor leader
  role of proregion in disulfide-linked dimer formation in IGF-1, 115,116t,117f
  signal for IGF-1 secretion, 107
α-Helix properties, *See* Amino acid α-helix propensities
Absorption spectroscopy, analysis of site-directed mutants of human growth hormone, 244,247,248f
Adenine nucleotides, GroEL susceptibility to trypsin digestion effect, 134,136f
Adenosine 5´triphosphate analogues, structures, 134,135f
Aggregation of proteins, scheme, 47,50f
Aggregation suppression, in vivo protein folding effect, 9–10
Aid for folding, requirements, 15–16
Alanine-scanning mutagenesis, application, 241
Amino acid α-helix propensities
  dimeric coiled-coil system, 168t,174
  host–guest system, 166–167,168t
  peptide systems, 168–174
  prediction by molecular dynamics and Monte Carlo simulation, 175–176
  preformed helix nucleus system, 168t,174
  protein studies, 175
Amino acid replacement
  alteration of insulin self-association, 254–259
  alteration of insulin stability, 259–268
Amino acid sequence
  determination of protein structure, 156
  role in protein folding, 47
  use to produce biologically functional proteins, 72

Amino acid sequence determinants of polypeptide chain folding and inclusion body formation
  folding pathways for P22 tailspike endorhamnosidase, 26,27–28f,29
  global suppressors of protein-folding defects and inclusion body formation, 30,31–32f,33
  mutational suppression of inclusion body formation in eukaryotic proteins, 33
  suppression of inclusion body formation by overexpression of chaperonins, 32,33t,35f
  temperature-sensitive folding mutations, 29–30
Antibody fragments
  in vivo expression from microorganisms, 203–215
  production from *Escherichia coli*, 204
Anti-p185^{HER2} antibody Fab fragments, humanized, *See* Humanized anti-p185^{HER2} antibody Fab fragments produced in *Escherichia coli*
AraBAD promoter, control of gene expression, 205–206
Assembly, definition, 73

B

Bacterial expression system
  advantages, 5–6
  characteristics, 1,4t,5
Binding protein
  cloning procedure, 104
  functions, 122–123
  levels in yeast cells secreting foreign proteins, 127,128t,f,129–130

Binding protein—*Continued*
  mathematical model of function, 123–127
  proofreading role in CHO cells, 122
  role in IGF-1 folding–unfolding, 108,109f,110t,f,111
Biologically functional proteins, production with amino acid sequence information, 72
Biotechnology, impact of protein folding, 1–18
Biotechnology industry, development, 1
Bis(8-anilino-1-naphthalenesulfonate), fluorescence, 136,137f
Breast cancer, role of p185[HER2], 220

C

CelD, endoglucanase, *See* Endoglucanase CelD
Cellular compartments, problem of understanding development and maintenance, 84
Cellulosome
  description, 38
  role of *Clostridium thermocellum* cellulase duplicated segment in structural organization, 41–42
Chaperones
  inclusion body formation, 25–26
  molecular, *See* Molecular chaperone(s)
  role in protein folding, 121
  types, 121
Chaperonin(s)
  comparison to detergent-assisted rhodanese folding, 160–161
  description, 76
  Escherichia coli chaperonins, 77–80
  role in biologically related rhodanese folding, 159–160
  suppression of inclusion body formation, 33,34,35f
  types, 76–77
Chaperonins 10 and 60, description, 76–77
Chemical stability, insulin, 260–263,267–268
Circular dichroism spectroscopy, analysis of site-directed mutants of human growth hormone, 242,244,245–246f

Citrate synthase, use as model protein for protein renaturation kinetic analysis, 93–94
Clostridium thermocellum, cellulase system, 38
Clostridium thermocellum cellulase duplicated segment
  alignment, 38–40,42–43
  ammonium sulfate precipitation curves, 42,43f,44
  formation of inclusion bodies containing active endoglucanase, 40–41
  structural organization of cellulosome, 41–42
Compact folding intermediates, definition, 159
Conformational stability, insulin, 262,264–268
Cpn60, comparison to detergent-assisted rhodanese folding, 159–161
Cyclophilin, role in protein folding, 111,112f,113t

D

4D5, monoclonal antibody, *See* Monoclonal antibody 4D5
Detergent micelles
  comparison to chaperonin-assisted folding of rhodanese, 160–161
  use for capture of folded rhodanese and intermediate studies, 158–159
Dimeric coiled-coil system, amino acid α-helix propensities, 168t,174
Disulfide bonds, folding of recombinant human granulocyte colony stimulating factor produced in *Escherichia coli*, 189–201
Duplicated segment, *Clostridium thermocellum* cellulases, 38–44

E

Endoglucanase CelD
  activity within inclusion bodies, 40–41
  ammonium sulfate precipitation curves, 42,43f,44

INDEX

Endoglucanase CelD—*Continued*
  purification from inclusion bodies, 41–42
Endorhamnosidase, P22 tailspike, *See* P22 tailspike endorhamnosidase
Equilibrium constant of unfolding for protein, definition, 260
Escherichia coli
  expression of Fab, Fab´, and F(ab´)$_2$, 205–211
  humanized anti-p185^{HER2} antibody Fab fragments, 218–237
  in vivo protein folding, 8–10
  role of disulfide bonds in folding of recombinant human granulocyte colony stimulating factor, 189–201
  single-step solubilization and folding of IGF-1 aggregates, 178–187
Escherichia coli chaperonins, in vitro protein assembly effect, 77–80
Escherichia coli expression system, advantages, 5–6
Eukaryotic cells, role of binding protein in secretion of foreign proteins, 121–130
Eukaryotic proteins, mutational suppression of inclusion body formation, 33
Expression of heterologous proteins in microorganisms
  advantages, 203
  antibody fragments, 204–215
Expression systems
  characteristics, 1,4t,5
  therapeutic proteins, 1,2–3t

F

Fab
  applications, 204
  construction of immunotoxins, 214
  expression in *Escherichia coli*, 206,207–208f
  expression in yeast, 211,212–213f,214
  fermentation, 209t,210f,211
  production from *Escherichia coli*, 204
  purification, 204

Fab´
  construction of immunotoxins, 214,215t
  description, 204
  expression in *Escherichia coli*, 206,209,210f
  expression in yeast, 211,212–213f,214
  fermentation, 209t,210f,211
  purification, 204
F(ab´)$_2$
  construction of immunotoxins, 214,215t
  expression in *Escherichia coli*, 206,209
  expression in yeast, 211,212–213f,214
  fermentation, 209t,210f,211
  purification, 204
Fd and Fd´ modules, construction, 205,207f
Fermentation media composition, in vivo protein folding, 8–9
FK506 binding protein, role in protein folding, 111
Fluorescence, bisAns, 136,137f
Fluorescence quenching, definition, 242
Folding accelerating proteins
  in vitro protein folding, 93
  in vivo protein folding, 94,95f,96
Folding aid, requirements, 15–16
Folding intermediates, function, 24–25
Folding of insulin-like growth factor 1 aggregates from *Escherichia coli*
  aggregate concentration effects, 186,187f
  characterization of folded IGF-1, 181,182f,184f
  initial folding, 179,181,182f
  kinetics, 183,185f
  optimization, 181,183–189
  pH effect, 181,183,184f
  solvent effects, 183,185f
  temperature effects, 183,186,187f
Folding of polypeptide chain under native conditions, amino acid sequence effect, 142
Folding of proteins
  fast and slow reactions, 143f,144
  kinetic control by detergent micelles, liposomes, and chaperonins, 156–161
  need for control, 84
  sources of complexity, 142–143
  See also Protein folding

Folding of recombinant human-insulin-like growth factor 1 in yeast
  binding expression cassette, 105
  binding protein cloning procedure, 104
  binding protein role, 108,109f,110t,f,111
  colony blot analytical procedure, 106
  covalent attachment of proregion to signal sequence for translocation, 113–117
  cytoplasmic cyclophilin expression cassette, 105
  cytoplasmic cyclophilin gene cloning procedure, 104
  experimental materials, 103–104
  immunoblot analytical procedure, 106
  immunophilin role, 111,112f,113t
  incapability of translocation via signal sequences, 107–108
  plasmids containing Kex2p variants, 105
  plasmids with expression cassettes containing different secretion signals, 104–105
  procedure for expression in yeast, 106
  quantitative determination procedure, 106
  role of signal sequences in secretion, 106
  secretion signals, 107
  strains, 104
  transformations, 104
Folding of ribonuclease T_1, 147–153
Foreign protein secretion in yeast, binding protein levels, 127,128t,f,129–130
Fv, description, 204

G

Global suppressors of protein-folding defects and inclusion body formation
  intracellular folding of suppressor and wild-type polypeptide chains, 30–31,32f
  intracellular folding of wild-type, tsf, and su–tsf polypeptide chains, 30,31f
Granulocyte colony stimulating factor, applications, 190
GroEL, complex formation, 133–134

GroEL-mediated protein folding
  bisAns fluorescence, 136,137f,138
  experimental description, 133
  GroES binding effect, 138,139f,140
  inhibition of adenosine triphosphatase activity of GroEL by adenine nucleotides, 134t
  susceptibility of GroEL to trypsin digestion, 134,135f,136
GroES
  binding effect on GroEL-mediated protein folding, 138,139f,140
  GroEL susceptibility to trypsin digestion effect, 134,135f
Growth conditions, in vivo protein folding, 8
Growth hormone family of proteins, function, 240

H

Helix properties, *See* Amino acid α-helix propensities
Heregulin, definition, 220
Host cells in expression systems
  glycosylation effect, 4t,5
  in vivo protein-folding efficiency, 5
Host–guest system, amino acid α-helix propensities, 166–167,168t
Hsp60, function and occurrence, 121–122
Hsp70, function and occurrence, 121–122
Human growth hormone, spectral analysis of site-directed mutants, 240–251
Human insulin, amino acid sequence, 254,255f
Human-insulin-like growth factor 1, folding in yeast, 103–118
Humanization, process, 218,219f,220
Humanized anti-p185[HER2] antibody Fab fragments produced in *Escherichia coli*
  antigen-binding determinants in human 4D5 Fab, 236,237f
  antigen-binding thermodynamics, 223,226,228–232
  antiproliferative activity of 4D5 fragments, 234,235f
  circular dichroism spectroscopic procedure, 223,225f
  conformation and stability of Fab, 226,227f,228

INDEX

Humanized anti-p185^{HER2} antibody Fab
  fragments produced in *Escherichia coli*—
  Continued
  differential scanning calorimetric
    procedure, 223
  experimental description, 218
  expression and purification of 4D5
    Fab, 226
  humanization of murine MAb4D5, 222
  procedure for expression and
    purification of 4D5 Fab fragments,
    222–223,224f
  properties of MAb4D5, 220,221t
  purification of extracellular domain of
    p185^{HER2}, 223
  recruitment of antiproliferative effect
    into humanized 4D5 Fab, 234,236t
  role of framework residues in antigen
    binding, 229,233t,234
  role of p185^{HER2} in breast cancer, 220
  structure of MAb4D5, 220,222,223f
Hydrophobic surfaces, importance to
  structure and function of
  rhodanese, 157

I

Immunophilins, role in protein folding,
  111,112f,113t
In vitro protein folding
  applications, 10
  citrate synthase as model, 93–94
  description, 10
  model, 91,92f,93–94
  molecular chaperone effect,
    10,11–14t,15
  requirements of folding aid, 15–16
  stabilizer effect, 16
In vivo expression of correctly folded
  antibody fragments from microorganisms
  construction of Fab, Fab′, and F(ab′)$_2$
    in immunotoxins, 214,215t
  construction of Fd and Fd′ molecules,
    205,207f
  expression of Fab, Fab′, and F(ab′)$_2$ in
    Escherichia coli, 205–211
  expression of Fab, Fab′, and F(ab′)$_2$
    in yeast, 211,212–213f,214

In vivo kinetics of inclusion body
  formation
  bacterial strains, 60
  capability of centrifugation to achieve
    complete recovery of insoluble
    protein, 67–68
  chloramphenicol pulse–chase experiments
    for CheY protein, 65,66f
  continued incorporation of radiolabel,
    67–68
  experimental description, 59–62
  implications for formation mechanism,
    68,70
  material balance, 65,67
  previous studies, 59
  protein partitioning, 68,69f
  proteolysis of CheY protein, 63
  pulse–chase experiments for CheY
    protein, 62–63,64f
  pulse–chase experiments for EGD protein,
    63,65,66f
In vivo protein folding, model,
  94,95f,96
In vivo protein folding in
  Escherichia coli
  aggregation-suppressing alteration
    effect, 9–10
  analysis, 16–17
  fermentation media composition
    effect, 8–9
Inclusion bodies, role of *Clostridium
  thermocellum* cellulase duplicated
  segment in formation, 40–41
Inclusion body formation
  advantages, 46
  amino acid sequence determinants, 24–35
  chaperones, 25–26
  in vitro thermal studies, 51–54,56
  in vivo kinetics, 59–70
  in vivo temperature effects, 48,49t
  influencing factors, 46–47
  kinetic studies, 53,54f
  mutational effects, 47,48t
  mutational suppression in eukaryotic
    proteins, 33
  position 97 replacement effect, 56
  problem for protein recovery, 46
  protein-folding intermediates, 25
  protein stability effect,
    49,50f,51,52f

Inclusion body formation—*Continued*
  suppression by overexpression of
    chaperonins, 33,34*t*,35*f*
  three-dimensional structure, 53,55*f*,56
Insect cell expression system,
  characteristics, 1,4*t*,5
Insulin
  alteration for insulin by amino acid
    replacement, 259
  human, amino acid sequence,
    254,255*f*
Insulin-like growth factor 1
  activity, 178
  human, folding in yeast, 103–118
  single-step solubilization and folding
    of aggregates from *Escherichia coli*,
    178–187
  structure, 178
Insulin-like growth factor 1 aggregate
  concentration, IGF-1 folding effect,
  186,187*f*
Intermediates, role in native protein
  folding, 156–157

K

KAR2 gene product, *See* Binding protein
Kinetic control, protein folding by
  detergent micelles, liposomes, and
  chaperonins, 156–161
Kinetics
  folding of IGF-1 aggregates from
    Escherichia coli, 183,185*f*
  in vivo, *See* In vivo kinetics of
    inclusion body formation

L

Liposomes, binding of rhodanese, 161

M

Major coat protein of phage P22,
  temperature-sensitive folding
  mutants, 30

Mammalian cell expression system,
  characteristics, 1,4*t*,5
Mammalian endoplasmic reticulum, role of
  molecular chaperones in protein
  transport, 84–99
Mammalian growth hormones, examples for
  studies, 240
Material balance, description, 65
Mathematical model of binding protein
  fraction in vivo, description,
  123–127
Microorganisms, in vivo expression of
  correctly folded antibody fragments,
  203–215
Misfolded chain generation,
  problem, 25
Molecular chaperone(s)
  chaperonin family, 77
  classes, 74,75*t*,76
  definition, 73,133
  development of concept, 73
  implications for biotechnology, 80
  in vitro protein folding,
    10,11–14*t*,15
  role in vivo, 73–74,75*t*
Molecular chaperone role in protein
  transport into mammalian endoplasmic
  reticulum
  ammonium sulfate fraction analysis,
    88,90*f*
  assay, 85,87*f*,88
  experimental objective, 85
  Hsc70 addition, 88
  Hsp90 effect on protein folding, 88,91
  in vitro protein-folding model,
    91,92*f*,93–94
  in vivo protein-folding model,
    94,95*f*,96
  procoat protein assembly analysis, 88,89*f*
  transport competence model,
    96,97*f*,98–99
Molecular dynamics simulation, prediction
  of helix propensity, 175–176
Molten globules
  definition, 159
  formation, 79
Monoclonal antibodies
  advantages for treatment of human
    disease, 218
  problem of human immune response, 218

INDEX

Monoclonal antibody 4D5
  humanization procedure, 222
  properties, 220,221t
  structure, 220,222,223f
Monte Carlo simulation, prediction of
  helix propensity, 176

N

Native protein folding, role of
  intermediates, 156–157
Newly synthesized polypeptide chains,
  role of folding intermediates, 24–25
Nucleoplasmin, description, 75t,76

P

P22 tailspike endorhamnosidase, folding
  pathways, 26,27–28f,29–30
P39G variant of ribonuclease T_1, role of
  prolyl isomerization in folding,
  151–153
p185^{HER2}, role in breast cancer, 220
Partitioning of recombinant protein,
  problem for product recovery, 59
Peptide systems, amino acid α-helix
  propensities, 168–174
Peptidyl prolyl isomerase, function, 5
pH, IGF-1 folding effect, 181,183,184f
Polypeptide chain folding, amino acid
  sequence determinants, 24–35
Preformed helix nucleus system, amino acid
  α-helix propensities, 168t,174
Prepro-α-factor, signal for IGF-1
  secretion, 107
Primary sequence analysis, protein
  folding, 17
Proinsulin, structure, 102–103
Prolyl isomerizations in ribonuclease T_1
  folding
  directed mutagenesis probing of role,
    149–153
  experimental basis of folding models,
    147–149
  experimental objective, 142
  fast and slow protein-folding reactions,
    143f,144

Prolyl isomerizations in ribonuclease T_1
  folding—*Continued*
  kinetic models for unfolding and slow
    refolding of ribonuclease T_1,
    145–147
  structure and stability of ribonuclease
    T_1, 144,145f
Proregion of α-factor leader
  role in disulfide-linked dimer formation
    in IGF-1, 115,116t,117f
  role in IGF-1 folding, 115
Protein(s), helix propensity studies, 175
Protein aggregation, scheme, 47,50f
Protein assembly, *Escherichia coli*
  chaperonins, 77–80
Protein cellular alteration, in vivo
  protein folding, 9
Protein disulfide isomerase, function, 5
Protein folding
  analysis of folding, 16–18
  concentrations required, 121
  GroEL mediated, *See* GroEL-mediated
    protein folding
  in vitro, *See* In vitro protein folding
  in vivo, *See* In vivo protein folding in
    Escherichia coli
  interest, 121
  potential pathways, 6,7f
  primary sequence analysis, 17
  scheme, 47,50f
Protein-folding effect on inclusion body
  formation
  in vitro thermal studies, 51–54,56
  in vivo temperature effects, 48,49t
  inclusion body screening, 47,48t
  kinetic studies, 53,54f
  mutational effects, 56–57
  protein stability effect, 49,50f,51,52f
  replacements at position 97, 56
  three-dimensional structure, 53,55f,56
Protein-folding intermediates, inclusion
  body formation, 25
Protein-folding methods for industrial
  production
  in vitro protein folding, 10–16
  in vivo folding in *Escherichia
    coli*, 8–10
Protein-folding pathway, measurement of
  structural properties of
  intermediates, 189

Protein partitioning, calculation, 68
Protein proteolysis, scheme, 47,50f
Protein refolding in vitro, schematic representation, 77–78
Protein transport, 85,86
Proteolysis of proteins, scheme, 47,50f

R

Recombinant human granulocyte colony stimulating factor produced in *Escherichia coli*
analogues without disulfide bonds, 196t,197
covalent structure, 190,191f
Cys^{17} at NH$_2$-terminal helix, 199,200f,201
disulfide bond role in structure, 199
disulfide-reduced intermediates, 192,195f,196t,197
experimental description, 190
properties, 196t,197–198,200f
reverse-phase high-performance liquid chromatography, 190,192–195
sequential folding pathway, 198–199
Refolding, *See* In vitro protein folding
Refolding of proteins in vitro, schematic representation, 77–78
Refolding of ribonuclease T$_1$, kinetic model, 145–147
Refractile particles, description, 178
Rhodanese
advantages as model for protein-folding kinetics, 157
binding to liposomes, 161
catalytic function, 157
detergent-assisted refolding, 158
importance of hydrophobic surfaces in structure and function, 157
refolding difficulties after denaturation, 158
sequence, 157
Rhodanese folding, 158–161
Ribonuclease T$_1$
kinetic models for folding, 145–147
structure, 144,145f
Ribonuclease T$_1$ folding, *See* Folding of ribonuclease T$_1$

S

S value, definition, 166
S54G,P55N variant of ribonuclease T$_1$, role of prolyl isomerizations in folding, 150
Self-assembly hypothesis, description, 72
Self-association, alteration for insulin by amino acid replacement, 254–259
Self-association of insulin, description, 254,256–259
Signal peptide
description, 85
role in protein transport, 98
Signal recognition peptide, role in protein transport, 98
Signal sequences, role in IGF-1 translocation, 107–108
Single-step solubilization of insulin-like growth factor 1 aggregates from *Escherichia coli*, 178–180
Site-directed mutagenesis, use in growth hormone studies, 240
Site-directed mutants of human growth hormone, spectral analysis, 240–251
Solubilization, single step, 178–180
Solubilization-achieving proteins
in vitro protein folding, 93
in vivo protein folding, 94,95f,96
protein transport, 97f,98
Solvent, IGF-1 folding effect, 183,185f
Spectral analysis of site-directed mutants of human growth hormone
absorption spectra, 244,247,248f
circular dichroism spectra, 242,244,245–246f
distribution of aromatic amino acids and mutants, 242,243f
experimental description, 240–242
reduction in binding affinity, 242
second-derivative spectra, 247,249f
Stern–Volmer analysis of tryptophan fluorescence, 247,250,251f
tryptophan indole structure, 247,248f,251t
Stability, ribonuclease T$_1$, 145
Stability of insulin
correlation between chemical and conformational stability, 260–268
importance, 259–260

INDEX

Stability of proteins, inclusion body formation effect, 49,50f,51,52f
Stabilizers, in vitro protein-folding effect, 16
Structure, ribonuclease T_1, 144,145f
Structure prediction, protein folding, 17–18

T

Temperature, IGF-1 folding effect, 183,186,187f
Temperature-sensitive folding mutants, major coat protein of phage P22, 30
Temperature-sensitive folding mutations, identification of critical residues for tailspike folding, 29–30
Therapeutic proteins, characteristics, 1,2–3t
Translocation, role in protein transport, 99
Translocation-mediating proteins
in vivo protein folding, 94,95f,96
protein transport, 97f,98–99
Transport competence model
cytosolic proteins, 96,97f,98
noncytosolic proteins, 97f,98–99
Transport of proteins into mammalian endoplasmic reticulum, model, 85,86f

U

Unfolding of ribonuclease T_1, kinetic model, 145–146

V

Vesicular transport, prerequisites, 122

W

W59Y variant of ribonuclease T_1, role of prolyl isomerization in folding, 151

Y

Yeast
expression of Fab, Fab´, and F(ab´)$_2$, 211,212–213f,214
folding of IGF-1, 102–110
Yeast cells secreting foreign proteins, binding protein levels, 127,128t,f,129–130
Yeast expression system, characteristics, 1,4t,5

Production: C. Buzzell-Martin
Indexing: Deborah H. Steiner
Acquisition: Barbara C. Tansill
Cover design: Tana Powell

Printed and bound by Maple Press, York, PA

Bestsellers from ACS Books

The ACS Style Guide: A Manual for Authors and Editors
Edited by Janet S. Dodd
264 pp; clothbound ISBN 0–8412–0917–0; paperback ISBN 0–8412–0943–X

The Basics of Technical Communicating
By B. Edward Cain
ACS Professional Reference Book; 198 pp;
clothbound ISBN 0–8412–1451–4; paperback ISBN 0–8412–1452–2

Chemical Activities (student and teacher editions)
By Christie L. Borgford and Lee R. Summerlin
330 pp; spiralbound ISBN 0–8412–1417–4; teacher ed. ISBN 0–8412–1416–6

Chemical Demonstrations: A Sourcebook for Teachers,
Volumes 1 and 2, Second Edition
Volume 1 by Lee R. Summerlin and James L. Ealy, Jr.;
Vol. 1, 198 pp; spiralbound ISBN 0–8412–1481–6;
Volume 2 by Lee R. Summerlin, Christie L. Borgford, and Julie B. Ealy
Vol. 2, 234 pp; spiralbound ISBN 0–8412–1535–9

Chemistry and Crime: From Sherlock Holmes to Today's Courtroom
Edited by Samuel M. Gerber
135 pp; clothbound ISBN 0–8412–0784–4; paperback ISBN 0–8412–0785–2

Writing the Laboratory Notebook
By Howard M. Kanare
145 pp; clothbound ISBN 0–8412–0906–5; paperback ISBN 0–8412–0933–2

Developing a Chemical Hygiene Plan
By Jay A. Young, Warren K. Kingsley, and George H. Wahl, Jr.
paperback ISBN 0–8412–1876–5

Introduction to Microwave Sample Preparation: Theory and Practice
Edited by H. M. Kingston and Lois B. Jassie
263 pp; clothbound ISBN 0–8412–1450–6

Principles of Environmental Sampling
Edited by Lawrence H. Keith
ACS Professional Reference Book; 458 pp;
clothbound ISBN 0–8412–1173–6; paperback ISBN 0–8412–1437–9

Biotechnology and Materials Science: Chemistry for the Future
Edited by Mary L. Good (Jacqueline K. Barton, Associate Editor)
135 pp; clothbound ISBN 0–8412–1472–7; paperback ISBN 0–8412–1473–5

For further information and a free catalog of ACS books, contact:
American Chemical Society
Distribution Office, Department 225
1155 16th Street, NW, Washington, DC 20036
Telephone 800–227–5558

Highlights from ACS Books

Good Laboratory Practice Standards: Applications for Field and Laboratory Studies
Edited by Willa Y. Garner, Maureen S. Barge, and James P. Ussary
ACS Professional Reference Book; 572 pp; clothbound ISBN 0–8412–2192–8

Silent Spring Revisited
Edited by Gino J. Marco, Robert M. Hollingworth, and William Durham
214 pp; clothbound ISBN 0–8412–0980–4; paperback ISBN 0–8412–0981–2

The Microkinetics of Heterogeneous Catalysis
By James A. Dumesic, Dale F. Rudd, Luis M. Aparicio, James E. Rekoske, and Andrés A. Treviño
ACS Professional Reference Book; 316 pp; clothbound ISBN 0–8412–2214–2

Helping Your Child Learn Science
By Nancy Paulu with Margery Martin; Illustrated by Margaret Scott
58 pp; paperback ISBN 0–8412–2626–1

Handbook of Chemical Property Estimation Methods
By Warren J. Lyman, William F. Reehl, and David H. Rosenblatt
960 pp; clothbound ISBN 0–8412–1761–0

Understanding Chemical Patents: A Guide for the Inventor
By John T. Maynard and Howard M. Peters
184 pp; clothbound ISBN 0–8412–1997–4; paperback ISBN 0–8412–1998–2

Spectroscopy of Polymers
By Jack L. Koenig
ACS Professional Reference Book; 328 pp;
clothbound ISBN 0–8412–1904–4; paperback ISBN 0–8412–1924–9

Harnessing Biotechnology for the 21st Century
Edited by Michael R. Ladisch and Arindam Bose
Conference Proceedings Series; 612 pp;
clothbound ISBN 0–8412–2477–3

From Caveman to Chemist: Circumstances and Achievements
By Hugh W. Salzberg
300 pp; clothbound ISBN 0–8412–1786–6; paperback ISBN 0–8412–1787–4

The Green Flame: Surviving Government Secrecy
By Andrew Dequasie
300 pp; clothbound ISBN 0–8412–1857–9

For further information and a free catalog of ACS books, contact:
American Chemical Society
Distribution Office, Department 225
1155 16th Street, NW, Washington, DC 20036
Telephone 800–227–5558